Networks and Netwars

The Future of Terror, Crime, and Militancy

Edited by

John Arquilla
and David Ronfeldt

Prepared for the Office of the Secretary of Defense

National Defense Research Institute

RAND

The research described in this report was sponsored by the Office of the Secretary of Defense (OSD). The research was conducted in RAND's National Defense Research Institute, a federally funded research and development center supported by the OSD, the Joint Staff, the unified commands, and the defense agencies under Contract DASW01-01-C-0004.

Library of Congress Cataloging-in-Publication Data

Networks and netwars : The future of terror, crime, and militancy / John Arquilla, David Ronfeldt.
    p. cm.
    MR-1382
    Includes bibliographical references.
    ISBN 0-8330-3030-2
    1. Information warfare. 2. Cyberterrorism. I. Arquilla, John. II. Ronfeldt, David F.

HV6773 .N47 2001
303.6'25—dc21

2001041739

Cover design by Peter Soriano

Published 2001 by RAND
1700 Main Street, P.O. Box 2138, Santa Monica, CA 90407-2138
1200 South Hayes Street, Arlington, VA 22202-5050
201 North Craig Street, Suite 102, Pittsburgh, PA 15213
RAND URL: http://www.rand.org/
To order RAND documents or to obtain additional information, contact Distribution Services: Telephone: (310) 451-7002; Fax: (310) 451-6915; Email: order@rand.org

To Dick O'Neill, for his stalwart support and keen eye

The fight for the future makes daily headlines. Its battles are not between the armies of leading states, nor are its weapons the large, expensive tanks, planes, and fleets of regular armed forces. Rather, the combatants come from bomb-making terrorist groups like Osama bin Laden's al-Qaeda, drug smuggling cartels like those in Colombia and Mexico, and militant anarchists like the Black Bloc that ran amok during the Battle of Seattle. Other protagonists are civil-society activists fighting for democracy and human rights—from Burma to the Balkans. What all have in common is that they operate in small, dispersed units that can deploy nimbly—anywhere, anytime. They know how to penetrate and disrupt, as well as elude and evade. All feature network forms of organization, doctrine, strategy, and technology attuned to the information age. And, from the Intifadah to the drug war, they are proving very hard to beat; some may actually be winning. This is the story we tell.

This book also provides a further step in the elaboration of our ideas about how and why the information revolution is affecting the whole spectrum of conflict. Our notion of *cyberwar* (1993) focused on the military domain, and our first study of *netwar*[1] (1996) on irregular modes of conflict, including terror, crime, and militant social activism.[2] The implications of these concepts for organization, doctrine, and technology across the spectrum of conflict were further elaborat-

---

[1]Our netwar concept predates, and should not be confused with, the U.S. military's "network warfare simulation system" (NETWARS).

[2]For full citations of these and our other studies, please see the bibliographies for Chapters One and Ten.

ed in our book, *In Athena's Camp* (1997). More recently, we noted that many activists who practice netwar are helping to create a new approach to strategy and diplomacy that we call *noopolitik* (1999). Next, we expanded on our idea that *swarming* (2000) will emerge as a 21st-century doctrine that will encompass and enliven both cyberwar and netwar. Here, we offer new analysis about netwar. The analysis includes case studies about terrorists, criminals, and gangs; social netwars in Burma, Mexico, and Seattle; and closing chapters on some of the technological, organizational, and doctrinal dynamics of netwar.

U.S. policymakers and strategists will be interested in this book. It should also interest analysts in academia and research institutes concerned with how the information revolution is altering the nature of conflict.

This book was prepared for a project on "Networks and Netwars," directed by John Arquilla and David Ronfeldt. The project was sponsored by the Office of the Assistant Secretary of Defense (Command, Control, Communications and Intelligence), OASD/C3I, and was conducted within the International Security and Defense Policy Center of RAND's National Defense Research Institute (NDRI). NDRI is a federally funded research and development center sponsored by the Office of the Secretary of Defense, the Joint Staff, the unified commands, and the defense agencies.

Comments are invited. We can be reached via email at arquilla@ rand.org and ronfeldt@rand.org.

CONTENTS

SUMMARY

Netwar is the lower-intensity, societal-level counterpart to our earlier, mostly military concept of cyberwar. Netwar has a dual nature, like the two-faced Roman god Janus, in that it is composed of conflicts waged, on the one hand, by terrorists, criminals, and ethnonationalist extremists; and by civil-society activists on the other. What distinguishes netwar as a form of conflict is the networked organizational structure of its practitioners—with many groups actually being leaderless—and the suppleness in their ability to come together quickly in swarming attacks. The concepts of cyberwar and netwar encompass a new spectrum of conflict that is emerging in the wake of the information revolution.

This volume studies major instances of netwar that have occurred over the past several years and finds, among other things, that netwar works very well. Whether the protagonists are civil-society activists or "uncivil-society" criminals and terrorists, their netwars have generally been successful. In part, the success of netwar may be explained by its very novelty—much as earlier periods of innovation in military affairs have seen new practices triumphant until an appropriate response is discovered. But there is more at work here: The network form of organization has reenlivened old forms of licit and illicit activity, posing serious challenges to those—mainly the militaries, constabularies, and governing officials of nation states—whose duty is to cope with the threats this new generation of largely nonstate actors poses.

Strategists and policymakers in Washington and elsewhere have already begun to discern the dark side of the netwar phenomenon, especially as manifested in terrorist and criminal organizations. This growing awareness is quite evident in recent official studies of this burgeoning problem: *Patterns of Global Terrorism: 1999* (State Department, 2000), *International Crime Threat Assessment* (Interagency Working Group, 2000), and *Global Trends 2015* (National Intelligence Council, 2000). But strategists and policymakers still have much work to do to harness the brighter, civil-society-building potential of networked nonstate actors. Thus, a fundamental challenge in the coming decade will be to focus on the opportunities that may arise from closer cooperation with nongovernmental organizations (NGOs) and other nonstate actors.

For the U.S. Department of Defense, a range of possibilities opens up, from encouraging the early involvement of appropriate NGO networks in helping to detect and head off a looming crisis, to working closely with them in the aftermath of conflicts to improve the effectiveness of U.S. forces still deployed, to reduce the residual hazards they face, and to strengthen the often fragile peace. In short, American policymakers and strategists must continue to keep an eye on the perils posed by criminal and terrorist networks. But they must enlarge their vision and their practices to encompass the tremendous opportunities likely to attend the rise of a network-based realm devoted to the protection of human rights, the spread of democratic values, and the formation of deep coalitions between states and civil-society NGOs. Netwar, the emergent mode of conflict of choice for networked nonstate actors, has two faces—and both matter very much.

In this volume, we and our colleagues examine various types of netwar, from the most violent to the most socially activist. In so doing, we find that, despite the variety, all networks that have been built for waging netwar may be analyzed in terms of a common analytic framework. There are five levels of theory and practice that matter: the technological, social, narrative, organizational, and doctrinal levels. A netwar actor must get all five right to be fully effective.

While a network's level of technological sophistication does make a difference—and people do tend to think that netwar depends heavily on technology—the other levels have just as much, if not more, of an effect on the potential power of a given group. One key level is the social basis for cooperation among network members. When social ties are strong, building mutual trust and identity, a network's effectiveness is greatly enhanced. This can be seen most clearly in ethnically based terror, crime, and insurgent groups in which clan ties bind together even the loosest, most dispersed organization.

Among civil-society netwarriors, the narrative level of analysis may matter most. Sharing and projecting a common story about their involvement in an activist network enliven and empower these groups, and attract their audiences. The narrative level is also important to practitioners of the dark side of netwar, but it may be more necessary for civil-society networks to emphasize this level and get it right because they are less likely to be held together by the kinds of ethnic or clan ties so common among crime and terror networks.

In trying to confront or cope with a networked adversary, it is important to assess the opponent's strengths and weaknesses at the technological, social, and narrative levels. Yet, the defining level of a netwar actor is its organizational design. Analysts must realize that the structures of networks may feature much variety—from simple chain or line networks, to less simple hub or star designs, to complex all-channel designs, any and all of which may be blended into sprawling multihub and spider's-web networks. To cope with a network, analysts must first learn what *kind* of network it is and then draw on the best methods for analysis. In the past, intelligence assessments of adversaries have tended to focus on their hierarchical leadership structures. This is insufficient for analyzing netwar actors—which, like some of today's terrorist networks, may well consist of various small, dispersed groups that are linked in odd ways and do not have a clear leadership structure.

Another important level of analysis is to parse just what sort of doctrine the netwar actor is employing. Most networks—of both the civil and uncivil variety—will have a great capacity for swarming. This does not mean that all will swarm all the time, or even that all will swarm well. Moreover, few netwar actors have an explicit doctrine for swarming. But most are moving in that direction. Swarming is the key doctrinal approach for which to prepare.

The most potent netwarriors will not only be highly networked and have a capacity to swarm, they will also be held together by strong social ties, have secure communications technologies, and project a common "story" about why they are together and what they need to do. These will be the most serious adversaries. But even those networks that are weak on some levels (e.g., technological) may pose stiff challenges to their nation-state adversaries. With this in mind, it is necessary to go beyond just diagnosing the nature of the networked nonstate opponent in a given conflict. It will become crucial for governments and their military and law enforcement establishments to begin networking themselves. Perhaps this will become the greatest challenge posed by the rise of netwar.

ACKNOWLEDGMENTS

We are deeply grateful for the thoughtful, generous participation of all the authors of the case studies in this volume: in order of their appearance, Michele Zanini, Sean Edwards, Phil Williams, John Sullivan, Tiffany Danitz, Warren Strobel, Paul de Armond, Dorothy Denning, and Luther Gerlach. We are also thankful for the RAND managers who showed interest in this project, Jeff Isaacson and Stuart Johnson, and for the constructive comments and criticisms provided by the two formal reviewers of the draft version of this volume, Robert Anderson and Greg Treverton. We also appreciate the efforts of people who occasionally send us comments or items of information pertaining to netwar: at RAND, Robert Anderson, Kevin McCarthy, James Mulvenon, William O'Malley, and Willis Ware; outside RAND, Stephen Borgatti, Robert Bunker, Steve Cisler, Harry Cleaver, Peter Monge, Joel Simon, Steven Strogatz, and especially Margarita Studemeister. A number of individuals in the activist communities also deserve our thanks, but they would undoubtedly be discomfited by our explicitly acknowledging them.

Furthermore, we are indebted to Richard O'Neill and other members of the Highlands Forum (including David Brin, Robert Scott, and Bruce Sterling) for discussions about netwar that we have had with them over the past several years.

We are also profoundly thankful for the continuing support and guidance provided to us in the Pentagon by former Assistant Secretary of Defense (C3I) Art Money and his principal deputy, Lin Wells, as well as Deputy Assistant Secretary of Defense William Leonard and Lieutenant Colonel Robert Walter, U.S. Army (ret.).

Finally, our heartfelt thanks extend to Christina Pitcher for the splendid effort she put into editing our manuscript.

Chapter One

THE ADVENT OF NETWAR (REVISITED)[1]

John Arquilla and David Ronfeldt

Editors' abstract. This introductory chapter provides a reprise of many of the points we have made about the netwar concept since 1993. In this book, we depict netwar as having two major faces, like the Roman god Janus—one dominated by terrorists and criminals that is quite violent and negative, and another evinced by social activists that can be militant but is often peaceable and even promising for societies. Indeed, the book is structured around this theme.

The information revolution is altering the nature of conflict across the spectrum. We call attention to two developments in particular. First, this revolution is favoring and strengthening network forms of organization, often giving them an advantage over hierarchical forms. The rise of networks means that power is migrating to nonstate actors, because they are able to organize into sprawling multiorganizational networks (especially "all-channel" networks, in which every node is connected to every other node) more readily than can traditional, hierarchical, state actors. This means that conflicts may increasingly be waged by "networks," perhaps more than by "hierarchies." It also means that whoever masters the network form stands to gain the advantage.

Second, as the information revolution deepens, the conduct and outcome of conflicts increasingly depend on information and communications. More than ever before, conflicts revolve around "knowledge"

---

[1]Our netwar concept predates, and should not be confused with, the U.S. military's network warfare simulation (NETWARS) system.

and the use of "soft power."[2] Adversaries are learning to emphasize "information operations" and "perception management"—that is, media-oriented measures that aim to attract or disorient rather than coerce, and that affect how secure a society, a military, or other actor feels about its knowledge of itself and of its adversaries. Psychological disruption may become as important a goal as physical destruction.

These propositions cut across the entire conflict spectrum. Major transformations are thus coming in the nature of adversaries, in the type of threats they may pose, and in how conflicts can be waged. Information-age threats are likely to be more diffuse, dispersed, multi-dimensional nonlinear, and ambiguous than industrial-age threats. Metaphorically, then, future conflicts may resemble the Oriental game of *Go* more than the Western game of chess. The conflict spectrum will be remolded from end to end by these dynamics.

A CONCEPT AND ITS BRIEF HISTORY

Back in 1992, while first wondering about such propositions and writing about *cyberwar* as a looming mode of military conflict, we thought it would be a good idea to have a parallel concept about information-age conflict at the less military, low-intensity, more social end of the spectrum. The term we coined was *netwar*, largely because it resonated with the surety that the information revolution favored the rise of network forms of organization, doctrine, and strategy. Through netwar, numerous dispersed small groups using the latest communications technologies could act conjointly across great distances. We had in mind actors as diverse as transnational terrorists, criminals, and even radical activists. Some were already moving from hierarchical to new information-age network designs.

We fielded the netwar concept in our first journal article, "Cyberwar Is Coming" (1993), then provided a full exposition in our RAND report, *The Advent of Netwar* (1996). Additional insights were advanced in the concluding chapter of our book, *In Athena's Camp* (1997). Elaborations appeared in multiauthored RAND volumes on *The Zapatista*

---

[2]The concept of soft power was introduced by Nye (1990), and further elaborated in Nye and Owens (1996).

"Social Netwar" in Mexico (Ronfeldt et al., 1998) and *Countering the New Terrorism* (Lesser et al., 1999). Our study *The Emergence of Noopolitik: Toward an American Information Strategy* (1999) observed that many socially minded nongovernmental organizations (NGOs) were already using netwar strategies to enhance their soft power. Our recent study *Swarming and the Future of Conflict* (2000) is mainly about developing a new military doctrine for wielding "hard" power, but it generally advances our view that swarming is likely to become the dominant approach to conflict across the spectrum, including among netwar actors. While the Zapatista study provided early evidence for this, short opinion pieces on the military war in Kosovo (1999) and the activist "Battle for Seattle"(1999) identified new cases.[3]

As these writings have spread, the netwar concept has struck a chord with a growing number of theorists, futurists, journalists, and practitioners. In forward-looking books, scholars as diverse as Manuel Castells (1997), Chris Hables Gray (1997), and David Brin (1998) have used the concept for discussing trends at the mostly nonmilitary end of the conflict spectrum. For several years, a web site maintained by Jason Wehling carried a wide range of articles about netwar, social activism, and information-age conflict, leading off with a paper he had written about the netwar concept (1995). Meanwhile, interesting flurries of discussion about netwar arose on email lists related to the Zapatista movement in Mexico following the armed uprising in January 1994. Harry Cleaver's writings (e.g., 1995, 1998, 1999) are particularly illuminating. They show that Mexico became a laboratory for the emergence of a new, non-Leninist model of radicalism. The Zapatista leader, Subcomandante Marcos, even averred in 1999 that netwar described the Zapatista movement, and that *counternetwar* instructed the strategy of its military and paramilitary opponents. For its part, the high command of the Mexican military also espoused admiration for the concept during 2000.[4] Also in 2000, a leader of the International Campaign to Ban Landmines (ICBL), Jody Williams, remarked in a

---

[3]John Arquilla and David Ronfeldt, "Need for Networked, High-Tech Cyberwar," *Los Angeles Times*, June 20, 1999, pp. A1, A6; John Arquilla and David Ronfeldt, "A Win for Netwar in Seattle," December 1999, posted on the web site for the Highlands Forum.

[4]Both the Zapatista and the Mexican army leadership had read the RAND report analyzing the Zapatista movement as a case of social netwar (Ronfeldt et al., 1998).

radio interview that she had heard that RAND researchers were developing the netwar concept to help governments control movements like the ICBL. Elsewhere, the concept cropped up in marginal rants and ruminations by militants associated with various left-wing, right-wing, and eclectic religious movements who posted on Usenet discussion groups.

Meanwhile, officials and analysts in U.S. and European government, military, and police circles began showing an interest in the concept. They were finding it difficult to deal with terrorists, criminals, and fanatics associated with militias and extremist single-issue movements, largely because these antagonists were organizing into sprawling, loose, "leaderless" networks, overcoming their former isolated postures as stand-alone groups headed by "great men." U.S. and European officials realized that these troublesome trends put a premium on interagency communication and coordination, for everything from intelligence sharing to tactical operations. But this implied a degree of cross-jurisdictional and international networking, especially for intelligence sharing, that is difficult for state hierarchies to accomplish. The concepts of netwar and counternetwar attracted some interest because they had a potential for motivating officials to build their own networks, as well as hybrids of hierarchies and networks, to deal with the networked organizations, doctrines, and strategies of their information-age adversaries. A special issue of the journal *Studies in Conflict and Terrorism* on "Netwar Across the Spectrum of Conflict" (1999) may have helped heighten awareness of this.[5]

Our formulation of the netwar concept has always emphasized the organizational dimension. But we have also pointed out that an organizational network works best when it has the right doctrinal, technological, and social dynamics. In our joint work, we have repeatedly insisted on this. However, writers enamored of the flashy, high-tech aspects of the information revolution have often depicted netwar (and cyberwar) as a term for computerized aggression waged via stand-off attacks in cyberspace—that is, as a trendy synonym for in-

---

[5]This special issue was partly assembled and edited by David Ronfeldt. Some text in this section comes from his introduction to that issue.

fowar, information operations, "strategic information warfare," Internet war, "hacktivism," cyberterrorism, cybotage, etc.[6]

Thus, in some quarters, the Serb hacks of NATO's web site in 1999 were viewed as netwar (or cyberwar). Yet, little was known about the perpetrators and the nature of their organization; if they amounted to just a few, clever, government-sponsored individuals operating from a site or two, then the netwar dimensions of this case were minimal, and it was just a clever instance of minor cybotage. This case also speaks to another distortion: These Serbs (presumably they were Serbs) aimed to bring a piece of "the Net" down. Yet, in a full-fledged ethnonationalist, terrorist, criminal, or social netwar, the protagonists may be far more interested in keeping the Net up. They may benefit from using the Internet and other advanced communications services (e.g., fax machines and cellular telephones) for purposes that range from coordinating with each other and seeking recruits, to projecting their identity, broadcasting their messages to target audiences, and gathering intelligence about their opponents.

With respect to Serbia, then, a better case of netwar as we define it was the effort by Serbia's reformist Radio B-92, along with a supportive network of U.S. and European government agencies and NGOs, to broadcast its reportage back into Serbia over the Internet, after B-92's transmitters were shut down by the Milosevic regime in 1998 and again in 1999. For a seminal case of a worldwide netwar, one need look no further than the ICBL. This unusually successful movement consists of a loosely internetted array of NGOs and governments, which rely heavily on the Internet for communications. Through the personage of one of its many leaders, Jody Williams, this netwar won a well-deserved Nobel peace prize.[7]

---

[6]For an interesting paper by a leading proponent of hacktivism, see Wray (1998).

[7]See speech by Jody Williams accepting the Nobel Peace Prize in 1997, www.waging peace.org/articles/nobel_lecture_97_williams.html; and the speech she gave at a gathering of recipients at the University of Virginia in 1998, www.virginia.edu/nobel/transcript/jwilliams.html, as well as Williams and Goose (1998).

DEFINING NETWAR[8]

To be precise, the term netwar refers to an emerging mode of conflict (and crime) at societal levels, short of traditional military warfare, in which the protagonists use network forms of organization and related doctrines, strategies, and technologies attuned to the information age. These protagonists are likely to consist of dispersed organizations, small groups, and individuals who communicate, coordinate, and conduct their campaigns in an internetted manner, often without a precise central command. Thus, netwar differs from modes of conflict and crime in which the protagonists prefer to develop formal, stand-alone, hierarchical organizations, doctrines, and strategies as in past efforts, for example, to build centralized movements along Leninist lines. Thus, for example, netwar is about the Zapatistas more than the Fidelistas, Hamas more than the Palestine Liberation Organization (PLO), the American Christian Patriot movement more than the Ku Klux Klan, and the Asian Triads more than the Cosa Nostra.[9]

The term *netwar* is meant to call attention to the prospect that network-based conflict and crime will become major phenomena in the decades ahead. Various actors across the spectrum of conflict and crime are already evolving in this direction. This includes familiar adversaries who are modifying their structures and strategies to take advantage of networked designs—e.g., transnational terrorist groups, black-market proliferators of weapons of mass destruction (WMD), drug and other crime syndicates, fundamentalist and ethnonationalist movements, intellectual-property pirates, and immigration and refugee smugglers. Some urban gangs, back-country militias, and militant single-issue groups in the United States have also been developing netwar-like attributes. The netwar spectrum also includes a new generation of revolutionaries, radicals, and activists who are beginning to create information-age ideologies, in which identities and

---

[8]This section reiterates but also updates our earlier formulations about the nature of netwar (notably those in Arquilla and Ronfeldt, 1996; Ronfeldt et al., 1998; and Arquilla, Ronfeldt, and Zanini, 1999). Readers who are already familiar with this work may prefer to skip this section.

[9]This is just a short exemplary statement. Many other examples could be noted. Instead of Hamas, for example, we might mention the Committee for the Defense of Legitimate Human Rights (CDLHR), an anti-Saudi organization based in London.

loyalties may shift from the nation state to the transnational level of "global civil society." New kinds of actors, such as anarchistic and nihilistic leagues of computer-hacking "cyboteurs," may also engage in netwar.

Many—if not most—netwar actors will be nonstate, even stateless. Some may be agents of a state, but others may try to turn states into *their* agents. Also, a netwar actor may be both subnational and transnational in scope. Odd hybrids and symbioses are likely. Furthermore, some bad actors (e.g., terrorist and criminal groups) may threaten U.S. and other nations' interests, but other actors (e.g., NGO activists in Burma or Mexico) may not—indeed, some actors who at times turn to netwar strategies and tactics, such as the New York–based Committee to Protect Journalists (CPJ), may have salutary liberalizing effects. Some actors may aim at destruction, but more may aim mainly at disruption and disorientation. Again, many variations are possible.

The full spectrum of netwar proponents may thus seem broad and odd at first glance. But there is an underlying pattern that cuts across all variations: *the use of network forms of organization, doctrine, strategy, and technology attuned to the information age.*

More About Organizational Design

In an archetypal netwar, the protagonists are likely to amount to a set of diverse, dispersed "nodes" who share a set of ideas and interests and who are arrayed to act in a fully internetted "all-channel" manner. In the scholarly literature (e.g., Evan, 1972), networks come in basically three types or topologies (see Figure 1.1):

- The *chain* or line network, as in a smuggling chain where people, goods, or information move along a line of separated contacts, and where end-to-end communication must travel through the intermediate nodes.

- The *hub*, star, or wheel network, as in a franchise or a cartel where a set of actors are tied to a central (but not hierarchical) node or actor, and must go through that node to communicate and coordinate with each other.

RAND *MR1382-1.1*

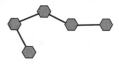

Chain network   Star or hub network   All-channel network

Figure 1.1—Three Basic Types of Networks

- The *all-channel* or full-matrix network, as in a collaborative network of militant peace groups where everybody is connected to
  everybody else.

Each node in the diagrams may refer to an individual, a group, an organization, part of a group or organization, or even a state. The nodes
may be large or small, tightly or loosely coupled, and inclusive or exclusive in membership. They may be segmentary or specialized—that
is, they may look alike and engage in similar activities, or they may
undertake a division of labor based on specialization. The boundaries
of the network, or of any node included in it, may be well-defined, or
blurred and porous in relation to the outside environment. Many
variations are possible.

Each type may be suited to different conditions and purposes, and all
three may be found among netwar-related adversaries—e.g., the
chain in smuggling operations; the hub at the core of terrorist and
criminal syndicates; and the all-channel type among militant groups
that are highly internetted and decentralized. There may also be hybrids of the three types, with different tasks being organized around
different types of networks. For example, a netwar actor may have an
all-channel council or directorate at its core but use hubs and chains
for tactical operations. There may also be hybrids of network and hierarchical forms of organization. For example, traditional hierarchies
may exist inside particular nodes in a network. Some actors may have
a hierarchical organization overall but use network designs for tactical operations; other actors may have an all-channel network design

overall but use hierarchical teams for tactical operations. Again, many configurations are possible, and it may be difficult for an analyst to discern exactly what type characterizes a particular network.

Of the three network types, the all-channel has been the most difficult to organize and sustain, partly because it may require dense communications. But it is the type that gives the network form its new, high potential for collaborative undertakings and that is gaining new strength from the information revolution. Pictorially, an all-channel netwar actor resembles a geodesic "Bucky ball" (named for Buckminster Fuller); it does not look like a pyramid. The organizational design is flat. Ideally, there is no single, central leadership, command, or headquarters—no precise heart or head that can be targeted. The network as a whole (but not necessarily each node) has little to no hierarchy; there may be multiple leaders. Decisionmaking and operations are decentralized, allowing for local initiative and autonomy. Thus the design may sometimes appear acephalous (headless), and at other times polycephalous (Hydra-headed).[10]

The capacity of this design for effective performance over time may depend on the existence of shared principles, interests, and goals—perhaps an overarching doctrine or ideology—which spans all nodes and to which the members subscribe in a deep way. Such a set of principles, shaped through mutual consultation and consensus-building, can enable members to be "all of one mind" even though they are dispersed and devoted to different tasks. It can provide a central ideational and operational coherence that allows for tactical decentralization. It can set boundaries and provide guidelines for decisions and actions so that the members do not have to resort to a hierarchy because "they know what they have to do."[11]

The network design may depend on having an infrastructure for the dense communication of functional information. This does not mean that all nodes must be in constant communication; that may not

---

[10]The structure may also be cellular. However, the presence of "cells" does not necessarily mean a network exists. A hierarchy can also be cellular, as is the case with some subversive organizations.

[11]The quotation is from a doctrinal statement by Beam (1992) about "leaderless resistance," which has strongly influenced right-wing white-power groups.

make sense for a secretive, conspiratorial actor. But when communication is needed, the network's members must be able to disseminate information promptly and as broadly as desired within the network and to outside audiences.

In many respects, then, the archetypal netwar design corresponds to what earlier analysts (Gerlach, 1987, p. 115, based on Gerlach and Hine, 1970) called a "segmented, polycentric, ideologically integrated network" (SPIN):

> By segmentary I mean that it is cellular, composed of many different groups. . . . By polycentric I mean that it has many different leaders or centers of direction. . . . By networked I mean that the segments and the leaders are integrated into reticulated systems or networks through various structural, personal, and ideological ties. Networks are usually unbounded and expanding. . . . This acronym [SPIN] helps us picture this organization as a fluid, dynamic, expanding one, spinning out into mainstream society.[12]

Caveats About the Role of Technology

Netwar is a result of the rise of network forms of organization, which in turn is partly a result of the computerized information revolution.[13] To realize its potential, a fully interconnected network requires a capacity for constant, dense information and communications flows, more so than do other forms of organization (e.g., hierarchies). This capacity is afforded by the latest information and communication technologies—cellular telephones, fax machines, electronic mail (email), web sites, and computer conferencing. Such technologies are highly advantageous for netwar actors whose constituents are geographically dispersed.

---

[12]The SPIN concept is a precursor of the netwar concept. Proposed by Luther Gerlach and Virginia Hine in the 1960s to depict U.S. social movements, it anticipates many points about network forms of organization, doctrine, and strategy that are now coming into focus in the analysis not only of social movements but also of some terrorist, criminal, ethnonationalist, and fundamentalist organizations.

[13]For explanation of this point, see Ronfeldt (1996), Arquilla and Ronfeldt (1996), and other sources cited in those documents.

But two caveats are in order. First, the new technologies, however enabling for organizational networking, are not absolutely necessary for a netwar actor. Older technologies, like human couriers, and mixes of old and new systems may do the job in some situations. The late Somali warlord, Mohamed Farah Aidid, for example, proved very adept at eluding those seeking to capture him while at the same time retaining full command and control over his forces by means of runners and drum codes (see Bowden, 1999). Similarly, the first Chechen War (1994–1996), which the Islamic insurgents won, made wide use of runners and old communications technologies like ham radios for battle management and other command and control functions (see Arquilla and Karasik, 1999). So, netwar may be waged in high-, low-, or no-tech fashion.

Second, netwar is not simply a function of "the Net" (i.e., the Internet); it does not take place only in "cyberspace" or the "infosphere." Some *battles* may occur there, but a *war's* overall conduct and outcome will normally depend mostly on what happens in the "real world"—it will continue to be, even in information-age conflicts, generally more important than what happens in cyberspace or the infosphere.[14]

Netwar is not solely about Internet war (just as cyberwar is not just about "strategic information warfare"). Americans have a tendency to view modern conflict as being more about technology than organization and doctrine. In our view, this is a misleading tendency. For example, social netwar is more about a doctrinal leader like Subcomandante Marcos than about a lone, wild computer hacker like Kevin Mitnick.

---

[14]This point was raised specifically by Paul Kneisel, "Netwar: The Battle over Rec.Music.White-Power," *ANTIFA INFO-BULLETIN*, Research Supplement, June 12, 1996, which is available on the Internet. He analyzes the largest vote ever taken about the creation of a new Usenet newsgroup—a vote to prevent the creation of a group that was ostensibly about white-power music. He concludes that "The *war* against contemporary fascism will be won in the 'real world' off the net; but *battles* against fascist netwar are fought and won on the Internet." His title is testimony to the spreading usage of the term *netwar*.

A Capacity for Swarming, and the Blurring of Offense and Defense

This distinctive, often ad-hoc design has unusual strengths, for both offense and defense. On the offense, networks tend to be adaptable, flexible, and versatile vis-à-vis opportunities and challenges. This may be particularly the case where a set of actors can engage in *swarming*. Little analytic attention has been given to swarming, [15] which is quite different from traditional mass- and maneuver-oriented approaches to conflict. Yet swarming may become the key mode of conflict in the information age (Arquilla and Ronfeldt, 2000, and Edwards, 2000), and the cutting edge for this possibility is found among netwar protagonists.

Swarming is a seemingly amorphous, but deliberately structured, co-ordinated, strategic way to strike from all directions at a particular point or points, by means of a sustainable pulsing of force and/or fire, close-in as well as from stand-off positions. This notion of "force and/or fire" may be literal in the case of military or police operations, but metaphorical in the case of NGO activists, who may, for example, be blocking city intersections or emitting volleys of emails and faxes. Swarming will work best—perhaps it will only work—if it is designed mainly around the deployment of myriad, small, dispersed, net-worked maneuver units. Swarming occurs when the dispersed units of a network of small (and perhaps some large) forces converge on a target from multiple directions. The overall aim is *sustainable pulsing*—swarm networks must be able to coalesce rapidly and stealthily on a target, then dissever and redisperse, immediately ready to re-combine for a new pulse. The capacity for a "stealthy approach" suggests that, in netwar, attacks are more likely to occur in "swarms" than in more traditional "waves." The Chechen resistance to the Russian army and the Direct Action Network's operations in the anti–World Trade Organization "Battle of Seattle" both provide excellent examples of swarming behavior.

---

[15]The first mention of "swarm networks" we encountered was in Kelly (1994). A recent discussion, really about "swarm intelligence" rather than swarm networks, is in Bonabeau, Dorigo, and Theraulaz (1999).

Swarming may be most effective, and difficult to defend against, where a set of netwar actors do not "mass" their forces, but rather engage in dispersion and "packetization" (for want of a better term). This means, for example, that drug smugglers can break large loads into many small packets for simultaneous surreptitious transport across a border, or that NGO activists, as in the case of the Zapatista movement, have enough diversity in their ranks to respond to any discrete issue that arises—human rights, democracy, the environment, rural development, whatever.

In terms of their defensive potential, networks tend to be redundant and diverse, making them robust and resilient in the face of attack. When they have a capacity for interoperability and shun centralized command and control, network designs can be difficult to crack and defeat as a whole. In particular, they may defy counterleadership targeting—a favored strategy in the drug war as well as in overall efforts to tamp organized crime in the United States. Thus, whoever wants to attack a network is limited—generally, only portions of a network can be found and confronted. Moreover, the deniability built into a network affords the possibility that it may simply absorb a number of attacks on distributed nodes, leading an attacker to believe the network has been harmed and rendered inoperable when, in fact, it remains viable and is seeking new opportunities for tactical surprise.

The difficulty of dealing with netwar actors deepens when the lines between offense and defense are blurred, or blended. When *blurring* is the case, it may be difficult to distinguish between attacking and defending actions, particularly where an actor goes on the offense in the name of self-defense. For example, the Zapatista struggle in Mexico demonstrates anew the blurring of offense and defense. The *blending* of offense and defense will often mix the strategic and tactical levels of operations. For example, guerrillas on the defensive strategically may go on the offense tactically, as in the war of the *mujahideen* in Afghanistan during the 1980s, and in both recent Chechen wars with the Russians.

Operating in the Seams

The blurring of offense and defense reflects another feature of netwar (albeit one that is exhibited in many other policy and issue areas): It tends to defy and cut across standard boundaries, jurisdictions, and distinctions between state and society, public and private, war and peace, war and crime, civilian and military, police and military, and legal and illegal. This makes it difficult if not impossible for a government to assign responsibility to any single agency—e.g., military, police, or intelligence—to be in charge of responding.

As Richard Szafranski (1994, 1995) illuminated in his discussions of how information warfare ultimately becomes "neo-cortical warfare," the challenge for governments and societies becomes "epistemological." A netwar actor may aim to confound people's fundamental beliefs about the nature of their culture, society, and government, partly to foment fear but perhaps mainly to disorient people and unhinge their perceptions. This is why a netwar with a strong social content—whether waged by ethnonationalists, terrorists, or social activists—may tend to be about disruption more than destruction. The more epistemological the challenge, the more confounding it may be from an organizational standpoint. Whose responsibility is it to respond? Whose roles and missions are at stake? Is it a military, police, intelligence, or political matter? When the roles and missions of defenders are not easy to define, both deterrence and defense may become problematic.

Thus, the spread of netwar adds to the challenges facing the nation state in the information age. Its sovereignty and authority are usually exercised through bureacracies in which issues and problems can be sliced up and specific offices can be charged with taking care of specific problems. In netwar, things are rarely so clear. A protagonist is likely to operate in the cracks and gray areas of a society, striking where lines of authority crisscross and the operational paradigms of politicians, officials, soldiers, police officers, and related actors get fuzzy and clash. Moreover, where transnational participation is strong, a netwar's protagonists may expose a local government to challenges to its sovereignty and legitimacy by arousing foreign governments and business corporations to put pressure on the local government to alter its domestic policies and practices.

NETWORKS VERSUS HIERARCHIES: CHALLENGES FOR COUNTERNETWAR

These observations and the case studies presented in this volume lead to four policy-oriented propositions about the information revolution and its implications for netwar and counternetwar (Arquilla and Ronfeldt, 1993, 1996):[16]

Hierarchies have a difficult time fighting networks. There are examples of this across the conflict spectrum. Some of the best are found in the failings of many governments to defeat transnational criminal cartels engaged in drug smuggling, as in Colombia. The persistence of religious revivalist movements, as in Algeria, in the face of unremitting state opposition, shows both the defensive and offensive robustness of the network form. The Zapatista movement in Mexico, with its legions of supporters and sympathizers among local and transnational NGOs, shows that social netwar can put a democratizing autocracy on the defensive and pressure it to continue adopting reforms.

It takes networks to fight networks. Governments that want to defend against netwar may have to adopt organizational designs and strategies like those of their adversaries. This does not mean mirroring the adversary, but rather learning to draw on the same design principles that he has already learned about the rise of network forms in the information age. These principles depend to some extent on technological innovation, but mainly on a willingness to innovate organizationally and doctrinally, perhaps especially by building new mechanisms for interagency and multijurisdictional cooperation.

Whoever masters the network form first and best will gain major advantages. In these early decades of the information age, adversaries who are advanced at networking (be they criminals, terrorists, or peaceful social activists, including ones acting in concert with states) are enjoying an increase in their power relative to state agencies. While networking once allowed them simply to keep from being suppressed, it now allows them to compete on more nearly equal terms with states and other hierarchically oriented actors. The histories of

---

[16]Also see Berger (1998) for additional observations about such propositions.

Hamas and of the Cali cartel illustrate this; so do the Zapatista movement in Mexico and the International Campaign to Ban Landmines.

Counternetwar may thus require very effective interagency approaches, which by their nature involve networked structures. It is not necessary, desirable, or even possible to replace all hierarchies with networks in governments. Rather, the challenge will be to blend these two forms skillfully, while retaining enough core authority to encourage and enforce adherence to networked processes. By creating effective hybrids, governments may become better prepared to confront the new threats and challenges emerging in the information age, whether generated by ethnonationalists, terrorists, militias, criminals, or other actors. (For elaboration, see Arquilla and Ronfeldt, 1997, Ch. 19.)

However, governments tend to be so constrained by hierarchical habits and institutional interests that it may take some sharp reverses before a willingness to experiment more seriously with networking emerges. The costs and risks associated with failing to engage in institutional redesign are likely to be high—and may grow ever higher over time. In the most difficult areas—crime and terrorism—steps to improve intra- and international networking are moving in the right direction. But far more remains to be done, as criminal and terrorist networks continuously remake themselves into ever more difficult targets.

RECENT CASES OF NETWAR

Since we first wrote about netwar over seven years ago, there have been at least ten prominent (i.e., front-page) instances of its employment, in conflicts ranging from social activist campaigns to violent ethnic insurgencies (see Table 1.1). The netwar record has been generally successful. In these ten cases, which feature networked non-state actors confronting states or groups of states, five netwars have achieved substantial success. Three have achieved limited success, while one (Burma) has yet to prove either a success or failure, and an-

Table 1.1

Prominent Cases of Netwar, 1994–2000

Campaign	Dates	Outcome	Type
Protracted Netwars			
EZLN[a]	1994–	Limited success	Autonomist
ICBL	1998–	Limited success	Globalist
Burma	1996–	Failing?	Mixed
Drug Cartels	1994–	Substantial success	Autonomist
Chechnya I	1994–1996	Substantial success	Autonomist
Chechnya II	1999–2000	Failure	Autonomist
Short-Duration Netwars			
Greenpeace	1994	Limited success	Globalist
Battle of Seattle	1999	Substantial success	Globalist
East Timor	1999	Substantial success	Autonomist
Serb Opposition	2000	Substantial success	Mixed

[a]Zapatista National Liberation Army.

other (Chechnya) must be judged, currently, as a failure.[17] Most of these cases, and the reasons for their success or the lack thereof, are discussed in detail in the following chapters.

The limits on some successes and the one failure imply a need to take a balanced view of netwar, analyzing the conditions under which it is most likely to succeed, fail, or fall somewhere in between. Clearly, there is enough success here to make netwar worth examining more closely. But it is important not to "tout" netwar, as Robert Taber (1970) once did guerrilla war. He was sharply rebutted by Lewis Gann (1970), who pointed out that guerrillas, far from being unstoppable, have of-

---

[17]Both Russo-Chechen conflicts are included as netwars, because of the extent to which the Chechens have relied upon networked forms of organization, both in field actions and in the struggle to win the "battle of the story." Arquilla and Karasik (1999) describe the Chechen victory in the 1994–1996 conflict as a clear triumph for networking but also posed concerns that the Russians would learn from this defeat—as they have learned from defeats throughout their history—and would improve, both in the field and in the arena of world perception. They have gotten better in the second conflict, driving the Chechens to their southern mountain redoubts and convincing state and nonstate actors around the world that Russian forces are fighting on behalf of a world community opposed to terrorism.

ten been defeated. Netwar will also have its ups and downs. Our purpose is to uncover and get a deeper understanding of its dynamics.

In Table 1.1, the cases are divided into those conflicts that were or have been drawn out, and those focused on specific crises—a useful distinction often made in studies of conflict. Interesting insights emerge. For example, the two most successful protracted campaigns were waged violently by ethnonationalists and criminals who sought freedom from state controls. The short-duration successes also included some use of violence (in two cases), and a global civil society reaction (that threatened a forceful response) to state violence in the other. And, though more muted, most of the other cases have violent aspects.

The table distributes netwars by type along a spectrum ranging from those that are globalist in orientation (e.g., the anti-landmine campaign), to those that are autonomist at the opposite end (e.g., the 1994 Chechen effort to secede from Russia). In the middle lie mixed cases where the objective is to gain power locally, but these netwars depend on the protagonists being able to open their societies to democratic, globalist influences.

The two unsuccessful netwar campaigns (in Russia and Burma) have featured networks confronting hierarchical authoritarian governments that have been willing to use substantial force to assert—in the case of Russia, to reassert—their hold on power. These networks' losses to hierarchies, combined with the fact that the principal successes to date have been gained by violent "uncivil society" actors, suggest being cautious about the claims for netwar. That said, the nonviolent International Campaign to Ban Landmines and the Greenpeace effort to curb nuclear testing both achieved reasonable measures of success without engaging in any violence whatsoever. This is a hopeful sign. And, while the civil society campaign to free Burma from authoritarian rule is a partial failure to date, this is a continuing campaign whose ultimate outcome is yet unknown.

Finally, these netwar conflicts feature an uneven split between those about globalist issues—aimed at fostering the rise of a rights- and ethics-based civil society—and the more frequent, somewhat darker "autonomist" variety of netwar, featuring nonstate actors trying to get out from under state controls. Most of the limited successes that have been achieved thus far are globalist in orientation, while most of the substantial successes (save for the Battle of Seattle and Serbia) have been autonomist. It will be interesting, as the instances of netwar increase over time, to see whether this pattern holds. The outcomes of the globalist cases suggest the prevalence of negotiated solutions, while the autonomist conflicts may, in general, have a much more inherently desperate character that drives them to greater violence and less willingness to reach accommodation. All this we will watch in the years to come. For now, these early cases have helped us to develop this taxonomy of netwar, further refining the concept.

Will netwar continue to empower nonstate actors, perhaps reducing the relative power advantage enjoyed by nation states? Civil society networks have already made much use of social netwar as a tool for advancing a globalist, ethics-based agenda focused on broadening and deepening human rights regimes—often in the context of an ongoing effort to foster movement from authoritarian rule to democracy (e.g., Burma). But there is another side of nonstate-actor-oriented netwar, characterized not by globalist impulses, but rather by the desire to avoid state control of a network's criminal, terrorist, or ethnic-separatist agenda (e.g., Hamas and Chechens). While the globalist netwars seem devoted to nonviolent tools of struggle, the autonomists may employ both means of engagement—often with a greater emphasis on violence.

VARIETIES OF NETWAR—DUAL PHENOMENA

Netwar is a deduced concept—it derives from our thinking about the effects and implications of the information revolution. Once coined, the concept helps show that evidence is mounting about the rise of network forms of organization, and about the importance of "information strategies" and "information operations" across the spectrum of conflict, including among ethnonationalists, terrorists, guerrillas, criminals, and activists.[18] Note that we do not equate ethnonational-

---

[18]These are not the only types of netwar actors; there are others. For example, corporations may also engage in netwars—or find themselves on the receiving end of netwar campaigns.

ists, terrorists, guerrillas, criminals, and activists with each other—each has different dynamics. Nor do we mean to tarnish social activism, which has positive aspects for civil society.[19] We are simply calling for attention to a cross-cutting meta-pattern about network forms of organization, doctrine, and strategy that we might not have spotted, by induction or deduction, if we had been experts focused solely on any one of those areas.

Netwar can be waged by "good" as well as "bad" actors, and through peaceful as well as violent measures. From its beginnings, netwar has appealed to a broad cross-section of nonstate actors who are striving to confront or cope with their state authorities. Ethnonationalists, criminals, and terrorists—all have found new power in networking. But so too have emerging global civil society actors who have emphasized nonviolent efforts to win the "battle of the story"—a more purely informational dimension of netwar—rather than the violent swarming characteristic of its darker side. Both categories of actors seem to realize, even if only implicitly, that, in the future, conflict will become even more "irregularized," with the set-piece confrontations and battles of earlier eras largely disappearing. While the U.S. military remains focused—in terms of budgetary emphasis, doctrine, and force structure—on the traditional forms of conflict, the rise of netwar should prompt a shift to a nimble "turn of mind," one far less attuned to fighting in the Fulda Gap or the Persian Gulf and more focused on engaging a range of odd new adversaries across a densely interconnected "global grid."

The duality of netwar in the real world—dark-side criminals and terrorists on the one hand, but enlightening civil society forces on the other—is mirrored in the virtual world of cyberspace, which is increasingly utilized for crime and terror (still embryonic), along with social activism. At present, social activism is far more robust and established in the cyber realm than is crime or terror. Will this continue to be the case? We think so. Activists will become more adept at integrating the mobilizing force of the Internet with the power and appeal of messages aimed at spreading and protecting human rights. Even

---

[19]See discussion in Ronfeldt (1996).

so, criminal and terrorist organizations will learn how to manipulate the infosphere with increasing skill.

Thus, netwar has two faces, like the Roman god Janus. Janus was the god of doors and gates, and thus of departures and returns, and new beginnings and initiatives. This, in a sense, meant he was the god of communications, too. His double face, one old and looking back, the other younger and peering forward, conveyed that he was an inherently dual god. At the beginning of creation, he partook in the separation of order from chaos. In Roman times, he was identified with the distinction between war and peace, for the gate to his temple at the Forum was kept ceremoniously closed in times of peace and open in times of war—which meant the gates were rarely closed. At the start of the 21st century, the world is again at a new beginning. It is uncertain whether it will be an era of peace or conflict; but how matters turn out will depend to some degree on which face of netwar predominates.

This volume explores the two faces of netwar, in three parts. The first part is composed of three chapters that chronicle the increasingly networked nature of major types of "uncivil-society" actors for whom violence is a principal mode of expression. The analyses by Michele Zanini and Sean Edwards of Arab terrorist groups, by Phil Williams of transnational criminal networks, and by John Sullivan of street-level gangs and hooligans, all speak to the increasingly sophisticated usage of the new information technologies to enhance both these groups' organizational and operational capabilities.

The second part of the book examines the rise of social netwar, again with three chapters. These chapters examine social netwars waged by networked civil society actors against various types of states. Tiffany Danitz and Warren Strobel show the limitations (but also some successful facets) of social netwar when waged against a resolute dictatorship that maintains a system virtually closed to civil society. Our own chapter on Mexico finds that an "NGO swarm" was quite effective in transforming a rural insurgency into a mostly peaceable netwar in a then rather authoritarian system. Paul de Armond provides insights into the full mobilizing potential of social netwar when conducted in a free society like the United States.

The final part considers the future of netwar, particularly regarding how technology, organization, and doctrine interact. Dorothy Denning assesses whether activists, hacktivists, or cyberterrorists may gain the most influence from exploiting the new information technologies. Luther Gerlach's chapter, though focused on environmental activism, identifies the dynamics of organizations that are segmentary, polycentric, and integrated as a network—from leaderlessness to operational fluidity. We think these dynamics apply, in varying degrees, to all the types of actors examined in the first two parts of the book. Our concluding chapter addresses likely trends in both the theory and practice of netwar—from how to draw on academic theories about networks, to how to think strategically about netwar itself. Thus, Part III should make the reader aware of both the perils and the promises of netwar, while also providing analytical guideposts for future studies of this phenomenon.

BIBLIOGRAPHY

Arquilla, John, and Theodore Karasik, "Chechnya: A Glimpse of Future Conflict?" *Studies in Conflict and Terrorism*, Vol. 22, No. 3, July–September 1999, pp. 207–230.

Arquilla, John, and David Ronfeldt, "Cyberwar Is Coming!" *Comparative Strategy*, Vol. 12, No. 2, Summer 1993, pp. 141–165. Available as RAND reprint RP-223.

Arquilla, John, and David Ronfeldt, *The Advent of Netwar*, Santa Monica, Calif.: RAND, MR-789-OSD, 1996.

Arquilla, John, and David Ronfeldt, *The Emergence of Noopolitik: Toward an American Information Strategy*, Santa Monica, Calif.: RAND, MR-1033-OSD, 1999.

Arquilla, John, and David Ronfeldt, *Swarming and the Future of Conflict*, Santa Monica, Calif.: RAND, DB-311-OSD, 2000.

Arquilla, John, and David Ronfeldt, eds., *In Athena's Camp: Preparing for Conflict in the Information Age*, Santa Monica, Calif.: RAND, MR-880-OSD/RC, 1997.

Arquilla, John, David Ronfeldt, and Michele Zanini, "Information-Age Terrorism and the U.S. Air Force," in Ian O. Lesser et al., *Countering the New Terrorism*, Santa Monica, Calif.: RAND, MR-989-AF, 1999.

Beam, Louis, "Leaderless Resistance," *The Seditionist*, Issue 12, February 1992 (text can also be located sometimes on the web).

Berger, Alexander, *Organizational Innovation and Redesign in the Information Age: The Drug War, Netwar, and Other Low-End Conflict*, master's thesis, Monterey, Calif.: Naval Postgraduate School, 1998.

Bonabeau, Eric, Marco Dorigo, and Guy Theraulaz, *Swarm Intelligence: From Natural to Artificial Systems*, Oxford: Oxford University Press, 1999.

Bowden, Mark, *Blackhawk Down: A Story of Modern War*, New York: Atlantic Monthly Press, 1999.

Brin, David, *The Transparent Society: Will Technology Force Us to Choose Between Privacy and Freedom?* Reading, Mass.: Addison-Wesley, 1998.

Castells, Manuel, *The Information Age: Economy, Society and Culture*, Vol. II, *The Power of Identity*, Malden, Mass.: Blackwell Publishers, 1997.

Cleaver, Harry, "The Zapatistas and the Electronic Fabric of Struggle," 1995, www.eco.utexas.edu/faculty/Cleaver/zaps.html, printed in John Holloway and Eloina Pelaez, eds., *Zapatista! Reinventing Revolution in Mexico*, Sterling, Va.: Pluto Press, 1998, pp. 81–103.

Cleaver, Harry, "The Zapatista Effect: The Internet and the Rise of an Alternative Political Fabric," *Journal of International Affairs*, Vol. 51, No. 2, Spring 1998, pp. 621–640.

Cleaver, Harry, *Computer-Linked Social Movements and the Global Threat to Capitalism*, July 1999, www.eco.utexas.edu/faculty/Cleaver/polnet.html.

Edwards, Sean J.A., *Swarming on the Battlefield: Past, Present and Future*, Santa Monica, Calif.: RAND, MR-1100-OSD, 2000.

Evan, William M., "An Organization-Set Model of Interorganizational Relations," in Matthew Tuite, Roger Chisholm, and Michael Radnor, eds., *Interorganizational Decisionmaking*, Chicago: Aldine Publishing Company, 1972, pp. 181–200.

Gann, Lewis, *Guerrillas in History*, Stanford, Calif.: Hoover Institution Press, 1970.

Gerlach, Luther P., "Protest Movements and the Construction of Risk," in B. B. Johnson and V. T. Covello, eds., *The Social and Cultural Construction of Risk*, Boston: D. Reidel Publishing Co., 1987, pp. 103–145.

Gerlach, Luther P., and Virginia Hine, *People, Power, Change: Movements of Social Transformation*, New York: The Bobbs-Merrill Co., 1970.

Gray, Chris Hables, *Postmodern War: The New Politics of Conflict*, New York: The Guilford Press, 1997.

Kelly, Kevin, *Out of Control: The Rise of Neo-Biological Civilization*, New York: A William Patrick Book, Addison-Wesley Publishing Company, 1994.

Lesser, Ian O., Bruce Hoffman, John Arquilla, David Ronfeldt, Michele Zanini, and Brian Jenkins, *Countering the New Terrorism*, Santa Monica, Calif.: RAND, MR-989-AF, 1999.

Nye, Joseph S., *Bound to Lead: The Changing Nature of American Power*, New York: Basic Books, 1990.

Nye, Joseph S., and William A. Owens, "America's Information Edge," *Foreign Affairs*, Vol. 75, No. 2, March/April 1996, pp. 20–36.

Ronfeldt, David, *Tribes, Institutions, Markets, Networks—A Framework About Societal Evolution*, Santa Monica, Calif.: RAND, P-7967, 1996.

Ronfeldt, David, John Arquilla, Graham Fuller, and Melissa Fuller, *The Zapatista "Social Netwar" in Mexico*, Santa Monica, Calif.: RAND, MR-994-A, 1998.

Szafranski, Colonel Richard, "Neo-Cortical Warfare? The Acme of Skill," *Military Review*, November 1994, pp. 41–55.

Szafranski, Colonel Richard, "A Theory of Information Warfare: Preparing for 2020," *Airpower Journal*, Spring 1995, pp. 56–65.

Taber, Robert, *The War of the Flea*, New York: Citadel, 1970.

Toffler, Alvin, and Heidi Toffler, *War and Anti-War: Survival at the Dawn of the Twenty-First Century*, Boston: Little, Brown and Company, 1993.

Van Creveld, Martin, *The Transformation of War*, New York: Free Press, 1991.

Wehling, Jason, *"Netwars" and Activists Power on the Internet*, March 25, 1995—as circulated on the Internet (and once posted at www.teleport.com/~jwehling/OtherNetwars.html).

Williams, Jody, and Stephen Goose, "The International Campaign to Ban Landmines," in Maxwell A. Cameron, Robert J. Lawson, and Brian W. Tomlin, eds., *To Walk Without Fear: The Global Movement to Ban Landmines*, New York: Oxford University Press, 1998, pp. 20–47.

Wray, Stefan, *Electronic Civil Disobedience and the World Wide Web of Hacktivism: A Mapping of Extraparliamentarian Direct Action Net Politics*, paper for a conference on The World Wide Web and Contemporary Cultural Theory, Drake University, November 1998, www.nyu.edu/projects/wray/wwwhack.html.

Part I

VIOLENCE-PRONE NETWARS

Chapter Two

THE NETWORKING OF TERROR IN THE INFORMATION AGE

Michele Zanini and Sean J.A. Edwards

Editors' abstract. Middle East Arab terrorists are on the cutting edge of organizational networking and stand to gain significantly from the information revolution. They can harness information technology to enable less hierarchical, more networked designs—enhancing their flexibility, responsiveness, and resilience. In turn, information technology can enhance their offensive operational capabilities for the war of ideas as well as for the war of violent acts. Zanini and Edwards (both at RAND) focus their analysis primarily on Middle East terrorism but also discuss other groups around the world. They conclude with a series of recommendations for policymakers. This chapter draws on RAND research originally reported in Ian Lesser et al., Countering the New Terrorism *(1999).*

INTRODUCTION

The information revolution has fueled the longest economic expansion in U.S. history and led to impressive productivity gains in recent years. Along with these benefits, however, has come the dark side of information technology—cyberterrorism. The idea of terrorists surreptitiously hacking into a computer system to introduce a virus, steal sensitive information, deface or swamp a web site, or turn off a crucial public service seriously concerns computer security personnel around the world. High profile attacks—such as the denial-of-service (DOS) attacks against major e-commerce sites Yahoo! and eBay in 1999 or the ongoing "cyber-jihad" against Israeli and American web sites being waged by Pakistani-based hackers in support of the Palestinian "al-Aqsa" Intifadah—continue to raise the specter of cyber-terrorism.

The information age is affecting not only the types of targets and weapons terrorists choose, but also the ways in which such groups operate and structure their organizations. Several of the most danger-ous terrorist organizations are using information technology (IT)—such as computers, software, telecommunication devices, and the In-ternet—to better organize and coordinate dispersed activities. Like the large numbers of private corporations that have embraced IT to operate more efficiently and with greater flexibility, terrorists are har-nessing the power of IT to enable new operational doctrines and forms of organization. And just as companies in the private sector are forming alliance networks to provide complex services to customers, so too are terrorist groups "disaggregating" from hierarchical bureau-cracies and moving to flatter, more decentralized, and often changing webs of groups united by a common goal.

The rise of networked terrorist groups is part of a broader shift to what Arquilla and Ronfeldt have called "netwar."[1] Netwar refers to an emerging mode of conflict and crime at societal levels, involving measures short of traditional war in which the protagonists are likely to consist of dispersed, small groups who communicate, coordinate, and conduct their campaigns in an internetted manner, without a precise central command. Netwar differs from modes of conflict in which the actors prefer formal, stand-alone, hierarchical organiza-tions, doctrines, and strategies, as in past efforts, for example, to build centralized revolutionary movements along Marxist lines.

This chapter assesses the degree to which—and how—networked ter-rorist groups are using IT, particularly in the Middle East. The analysis reviews past trends and offers a series of educated guesses about how such trends will evolve in the future. The first section discusses the or-ganizational implications of netwar, especially the degree to which IT is enabling different forms of terrorist structures and command, con-trol, and communications (C3). The second section examines past ev-

---

[1]The netwar concept is explained and discussed more thoroughly in Chapter One of this volume.

idence of terrorist use of IT for offensive netwar, such as destructive and disruptive attacks on information systems and for perception management. The third section contains a speculative look at how future terrorist uses of IT could develop in the near to medium term. The final section concludes with implications for counterterrorism policy.

ORGANIZATIONAL NETWORKING AND TECHNOLOGY ACQUISITION

In an archetypal netwar, the protagonists are likely to amount to a set of diverse, dispersed "nodes" who share a set of ideas and interests and who often are arrayed to act in a fully internetted "all-channel" manner. The potential effectiveness of the networked design compared to traditional hierarchical designs attracted the attention of management theorists as early as the 1960s.[2] Today, in the business world, virtual or networked organizations are heralded as effective alternatives to traditional bureaucracies because of their inherent flexibility, adaptiveness, and ability to capitalize on the talents of all of their members.

Networked organizations share three basic sets of features. First, communication and coordination are not formally specified by horizontal and vertical reporting relationships, but rather emerge and change according to the task at hand. Similarly, relationships are often informal and marked by varying degrees of intensity, depending on the needs of the organization. Second, internal networks are usually complemented by linkages to individuals outside the organization, often spanning national boundaries. Like internal connections, external relationships are formed and wind down according to the life cycle of particular joint projects. Third, both internal and external ties are enabled not by bureaucratic fiat, but rather by shared norms and

---

[2]In 1961, Burns and Stalker referred to the *organic* form as "a network structure of control, authority, and communication," with "lateral rather than vertical direction of communication." In organic structure,

> omniscience [is] no longer imputed to the head of the concern; knowledge about the technical or commercial nature of the here and now task may be located anywhere in the network; [with] this location becoming the ad hoc centre of control authority and communication.

values, as well as by reciprocal trust. Internally, the bulk of the work is conducted by self-managing teams, while external linkages compose "a constellation involving a complex network of contributing firms or groups" (Monge and Fulk, 1999, pp. 71–72).

The Emergence of Networked Terrorist Groups in the Greater Middle East

What has been emerging in the business world is now becoming apparent in the organizational structures of the newer and more active terrorist groups, which appear to be adopting decentralized, flexible network structures. The rise of networked arrangements in terrorist organizations is part of a wider move away from formally organized, state-sponsored groups to privately financed, loose networks of individuals and subgroups that may have strategic guidance but that, nonetheless, enjoy tactical independence.

For example, in the Greater Middle East, terrorist organizations have diverse origins, ideologies, and organizational structures but can be categorized roughly into traditional and new-generation groups. Traditional groups date to the late 1960s and early 1970s, and the majority were (and some still are) formally or informally linked to the Palestine Liberation Organization (PLO). Typically, they are also relatively bureaucratic and maintain a nationalist or Marxist agenda.[3] These groups have utilized autonomous cells as part of their organizational structure, but the operation of such cells is guided by a hierarchy through clear reporting relationships and virtually little horizontal coordination.

In contrast, the newer and less hierarchical groups (such as Hamas; the Palestinian Islamic Jihad; Hizbollah; Algeria's Armed Islamic Group; the Egyptian Islamic Group; and Osama bin Laden's terrorist

---

[3]The traditional, more bureaucratic groups have survived partly through support from states such as Syria, Libya, and Iran. These groups—such as the Abu Nidal Organization, the Popular Front for the Liberation of Palestine (PFLP), and three PFLP-related splinters (the PFLP-General Command, the Palestine Liberation Front, and the Democratic Front for the Liberation of Palestine)—retain an ability to train and prepare for terrorist missions; however, their involvement in actual operations has been limited in recent years, partly because of counterterrorism campaigns by Israeli and Western agencies and the ongoing peace process.

network, al-Qaeda) have become the most active organizations (Office of the Coordinator for Counterterrorism, 2000). In these loosely organized groups with religious or ideological motives, operatives are part of a network that relies less on bureaucratic fiat and more on shared values and horizontal coordination mechanisms to accomplish its goals.

The new and more active generation of Middle Eastern groups has operated both inside and outside the region. For instance, in Israel and the occupied territories, Hamas and to a lesser extent the Palestinian Islamic Jihad have demonstrated their strength over the last five years with a series of suicide bombings that have killed more than 100 people. In Egypt, the Islamic Group (also known as al-Gama'a al-Islamiya) carried out a 1997 attack at Luxor, killing 58 tourists and four Egyptians. Another string of terrorist attacks (and foiled attempts) has focused attention on a loosely organized group of "Arab Afghans"—radical Islamic fighters from several North African and Middle Eastern countries who have forged ties while resisting the Soviet occupation of Afghanistan. One of the leaders and founders of the Arab Afghan movement is Osama bin Laden, a Saudi entrepreneur based in Afghanistan.[4]

To varying degrees, these groups share the principles of the networked organization—relative flatness, decentralization and delegation of decisionmaking authority, and loose lateral ties among dispersed groups and individuals. Hamas, for example, is loosely structured with

> some elements working clandestinely and others working openly
> through mosques and social service institutions to recruit members,
> raise money, organize activities, and distribute propaganda (Office
> of the Coordinator for Counterterrorism, 2000, p. 74).

---

[4]Bin Laden allegedly sent operatives to Yemen to bomb a hotel used by American soldiers on their way to Somalia in 1992, plotted to assassinate President Bill Clinton in the Philippines in 1994 and Egyptian President Hosni Mubarak in 1995, and played a role in the Riyadh and Khobar blasts in Saudi Arabia that resulted in the deaths of 24 Americans in 1995 and 1996. U.S. officials have also pointed to bin Laden as the mastermind behind the American embassy bombings in Kenya and Tanzania in 1998, which claimed the lives of more than 260 people, including 12 Americans, and in the bombing of the *U.S.S. Cole* in Yemen, in which 17 American sailors were killed.

The pro-Iranian Hizbollah in southern Lebanon acts as an umbrella organization of radical Shiite groups and in many respects is a hybrid of hierarchical and network arrangements—although the organizational structure is formal, interactions among members are volatile and do not follow rigid lines of control (Ranstorp, 1994, p. 304).

Perhaps the most interesting example of a terrorist netwar actor is Osama bin Laden's complex network of relatively autonomous groups that are financed from private sources. Bin Laden uses his wealth and organizational skills to support and direct al-Qaeda (The Base), a multinational alliance of Islamic extremists. Al-Qaeda seeks to counter any perceived threats to Islam—wherever they come from— as indicated by bin Laden's 1996 declaration of a holy war against the United States and the West in general. In the declaration, bin Laden specified that such a holy war was to be waged by irregular, light, highly mobile forces. Although bin Laden finances al-Qaeda (exploiting a fortune of several million dollars, according to U.S. State Department estimates) and directs some operations, he apparently does not play a direct command-and-control role over all operatives. Rather, he is a key figure in the coordination and support of several dispersed nodes.[5]

There are reports that communications between al-Qaeda's members combine elements of a "hub-and-spoke" structure (where nodes of operatives communicate with bin Laden and his close advisers in Afghanistan) and a wheel structure (where nodes in the network communicate with each other without reference to bin Laden) (Simon and Benjamin, 2000, p. 70). Al-Qaeda's command-and-control structure includes a consultation council ("majlis al shura"), which discusses and approves major undertakings, and possibly a military committee.[6] At the heart of al-Qaeda is bin Laden's inner core group, which sometimes conducts missions on its own. Most of the other

---

[5]It is important to avoid equating the bin Laden network solely with bin Laden. He represents a key node in the Arab Afghan terror network. But the network conducts many operations without his involvement, leadership, or financing and will continue to be able to do so should he be killed or captured.

[6]See indictment testimony from U.S. District Court, Southern District of New York, *United States of America vs. Osama bin Laden et al.*, 98 Cr. and S(2) 98 Cr. 1023 (LBS) (www.library.cornell.edu/colldev/mideast/usavhage.htm).

member organizations remain independent, although the barriers between them are permeable. According to U.S. District Court testimony in New York, al-Qaeda has forged alliances with Egypt's Islamic Group (leading to an alleged influx of bin Laden operatives into its structure), the National Front in the Sudan, the government of Iran, and Hizbollah. Media reports also indicate that bin Laden has ties with other far-flung Islamic armed groups, such as Abu Sayyaf in the Philippines, as well as with counterparts in Somalia, Chechnya, and Central Asia.[7]

Command, Control, Communications, and the Role of IT

Lateral coordination mechanisms facilitate the operations of networked groups. In turn, such coordination mechanisms are enabled by advances in information technology—including increases in the speed of communication, reductions in the costs of communication, increases in bandwidth, vastly expanded connectivity, and the integration of communication and computing technologies (see Heydebrand, 1989). More specifically, new communication and computing technologies allow the establishment of networks in three critical ways (Monge and Fulk, 1999, p. 84).

First, new technologies have greatly reduced transmission time, enabling dispersed organizational actors to communicate and coordinate their tasks. This phenomenon is not new—in the early 20th century, the introduction of the telephone made it possible for large corporations to decentralize their operations through local branches.

Second, new technologies have significantly reduced the cost of communication, allowing information-intensive organizational designs such as networks to become viable.[8] As Thompson (1967) observed,

---

[7]See, for instance, Kurlantzick, 2000, and FBIS, 1997a and 1997b.

[8]The current IT revolution has not only increased the capacity and speed of communications networks, it has driven down telephone communication costs as well. The value and benefit of the Internet also rise as more servers and users link together online. Because the value of a network grows roughly in line with the square of the number of users, the benefit of being online increases exponentially with the number of connections (called Metcalfe's Law, attributed to Robert Metcalfe, a pioneer of computer networking). The number of users worldwide has already climbed to more than 350 million and may reach 1 billion within four years. See "Untangling e-conomics," *The Economist*, September 23, 2000.

in the past, organizations sought to reduce coordination and communications costs by centralizing and colocating those activities that are inherently more coordination-intensive. With the lowering of coordination costs, it is becoming increasingly possible to further disaggregate organizations through decentralization and autonomy.

Third, new technologies have substantially increased the scope and complexity of the information that can be shared, through the integration of computing with communications. Such innovations as tele- and computer conferencing, groupware, Internet chat, and web sites allow participants to have "horizontal" and rich exchanges without requiring them to be located in close proximity.

Thus, information-age technologies are highly advantageous for a netwar group whose constituents are geographically dispersed or carry out distinct but complementary activities.[9] IT can be used to plan, coordinate, and execute operations. Using the Internet for communication can increase speed of mobilization and allow more dialogue between members, which enhances the organization's flexibility, since tactics can be adjusted more frequently. Individuals with a common agenda and goals can form subgroups, meet at a target location, conduct terrorist operations, and then readily terminate their relationships and redisperse.

The bin Laden network appears to have adopted information technology to support its networked mode of operations. According to reporters who visited bin Laden's headquarters in a remote mountainous area of Afghanistan, the terrorist financier has modern computer and communications equipment. Bin Laden allegedly uses satellite phone terminals to coordinate the activities of the group's dispersed operatives and has even devised countermeasures to ensure his safety while using such communication systems.[10] Satellite phones reportedly travel in separate convoys from bin Laden's; the Saudi financier

---

[9]This is not to say that hierarchical terrorist groups will not also adopt IT to improve support functions and internal command, control, and communications. Aum Shinrikyo was highly centralized around the figure of Shoko Asahara and its structure was cohesive and extremely hierarchical; yet the use the IT was widespread within the group. See Cameron, 1999, p. 283.

[10]Afghanistan's ruling Taliban leaders have repeatedly claimed that bin Laden's movements and access to communications have been severely restricted.

also refrains from direct use, often dictating his message to an assistant, who then relays it telephonically from a different location. Bin Laden's operatives have used CD-ROM disks to store and disseminate information on recruiting, bomb making, heavy weapons, and terrorist operations.[11] Egyptian computer experts who fought alongside bin Laden in the Afghan conflict are said to have helped him devise a communications network that relies on the web, email, and electronic bulletin boards so that members can exchange information (FBIS, 1995).

This is a trend found among other terrorist actors in the Middle East. Counterterrorist operations targeting Algerian Armed Islamic Group (GIA) bases in the 1990s uncovered computers and diskettes with instructions for the construction of bombs (FBIS, 1996a). In fact, it has been reported that the GIA makes heavy use of floppy disks and computers to store and process orders and other information for its members, who are dispersed in Algeria and Europe (FBIS, 1996b). The militant Islamic group Hamas also uses the Internet to share and communicate operational information. Hamas activists in the United States use chat rooms to plan operations and activities. Operatives use email to coordinate actions across Gaza, the West Bank, and Lebanon. Hamas has realized that information can be passed relatively securely over the Internet because counterterrorism intelligence cannot accurately monitor the flow and content of all Internet traffic. In fact, Israeli security officials cannot easily trace Hamas messages or decode their content (more on this below).

In addition, terrorist networks can protect their vital communication flows through readily available commercial technology, such as encryption programs. Examples from outside the Middle East point in this direction—according to one report, Animal Liberation Front (ALF) cells in North America and Europe use the encryption program Pretty Good Privacy (PGP) to send coded email and share intelligence (Iuris, 1997, p. 64). New encryption programs emerging on the commercial market are becoming so sophisticated that coded emails may soon be extremely difficult to break. In fact, strong encryption pro-

---

[11]U.S. intelligence agencies recently obtained computer-disk copies of a six-volume training manual used by bin Laden to train his recruits (Kelley, 2000).

grams are being integrated into commercial applications and network protocols so that soon encryption will be easy and automatic (see Denning and Baugh, 1997). Rumors persist that the French police have been unable to decrypt the hard disk on a portable computer belonging to a captured member of the Spanish/Basque organization ETA (Fatherland and Liberty) (Denning and Baugh, 1997). It has also been suggested that Israeli security forces were unsuccessful in their attempts at cracking the codes used by Hamas to send instructions for terrorist attacks over the Internet (Whine, 1999, p. 128). Terrorists can also use steganography—a method of hiding secret data in other data such as embedding a secret message within a picture file (Denning and Baugh, 1997). Terrorists can also encrypt cell phone transmissions, steal cell phone numbers and program them into a single phone, or use prepaid cell phone cards purchased anonymously to keep their communications secure.[12]

The latest communications technologies are thus enabling terrorists to operate from almost any country in the world, provided they have access to the necessary IT infrastructure; and this affects the ways in which groups rely on different forms of sponsorship. Some analysts have argued that networked terrorists may have a reduced need for state support—indeed, governmental protection may become less necessary if technologies such as encryption allow a terrorist group to operate with a greater degree of stealth and safety (Soo Hoo, Goodman, and Greenberg, 1997, p. 142). Others point to the possibility that groups will increasingly attempt to raise money on the web, as in the case of Pakistan's Lashkar-e-Taiba ("Army of the Pure").[13]

---

[12]Cloned cell phones can either be bought in bulk (the terrorist discards the phone after use) or a phone number can be stolen and programmed into a single cell phone just before using it. A special scanner is used to "snatch" legitimate phone numbers from the airwaves, i.e., the Electronic Serial Number (ESN) and Mobile Identification Number (MIN). See Denning and Baugh, 1997.

[13]Lashkar and its parent organization, Markaz-e-Dawa wal Irshad (Center for Islamic Invitation and Guidance), have raised so much money, mostly from sympathetic Wahhabis in Saudi Arabia, that they are reportedly planning to open their own bank. See Stern, 2000.

Networked Organizations and IT: Mitigating Factors

To be sure, there are limits to how much reliance terrorist networks will place on information-age technology. For the foreseeable future, electronically mediated coordination will not be able to entirely supplant face-to-face exchanges, because uncertainty and risk will continue to characterize most organizational choices and interactions among individuals.[14] Moreover, informal linkages and the shared values mentioned above—which are critical enablers of networked designs—can only be fostered through personal contact. As Nohria and Eccles argue,

> electronically mediated exchange can increase the range, amount, and velocity of information flow in a network organization. But the viability and effectiveness of this electronic network will depend critically on an underlying network of social relationships based on face-to-face interaction (Nohria and Eccles, "Face-to-Face: Making Network Organizations Work," in Nohria and Eccles, 1992, pp. 289–290).

Moreover, while IT-enabled communication flows can greatly help a network coordinate dispersed activities (thus increasing its flexibility and responsiveness), they can also present a security risk. Communication over electronic channels can become a liability, since it leaves digital "traces." For instance, FBI officials have recently acknowledged that they used an Internet wiretap program called "Carnivore" to track terrorist email correspondence at least 25 times. According to *Newsweek*, Carnivore's ability to track Osama bin Laden's email was critical in thwarting several of his strikes.[15]

The case of Ramzi Yousef, the World Trade Center bomber, also provides a revealing example of how information-age technology can represent a double-edged sword for terrorists. Yousef's numerous calls to fellow terrorists during his preparation for the strike were registered in phone companies' computer databases, providing law en-

---

[14]In fact, ambiguous and complex situations are still better tackled through direct communications, because face-to-face interaction is generally faster at resolving outstanding issues and leaves less room for misunderstandings.

[15]"Tracking Bin Laden's E-mail," *Newsweek*, August 21, 2000.

forcement officials with a significant set of leads for investigating terrorists in the Middle East and beyond. Prior to his arrest, Yousef unintentionally offered the FBI another source of information when he lost control of his portable computer in the Philippines. In that laptop, U.S. officials found incriminating data, including plans for future attacks, flight schedules, projected detonation times, and chemical formulae (Reeve, 1999, pp. 39 and 97).

There are other examples of how electronic information belonging to terrorist groups has fallen into the hands of law enforcement personnel. In 1995, Hamas's Abd-al-Rahman Zaydan was arrested and his computer seized—the computer contained a database of Hamas contact information that was used to apprehend other suspects (Soo Hoo, Goodman, and Greenberg, 1997, p. 139). In December 1999, 15 terrorists linked to Osama bin Laden were arrested in Jordan; along with bomb-making materials, rifles, and radio-controlled detonators, a number of computer disks were seized. Intelligence analysts were able to extract information about bomb building and terrorist training camps in Afghanistan.[16] In June 2000, the names of 19 suspects were found on computer disks recovered from a Hizbollah-controlled house (see FBIS, 2000). Finally, several encrypted computer records belonging to the millennialist Aum Shinrikyo cult were retrieved by Japanese authorities after an electronic key was recovered (Denning and Baugh, 1997).

Thus, the organizational benefits associated with greater IT must be traded off against the needs for direct human contact and improved security. This makes it likely that terrorist groups will adopt designs that fall short of fully connected, all-channel networks. Hybrids of hierarchies and networks may better reflect the relative costs and benefits of greater IT reliance—as well as further the group's mission.[17] Another important factor determining the adoption of IT by terrorist groups involves the relative attractiveness of high-tech offensive information operations, to which we turn next.

---

[16]"Terrorist Threats Target Asia," *Jane's Intelligence Review,* Vol. 12, No. 7, July 1, 2000.

[17]In fact, strategy is likely to be an important driver of organizational form and therefore of the density and richness of communications among group members. For instance, any mission calling for quick, dispersed, and simultaneous actions by several nodes could simply not be achieved without some IT support.

NETWAR, TERRORISM, AND OFFENSIVE INFORMATION OPERATIONS[18]

In addition to enabling networked forms of organization, IT can also improve terrorist intelligence collection and analysis, as well as offensive information operations (IO).[19] The acquisition by terrorist groups of an offensive IO capability could represent a significant threat as the world becomes more dependent on information and communications flows.[20] We argue that information-age technology can help terrorists conduct three broad types of offensive IO. First, it can aid them in their perception management and propaganda activities. Next, such technology can be used to attack virtual targets for disruptive purposes. Finally, IT can be used to cause physical destruction.[21]

Perception Management and Propaganda

Given the importance of knowledge and soft power to the conduct of netwar, it is not surprising that networked terrorists have already begun to leverage IT for perception management and propaganda to influence public opinion, recruit new members, and generate funding. Getting a message out and receiving extensive news media exposure are important components of terrorist strategy, which ultimately seeks to undermine the will of an opponent. In addition to such traditional media as television or print, the Internet now offers terrorist

---

[18]The formal Joint Staff and Army definition of information operations is "actions taken to affect adversary information and information systems and defend one's own." See Chairman of the Joint Chiefs of Staff, 1998, 1996a, and 1996b; and Department of the Army, 1997.

[19]For example, IT improves intelligence collection because potential targets can be researched on the Internet. Commercial satellite imagery is now offered by several firms at 1-meter resolution, and in January 2001, the U.S. government granted at least one commercial firm a license to sell 0.5-m imagery. Satellite photos can be used to identify security vulnerabilities in large targets like nuclear reactors. See Koch, 2001.

[20]For more on the importance of information across the spectrum of conflict, see John Arquilla and David Ronfeldt, "Cyberwar Is Coming," in Arquilla and Ronfeldt, 1997, p. 28; also, Arquilla and Ronfeldt, 1993.

[21]The following discussion draws from a variety of terrorist cases, some of which do not necessarily fit the netwar actor description (that is, they may not be networked, as in the case of Aum Shinrikyo). However, we believe they are all indicative of the trends that are starting to shape netwar terrorist offensive operations and that will continue to do so in the coming years.

groups an alternative way to reach out to the public, often with much more direct control over their message.

The news media play an integral part in a terrorist act because they are the conduit for news of the violence to the general population. As Bruce Hoffman has noted, "[t]errorism . . . may be seen as a violent act that is conceived specifically to attract attention and then, through the publicity it generates, to communicate a message" (Hoffman, 1998, p. 131). Terrorists have improved their media management techniques to the point of using "spin doctoring" tactics (Hoffman, 1998, p. 134). In fact, some groups have even acquired their own television and radio stations to take direct control of the reporting of events. Hizbollah, through its television station, has broadcast footage of strikes carried out by its operatives and has a sophisticated media center that regularly—and professionally—briefs foreign journalists. Hizbollah field units have even included specially designated cameramen to record dramatic video footage of Israeli casualties that was then aired in Lebanon and usually rebroadcast by Israeli television. (On these points, see Nacos, 1994.)

The Internet now expands the opportunities for publicity and exposure beyond the traditional limits of television and print media. Before the Internet, a bombing might be accompanied by a phone call or fax to the press by a terrorist claiming responsibility. Now, bombings can be followed—should terrorists so desire—by an immediate press release from their own web sites (at little cost). (For a hypothetical example, see Devost, Houghton, and Pollard, 1997.) The fact that many terrorists now have direct control over the content of their message offers further opportunities for perception management, as well as for image manipulation, special effects, and deception.

An Internet presence could prove advantageous for mobilizing "part-time cyberterrorists"—individuals not directly affiliated with a given terrorist group who nonetheless support its agenda and who use malicious software tools and instructions available at a terrorist web site. This scenario would closely resemble the initiatives taken by both the Israeli and Palestinian governments, which have encouraged private citizens to download computer attack tools and become involved in the conflict surrounding the al-Aqsa Intifadah (more on this below).

It appears that nearly all terrorist groups have a web presence (see Table 2.1 for a selection). As the table indicates, Hizbollah even manages multiple sites—each with a different purpose (for instance, www.hizbollah.org is the site of the central press office, www.moqawama.org describes attacks on Israeli targets, and www.almanar.com.lb contains news and information).

Web sites can also be used to refine or customize recruiting techniques. Recording which types of propaganda receive the most browser hits could help tailor a message for a particular audience. Using some of the same marketing techniques employed by commercial enterprises, terrorist servers could capture information about the users who browse their web sites, and then later contact those who seem most interested. Recruiters may also use more interactive Internet technology to roam online chat rooms and cyber cafes looking for

Table 2.1

Sample of Web Sites Belonging to Militant Islamist Groups

Group Name	Country of Origin	Web Address
Almurabeton	Egypt	www.almurabeton.org
Al-Jama'ah Al-Islamiyyah	Egypt	www.webstorage.com/~azzam/
Hizb Al-Ikhwan Al-Muslimoon (Muslim Brotherhood Movement)	Egypt	www.ummah.org.uk/ikhwan/
Hizbollah	Lebanon	www.hizbollah.org www.moqawama.org/page2/main.htm www.almanar.com.lb http://almashriq.hiof.no/lebanon/300/320/324/324.2/hizballah http://almashriq.hiof.no/lebanon/300/320/324/324.2/hizballah/emdad
Hamas (Harakat Muqama al-Islamiyya)	Palestianian Authority	www.palestine-info.net/hamas/

receptive members of the public, particularly young people. Electronic bulletin boards and user nets can also serve as vehicles for reaching out to potential recruits. Interested computer users around the world can be engaged in long-term "cyber relationships" that could lead to friendship and eventual membership.

Disruptive Attacks

Netwar-oriented terrorists can also use IT to launch disruptive attacks—that is, electronic strikes that temporarily disable, but do not destroy, physical and/or virtual infrastructure. If the ultimate goal of a terrorist is to influence his opponent's will to fight, IO offer additional means to exert influence beyond using simple physical attacks to cause terror. Disruptive attacks include "choking" computer systems through such tools as e-bombs, fax spamming, and hacking techniques to deface web sites. These strikes are usually nonlethal in nature, although they can wreak havoc and cause significant economic damage.

To date, disruptive strikes by terrorists have been relatively few and fairly unsophisticated—but they do seem to be increasing in frequency. For example, in 1996, the Liberation Tigers of Tamil Eelam (LTTE) launched an email bomb attack against Sri Lankan diplomatic missions. Using automated tools, the guerilla organization flooded Sri Lankan embassies with thousands of messages, thus establishing a "virtual blockade."[22] Japanese groups have allegedly attacked the computerized control systems for commuter trains, paralyzing major cities for hours (Devost, Houghton, and Pollard, 1997, p. 67). In 2000, a group of Pakistani hackers who call themselves the Muslim Online Syndicate (MOS) defaced more than 500 web sites in India to protest the conflict in Kashmir (see Hopper, 2000). Finally, Pakistan's Lashkar-e-Taiba claimed to have attacked Indian military web sites in early 2000.[23]

---

[22]See Dorothy Denning's discussion of virtual sit-ins and email bombs in Chapter Eight of this volume.

[23]Jessica Stern, telephone interview with author Michele Zanini, September 2000.

Disruptive rather than destructive actions take place for several reasons. For example, terrorists who rely on the Internet for perception management and communication purposes may prefer not to take "the Net" down, but rather to slow it down selectively. In addition, groups may want to rely on nonlethal cyber strikes to pressure governments without alienating their own constituent audiences. Terrorist groups may also follow the lead of criminal hackers and use the threat of disruptive attacks to blackmail and extort funds from private-sector entities (e.g., the ongoing "cyber jihad" against Israel may come to target commercial enterprises that do business with the Israelis).[24] For instance, in the early 1990s, hackers and criminals blackmailed brokerage houses and banks for several million British pounds. Money can also be stolen from individual users who visit terrorist web sites.[25]

Destructive Attacks

As mentioned earlier, IT-driven information operations can lead to the actual destruction of physical or virtual systems. Malicious viruses and worms can be used to permanently destroy (erase) or corrupt (spoof) data and cause economic damage. In the worst case, these same software tools can be used to cause destructive failure in a critical infrastructure like air traffic control, power, or water systems, which can lead to casualties. Indeed, it is likely that information operations that result in the loss of life may offer the same level of drama as physical attacks with bombs. Also, striking targets through electronic means does not carry the risks associated with using conventional weapons—such as handling explosives or being in close proximity to the target.

---

[24]A survey conducted by the Science Applications International Corp. in 1996 found that 40 major corporations reported losing over $800 million to computer break-ins. This example is cited on several web sites including Don Gotterbarn's web site at www-cs.etsu.edu/gotterbarn/stdntppr.

[25]A related criminal case reveals the potential for this threat. In 1997, a group known as the Chaos Computer Club created an Active X Control, which, when downloaded and run on the user's home computer, could trick the Quicken accounting program into removing money from a user's bank account. See "ActiveX Used as Hacking Tool," CNET News.com, February 7, 1997, http://news.cnet.com/news/0,10000,0-1005-200-316425,00.html.

Offensive IO: Mitigating Factors

The extreme case where the use of IT results in significant human losses has yet to occur. The lack of destructive information attacks is arguably influenced by the relative difficulty of electronically destroying (rather than disrupting) critical infrastructure components—the level of protection of existing infrastructure may be too high for terrorists to overcome with their current IT skill set. In fact, a terrorist organization would first have to overcome significant technical hurdles to develop an electronic attack capability. Concentrating the necessary technical expertise and equipment to damage or destroy targeted information systems is no easy task, given the computer security risks involved. In developing and increasing their reliance on electronic attacks, terrorist organizations may be assuming risks and costs associated with the relative novelty of the technology. Terrorists wishing to expand the scope of their offensive IO activities would have to continue upgrading and researching new technologies to keep up with the countermeasures available to computer security experts and systems administrators. This technology "treadmill" would demand constant attention and the diversion of scarce organizational resources.[26]

Another important determinant in netwar terrorists choosing low-level IT is that such conventional weapons as bombs remain more cost-effective. In fact, most terrorism experts believe that existing groups see their current tactics as sufficient and are not interested in branching into computer network attacks. Since current tactics are simple and successful, there is no built-in demand to innovate: bombing still works.[27] As long as current tactics enable these groups to accomplish their short-term goals and move toward their long-term goals, there will be no strong incentives to change behavior. In addition, the fragility of computer hardware may make a physical attack on these targets more attractive because such an attack is signifi-

---

[26]These points are also elaborated considerably in unpublished RAND research by Martin Libicki, James Mulvenon, and Zalmay Khalilzad on information warfare.

[27]As one article puts it, "the gun and the bomb continue to be the terrorists' main weapon of choice, as has been the case for more than a century." See Hoffman, Roy, and Benjamin, 2000, p. 163.

cantly less challenging from a technical standpoint than attempting a virtual attack (Soo Hoo, Goodman, and Greenberg, 1997, p. 146).

Disruptive attacks may be easier to carry out, but because of their very nature they do not produce the same kind of visceral and emotional reaction that the loss of human life does. Indeed, some terrorism analysts argue that it is unlikely that terrorist groups will turn to disruptive attacks as the primary tactic. Brian Jenkins points out that IT-enabled disruptive strikes

> do not produce the immediate, visible effects. There is no drama. No lives hang in the balance . . . . Terrorist intentions regarding cyberterrorism are even more problematic. Linking the objectives of actual terrorist groups to scenarios of electronic sabotage that would serve those objectives is conjecture.[28]

In addition, many computer security experts believe that even disruptive attacks remain technologically challenging for most terrorist groups and too undervalued by the media to make them attractive for terrorists (Soo Hoo, Goodman, and Greenberg, 1997, pp. 145–146).

EVALUATING PAST, PRESENT, AND FUTURE TRENDS

Given that information technology brings drawbacks as well as benefits, the terrorist groups examined here have not chosen to rely exclusively on IT to coordinate their operations and execute attacks. The available evidence suggests that netwar terrorists have embraced IT for organizational purposes, especially to facilitate C3, but they have been either unable or unwilling to attempt more ambitious offensive IO. However, the benefits clearly outweigh the risks when it comes to utilizing IT for perception management and propaganda. See Table 2.2 for a summary.

---

[28]Email correspondence from Brian Jenkins (at RAND) October 2000, who is quoting a forthcoming manuscript by Paul Pillar.

Table 2.2

**Benefits and Drawbacks of IT Use for Netwar Terrorists
(facilitating and mitigating factors)**

IT Use	Facilitating	Mitigating
Organizational	Enables dispersed activities with reasonable secrecy, anonymity	Susceptibility to wire and wireless tapping
	Helps maintain a loose and flexible network	Digitally stored information can be easily retrievable unless well protected
	Lessens need for state sponsorship	Cannot by itself energize a network; common ideology and direct contact still essential
Offensive	Generally lower entry costs	Current bombing techniques already effective
	Eradication of national boundaries	Significant technical hurdles for disruptive and destructive IO
	Physically safer	Unique computer security risks impose recurring costs of "technology treadmill"
	Spillover benefits for recruitment/fundraising	

Future Developments in Information-Age Terrorism

Were the trends described above to persist, one could speculate that future netwar actors will continue to consolidate their IT use primarily for organizational purposes, with some emphasis on perception management on the offensive IO side. Under these conditions, networked terrorists would still rely on such traditional weapons as conventional bombs to cause physical violence. But they will also transmit information on how to build such weapons via CD-ROMs or email, use chat rooms to coordinate their activities, and use web sites to publicize and justify their strikes to a global audience.

The al-Aqsa Intifadah in the West Bank and Gaza highlights how protracted IO campaigns could be waged in conjunction with a campaign of conventional violence. Mirroring the real-world violence that has resulted in hundreds of casualties, a conflict has also been waged in cyberspace over economic and propaganda stakes. Palestinian hackers who support the al-Aqsa Intifadah have been waging a cyberjihad against Israeli government and commercial targets, defacing web sites and conducting DOS attacks. More than 40 Israeli sites have been hit, including the Tel Aviv Stock Exchange and the Bank of Israel. Israeli hackers have counterattacked, hitting more than 15 different Palestinian targets, including Hizbollah, Hamas, and the Palestinian National Authority. As the disruptive attacks have escalated, individuals and groups have joined both sides, from professional hackers to "script-kiddies" (relative amateurs who rely on off-the-shelf and easy to use tools). (See Lemos, 2000.)

That said, the swift and unpredictable changes associated with technology suggest that other outcomes are possible. The question is, will terrorists have the desire and opportunity to significantly increase their reliance on IT—primarily for offensive purposes—in the future? Several factors could influence such a shift, including the degree to which new technology will serve their main strategic goals in a safe and effective manner.[29] For instance, the introduction of easy-to-use, "unbreakable" encryption programs to support email and file exchange will encourage netwar terrorists to adopt such techniques. Moreover, terrorist access to technologies that can be readily employed without extensive internal development efforts[30]—by group members and part-time "volunteers" or through "hackers for hire"[31]—will be a critical facilitating factor. Equally important, the relative vulnerability of the information infrastructure plays a role in this calculus (more on this below).

These possible developments would likely prompt the evolution of current netwar terrorist groups toward greater reliance on IT for offensive purposes and could also encourage the emergence of new and completely virtual groups that exclusively operate in cyberspace. Each possibility is described briefly below.

---

[29]From a strategic perspective, the more that terrorist groups emphasize swarming doctrines to conduct dispersed and simultaneous operations, the greater the need for a sophisticated IT infrastructure.

[30]One example is Netcat, a free hacking tool made available in 1996. See Soo Hoo, Goodman, and Greenberg, 1997, p. 141.

[31]Rumors persist that people proficient in network attacks are available for hire. Press reports indicate that hacker groups have been approached by anonymous users claiming to be terrorists who have requested help gaining access to government classified information networks such as SIPRNET. For example, one teenage hacker was said to have received a $1,000 check. See McKay, 1998.

The Evolution of Current Groups

As Brian Jackson notes, the introduction of new technologies in an organization follows a complex and often lengthy process. Not only do innovative systems have to be developed or acquired, but organizational actors have to become familiar with new systems and be able to use them effectively (Jackson, unpublished). Given the challenge, terrorist groups are likely to channel their scarce organizational resources to acquire those IT skills that have the greatest leverage for the least amount of cost and effort.

This line of reasoning can help explain terrorists' recent emphasis on using communications technology for organizational purposes: Having access to the Internet and cellular telephones is not overly complicated, and it plays a significant role in enabling dispersed operations, a key goal of netwar groups. This reasoning also suggests that over time terrorist groups might begin to experiment more aggressively with information-age technologies for offensive information operations, as they become more familiar with such innovations. Indeed, some may follow a "migration" pattern as illustrated in Figure 2.1: The knowledge of IT issues gained by relying on technology to fa-

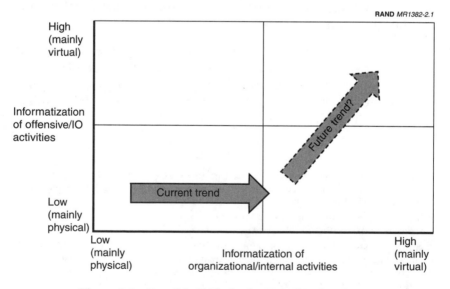

Figure 2.1—Possible Shifts in the Use of Technology

cilitate interactions among group members, or to gain a web presence, might eventually be expanded and harnessed for increasingly offensive uses.

The pace with which current groups move through such a path is also dependent on the degree of cooperation and information exchange among netwar terrorists. Such cooperation has often occurred in the past—for instance, Islamic radicals have organized "terror conferences"(Kushner, 1998, p. 41), while European terrorist groups such as the Irish Republican Army entered into joint ventures with counterparts across the globe to learn from one another and disseminate knowledge (such as designs of booby traps and radio-controlled bombs). (See Wilkinson, 1986, p. 40.) Given the loose and reciprocal nature of ties between actors in networks such as al-Qaeda, it is entirely possible that those with IT skills would be leveraged globally and placed at the disposal of the organization's various members.

Lastly, as leading-edge groups begin to move toward the upper-right quadrant of Figure 2.1, other groups may be tempted to follow suit: Terrorists that hitherto had decided to adopt a low-technology profile for their offensive operations could be emboldened by successful instances of IT attacks by others (Jackson, unpublished).

The Emergence of New Groups

An alternative hypothesis to the notion that existing terrorist groups should be watched for signs of movement toward cyberterror is that qualitative improvements in the informatization of networked terrorists will only be witnessed with the emergence of newer, and more technologically savvy, groups. Just as Hamas and al-Qaeda have overshadowed the PFLP and other Marxist groups formed in the 1960s, new-generation groups may further advance the trend toward networked and IT-reliant organizations. New groups could even be led by individuals who are technically skilled, suggesting the rise of a hybrid breed of "terrorists cum hackers." Like hackers, they would undertake most of their attacks in cyberspace. Like terrorists, they would seek to strike targets by both disruptive and destructive means to further a political or religious agenda.

The possibility that innovation will take place only with the advent of new groups finds support in previous work by such terrorism experts as Hoffman, who describes most groups today as operationally conservative (Hoffman, "Terrorism Trends and Prospects" in Lesser et al., 1999, p. 36). Aside from organizational inertia, current groups may also be hesitant to further rely on IT for offensive purpose because of large, "sunk costs" in traditional tactics, training, and weapon stockpiles. Existing groups wishing to "amortize" such capital cost may be unwilling to direct scarce resources toward the development of new and radically different offensive techniques.

POLICY IMPLICATIONS

The acquisition and use of information-age technology by terrorist groups are far from certain. Indeed, the scenarios painted above are not mutually exclusive. It is conceivable that current groups will acquire new IT skills over time and adopt more-offensive IT strategies. New hacker/terrorist groups may also emerge to compound this problem. Some terrorist networks may even become sophisticated enough to sustain and coordinate offensive campaigns in both the virtual and physical realms.

What is certain, however, is that counterterrorism policy will be able to counter the dangers associated with terrorist IT use only if it becomes attuned to the information age. Counterterrorist policies and tactics could even alter the speed with which terrorists become informatized—groups facing a robust counterterrorism campaign may have less time and resources to acquire new technologies (see Jackson, unpublished). For such reasons, it seems advisable that counterterrorism policymakers and strategists bear in mind the following recommendations.

First, monitor changes in the use of IT by terrorist groups, differentiating between organizational and offensive capabilities. Counterterrorism policies will have to take into account the type of IT capabilities developed by each group, targeting their specific technological vulnerabilities. Evaluating how IT shapes a group's organizational processes and offensive activities will remain a critical component of the threat assessment. Monitoring the shift in capabilities for each type of

IT use and then examining trends in the aggregate can also help forecast future terrorist behavior.

Among the most significant trends to be carefully examined is the possible emergence of a new, and potentially dangerous, breed of terrorists—groups that are highly informatized along both the organizational and offensive axes. In this regard, a number of "signposts" should be identified and tracked. These would include significant increases in the level of technical expertise of known leaders and their subordinates, increases in the frequency of disruptive attacks, increases in the seizures of IT equipment owned by terrorists, the presence—and successful recruiting—of "hackers for hire," and the availability of effective and relatively secure off-the-shelf information technologies (including those that facilitate hacking).

Second, target information flows. Since network designs are inherently information intensive, counterterrorism efforts should target the information flows of netwar groups. Intercepting and monitoring terrorist information exchanges should remain a top priority, and the implementation of "Project Trailblazer" by the National Security Agency—to develop a system that can crack new encryption software, fiber-optic cables, and cellular phone transmissions—represents a useful addition to America's signals intelligence capability (Kitfield, 2000).

Equally important, policymakers should consider going beyond the passive monitoring of information flows and toward the active disruption of such communications. To the degree that erroneous or otherwise misleading information is planted into a network's information flows by what are seemingly credible sources, over time the integrity and relevance of the network itself will be compromised. This in turn could breed distrust and further cripple a group's ability to operate in a dispersed and decentralized fashion—essentially eliminating a netwar group's key competitive advantage.

Increased emphasis on targeting information flows should not exclude nonelectronic efforts to gather intelligence and undermine the network. Indeed, human intelligence will remain an important tool for intercepting (and injecting) information not transmitted through electronic means of communication.[32] This is an especially pressing concern, given that several intelligence observers have pointed to a lack of U.S. capability in this area.

Third, deter IT-based offensive IO through better infrastructure protection. Changes in the vulnerability of critical infrastructures can significantly alter a terrorist's IT calculus. If such infrastructures, such as those that manage air traffic control, were to become relatively more vulnerable, they might become more attractive as targets: Terrorists could strike at a distance, generating as much—if not more—destruction as would have been caused by the use of traditional weapons. U.S. policy should identify specific vulnerabilities to expected threats and develop security techniques that mitigate each. An analysis of these issues is beyond the scope of the current chapter, but there are numerous studies that explore this process, including RAND's *Securing the U.S. Defense Information Infrastructure: A Proposed Approach* (Anderson et al., 1999). The FBI's National Infrastructure Protection Center and other newly created organizations represent useful steps in this direction. Counterterrorist agencies may also want to consider the option of employing a large number of hackers and leveraging their knowledge for defensive and possibly even retaliatory purposes.

Fourth, beat networked terrorists at their own game: "It takes networks to fight networks." Governments wishing to counter netwar terrorism will need to adopt organizational designs and strategies like those of their adversaries. This does not mean mirroring the opponent, but rather learning to draw on the same design principles of network forms. These principles depend to some extent upon technological innovation, but mainly on a willingness to innovate organizationally and doctrinally and by building new mechanisms for interagency and multijurisdictional cooperation. The Technical Support Working Group (TSWG) is a good example of a nontraditional government interagency group with more than 100 member organizations from at least 13 federal agencies and a growing number of local and state agencies. Its principal aim is to help develop and deploy technologies

---

[32]After Osama bin Laden noticed that his satellite phone connection was no longer secure, he began to use human couriers to pass information and instructions to his operatives.

to combat terrorism.[33] Another example is the Counter-Intelligence 21 (CI-21) plan, a set of reforms that seek to increase the level of cooperation between counterintelligence personnel at the CIA, FBI, and the Pentagon (Kitfield, 2000). If counterterrorism agencies become ready and willing to rely on networks of outside "ethical hackers" in times of crisis, the need to coordinate beyond the boundaries of government will increase.[34]

Supporters of these initiatives rightly recognize that the information age and the consequent advent of netwar have blurred the boundary between domestic and international threats, as well as between civilian and military threats. This in turn demands greater interagency coordination within the counterterrorism community. As terrorist groups evolve toward loose, ad-hoc networks that form and dissipate unpredictably, so must counterterrorism forces adopt a more flexible approach that crosses bureaucratic boundaries to accomplish the mission at hand. While militaries and governments will never be able to do away with their hierarchies entirely, there is nevertheless much room for them to develop more-robust organizational networks than they currently have—a change that may offset some, if not all, of the advantage now accruing mostly to networked terrorist groups.

BIBLIOGRAPHY

Anderson, Robert H., Phillip M. Feldman, Scott Gerwehr, Brian Houghton, Richard Mesic, John D. Pinder, Jeff Rothenberg, and James Chiesa, *Securing the U.S. Defense Information Infrastructure:*

---

[33]TSWG received $48 million in fiscal year 2000. Traditional terrorist threats such as bombs still generate the greatest concern, and most of TSWG's budget covers needs such as blast mitigation. See Stanton, 2000, p. 24.

[34]In some cases, hackers may be spontaneously driven to aid law enforcement officials in defending against particularly objectionable crimes. For instance, a group called Ethical Hackers Against Pedophilia has been created to identify and urge punishment of people who publish child pornography on the Internet (see www.ehap.org). The government could take the lead in mobilizing existing ethical hackers in the private sector to help in times of crisis. This would be different from deliberately organizing a virtual militia composed of relatively unsophisticated citizens armed with off-the-shelf hacking tools—something the Israeli government *has* experimented with during the al-Aqsa Intifadah. Given this option's potential to become a double-edged sword—as well as lack of information on its efficacy—more research on this topic is warranted.

A Proposed Approach, Santa Monica, Calif.: RAND, MR-993-OSD/NSA/DARPA, 1999.

Arquilla, John, and David F. Ronfeldt, *The Advent of Netwar,* Santa Monica, Calif.: RAND, MR-789-OSD, 1996.

Arquilla, John, and David F. Ronfeldt, "Cyberwar Is Coming," *Comparative Strategy,* Vol. 12, No. 2, Summer 1993, pp. 141–165. Available as RAND reprint RP-223.

Arquilla, John, and David Ronfeldt , eds., *In Athena's Camp: Preparing for Conflict in the Information Age,* Santa Monica, Calif.: RAND, MR-880-OSD/RC, 1997.

Bremer, Paul L. III, et al., *Countering the Changing Threat of International Terrorism,* Washington, D.C.: National Commission on Terrorism, 2000, www.fas.org/irp/threat/commission.html (August 28, 2000).

Burns, Tom, and G. M. Stalker, *The Management of Innovation,* London: Tavistock, 1961.

Cameron, Gavin, "Multi-Track Microproliferation: Lessons from Aum Shinrikyo and Al Qaida," *Studies in Conflict & Terrorism,* Vol. 22, 1999.

Chairman of the Joint Chiefs of Staff, *Doctrine for Joint Psychological Operations,* Joint Pub 3-53, Washington, D.C.: United States Government Printing Office, July 10, 1996a.

Chairman of the Joint Chiefs of Staff, *Joint Doctrine for Command and Control Warfare (C2W),* Joint Pub 3-13.1, Washington, D.C.: United States Government Printing Office, February 7, 1996b.

Chairman of the Joint Chiefs of Staff, *Joint Doctrine for Information Operations,* Joint Pub 3-13, Washington, D.C.: United States Government Printing Office, October 9, 1998.

Denning, Dorothy E., "Activism, Hacktivism, and Cyberterrorism: The Internet As a Tool for Influencing Foreign Policy," 1999, www.nautilus.org/info-policy/workshop/papers/denning.html (January 23, 2001).

Denning, Dorothy E., and William E. Baugh, Jr., *Cases Involving Encryption in Crime and Terrorism*, 1997, www.cs.georgetown.edu/~denning/crypto/cases.html (January 23, 2001).

Denning, Dorothy E., and William E. Baugh, Jr., *Encryption and Evolving Technologies As Tools of Organized Crime and Terrorism*, Washington, D.C.: U.S. Working Group on Organized Crime (WGOC), National Strategy Information Center, 1997.

Department of the Army, *Public Affairs Operations*, Field Manual FM 46-1, Washington, D.C.: United States Government Printing Office, May 30, 1997.

Devost, Matthew G., Brian K. Houghton, and Neal A. Pollard, "Information Terrorism: Can You Trust Your Toaster?" in Robert E. Neilson, ed., *Sun Tzu and Information Warfare*, Washington, D.C.: National Defense University Press, 1997.

Drogin, Bob, "State Dept. Report Cites Growing Reach of Bin Laden," *Los Angeles Times*, May 2, 2000.

FBIS, "Afghanistan, China: Report on Bin-Laden Possibly Moving to China," *Paris al-Watan al-'Arabi*, 23 May 1997a, pp. 19–20, FBIS-NES-97-102.

FBIS, "Arab Afghans Reportedly Transfer Operations to Somalia," *Cairo Al-Arabi*, 10 March 1997b, p. 1, FBIS-TOT-97-073.

FBIS, "Arab Afghans Said to Launch Worldwide Terrorist War," *Paris al-Watan al-'Arabi*, December 1, 1995, pp. 22–24, FBIS-TOT-96-010-L.

FBIS, "Italy: Security Alter Following Algerian Extremists' Arrests," *Milan Il Giornale*, November 12, 1996a, p. 10, FBIS-TOT-97-002-L.

FBIS, "Italy, Vatican City: Daily Claims GIA 'Strategist' Based in Milan," *Milan Corriere della Sera*, December 5, 1996b, p. 9, FBIS-TOT-97-004-L.

FBIS, "Trial of 19 Hizbullah Members Begins in Istanbul," June 2000, FBIS-WEU-2000-0612.

Heydebrand, Wolf V., "New Organizational Forms," *Work and Occupations*, Vol. 16, No. 3, August 1989, pp. 323–357.

Hoffman, Bruce, *Inside Terrorism*, New York: Columbia University Press, 1998.

Hoffman, Bruce, Olivier Roy, and Daniel Benjamin, "America and the New Terrorism: An Exchange," *Survival*, Vol. 42, No. 2, Summer 2000.

Hopper, D. Ian, "Kashmir Conflict Continues to Escalate—Online," *CNN*, March 20, 2000, www.cnn.com (September 5, 2000).

Iuris, Andre Pienaar, "Information Terrorism," in Amelia Humphreys, ed., *Terrorism: A Global Survey: A Special Report for Jane's Intelligence Review and Jane's Sentinel*, Alexandria, Va.: Jane's Information Group, 1997.

Jackson, Brian, unpublished RAND research on technology acquisition by terrorist groups.

Kelley, Jack, "U.S. Acquires Reputed Terrorism Guide," *USA Today*, September 18, 2000.

Kitfield, James, "Covert Counterattack," *National Journal*, September 16, 2000.

Koch, Andrew, "Space Imaging Gets .5m Go Ahead," *Jane's Defence Weekly*, January 10, 2001.

Kurlantzick, Joshua, "Muslim Separatists in Global Network of Terrorist Groups," *Washington Times*, May 2, 2000.

Kushner, H. W., *Terrorism in America: A Structured Approach to Understanding the Terrorist Threat*, Springfield, Ill.: Charles C. Thomas, 1998.

Laqueur, Walter, *The New Terrorism: Fanaticism and the Arms of Mass Destruction*, New York: Oxford University Press, 1996.

Lemos, Robert, "Hacktivism: Mideast Cyberwar Heats Up," *ZDNet News*, November 6, 2000, www.zdnet.com (December 4, 2000).

Lesser, Ian O., Bruce Hoffman, John Arquilla, David Ronfeldt, Michele Zanini, and Brian Jenkins, *Countering the New Terrorism*, Santa Monica, Calif.: RAND, MR-989-AF, 1999.

McKay, Niall, "Do Terrorists Troll the Net?" *Wired News*, November 1998, www.wired.com (September 15, 2000).

Monge, Peter, and Janet Fulk, "Communication Technology for Global Network Organizations," in Gerardine Desanctis and Janet Fulk, eds., *Shaping Organizational Form: Communication, Connection, and Community*, Thousand Oaks, Calif.: Sage, 1999.

Nacos, Brigitte, *Terrorism and the Media*, New York: Columbia University Press, 1994.

Nohria, Nitin, and Robert Eccles, eds., *Networks and Organizations*, Boston, Mass.: Harvard Business School Press, 1992.

Office of the Coordinator for Counterterrorism, *Patterns of Global Terrorism: 1999*, Washington, D.C.: U.S. Department of State, Publication #10687, 2000.

Ranstorp, Magnus, "Hizbollah's Command Leadership: Its Structure, Decision-Making and Relationship with Iranian Clergy and Institutions," *Terrorism and Political Violence*, Vol. 6, No. 3, Autumn 1994.

Reeve, Simon, *The New Jackals: Ramzi Yousef, Osama Bin Laden and the Future of Terrorism*, Boston, Mass.: Northeastern University Press, 1999.

Simon, Steven, and Daniel Benjamin, "America and the New Terrorism," *Survival*, Vol. 42, No. 1, Spring 2000.

Soo Hoo, Kevin, Seymour Goodman, and Lawrence Greenberg, "Information Technology and the Terrorist Threat," *Survival*, Vol. 39, No. 3, Autumn 1997, pp. 135–155.

Stanton, John J., "A Typical Pentagon Agency Waging War on Terrorism," *National Defense*, May 2000.

Stern, Jessica, "Pakistan's Jihad Culture," *Foreign Affairs*, November/December 2000, pp. 115–126.

"Terrorist Threats Target Asia," *Jane's Intelligence Review*, Vol. 12, No. 7, July 1, 2000.

Thompson, James D., *Organizations in Action*, New York: McGraw-Hill, 1967.

Whine, Michael, "Islamist Organizations on the Internet," *Terrorism and Political Violence*, Vol. 11, No. 1, Spring 1999.

Wilkinson, P., "Terrorism: International Dimensions," in W. Gutteridge, ed., *Contemporary Terrorism*, New York: Facts on File Publications, 1986.

Chapter Three

TRANSNATIONAL CRIMINAL NETWORKS[1]

Phil Williams

Editors' abstract. Many old-style criminal hierarchies (e.g., the Italian Mafia) are reorganizing into sprawling transnational networks. Williams (University of Pittsburgh) analyzes this trend, with an emphasis on developments unfolding in Russian criminal organizations. He draws on the academic literatures about social and business networks to deepen our understanding of this phenomenon. The chapter builds upon earlier articles in which he pioneered the study of transnational criminal organizations from network perspectives, notably "The Nature of Drug-Trafficking Networks," Current History, April 1998.

In a recent analysis of global trends, the U.S. National Intelligence Council included a short section on criminal organizations and networks. It noted that

> criminal organizations and networks based in North America, Western Europe, China, Colombia, Israel, Japan, Mexico, Nigeria, and Russia will expand the scale and scope of their activities. They will form loose alliances with one another, with smaller criminal entrepreneurs, and with insurgent movements for specific operations. They will corrupt leaders of unstable, economically fragile, or failing states, insinuate themselves into troubled banks and businesses,

---

[1]The author thanks John Picarelli, Bill Koenig, and Paul N. Woessner for a series of helpful discussions on the role of network analysis in intelligence; and Gregory O'Hayon, William Cook, Jeremy Kinsell, and Brian Joyce for their work at the University of Pittsburgh's Ridgway Center in mapping Russian and other criminal networks. He also thanks the I2 Company for providing the software used for these activities.

and cooperate with insurgent political movements to control substantial geographic areas.[2]

In other words, there is a growing recognition that organized crime is increasingly operating through fluid network structures rather than more-formal hierarchies.

However, the traditional paradigm for studying organized crime emphasized identifying the hierarchical or pyramidal structures of criminal organizations. Finding its fullest expression in the 1967 report on organized crime by the President's Commission on Law Enforcement and Administration of Justice and in Donald Cressey's famous analysis in *Theft of the Nation*, this interpretation of organized crime was based on the example provided by La Cosa Nostra in the United States. It emphasized the existence of a "nationwide illicit cartel and confederation," the governing role of a national commission, hierarchical structure, and the clear division of labor between local branches.[3]

Cressey's analysis, though, provoked several pioneering studies that challenged the mainstream interpretation of organized crime as having a rational corporate structure, arguing not only that organized crime was much more fluid than portrayed by the conventional wisdom, but also that patron-client relations and network structures played a pivotal role. Francis Ianni looked at the role of African American and Puerto Rican criminal networks in New York, while Joseph Albini contended that even Italian organized crime in the United States could best be understood through patron-client relations rather than formal hierarchies.[4] In an important historical study of organized crime in New York, Alan Block discovered that it was not only more fragmented and chaotic than believed, but also that it involved

---

[2]National Intelligence Council, *Global Trends 2015* (Washington: National Intelligence Council, December 2000) p. 41.

[3]For an excellent excerpt summarizing Cressey's views, see Donald R. Cressey, "The Functions and Structure of Criminal Syndicates," in Patrick J. Ryan and George E. Rush, eds., *Understanding Organized Crime in Global Perspective* (London: Sage, 1997) pp. 3–15, especially p. 3.

[4]Francis J. Ianni, *Black Mafia: Ethnic Succession in Organized Crime* (New York: Simon and Schuster, 1974), and Joseph Albini, *The American Mafia: Genesis of a Legend* (New York: Appleton, Crofts, 1971).

"webs of influence" that linked criminals with those in positions of power in the political and economic world. These patterns of affiliation and influence were far more important than any formal structure and allowed criminals to maximize opportunities.[5]

More recently, Gary Potter has suggested that organized crime in the United States can best be understood in network terms, while Finckenauer and Waring, in a study on Russian émigré criminals in the United States, concluded that they operate largely through network structures.[6] Of particular importance is the work of Malcolm Sparrow, who not only applies concepts from social network analysis to the operation of criminal networks, but also offers innovative insights into ways in which the vulnerabilities of these networks might be identified and exploited.[7] My own work has also moved in this direction.[8]

This emphasis on criminal networks reflects a growing acknowledgment, among researchers into crime, that there is no single, dominant organizational structure with universal applicability, and the realization by law enforcement agencies that they arc seeing patterns of organized crime that do not fit the traditional hierarchical structure. The German BKA (the equivalent of the American FBI), for example, has observed that most of the criminal organizations it investigates are "loose, temporary networks" and that "lastingly established, hierarchical structures" are rather in the minority. In the BKA's view, the evidence in Germany suggests that even hierarchical organizations

---

[5]Alan Block, *East Side–West Side: Organizing Crime in New York 1939–1959* (Swansea U.K.: Christopher Davis, 1979).

[6]Gary Potter, *Criminal Organizations: Vice Racketeering and Politics in an American City* (Prospect Heights, Ill.: Waveland Press, 1993), and James O. Finckenauer and Elin J. Waring, *Russian Mafia in America* (Boston: Northeastern University Press, 1998).

[7]Malcolm K. Sparrow, "The Application of Network Analysis to Criminal Intelligence: An Assessment of the Prospects," *Social Networks*, Vol. 13, No. 3, September 1991, and "Network Vulnerabilities and Strategic Intelligence in Law Enforcement," *International Journal of Intelligence and Counterintelligence*, Vol. 5, No. 3, Fall 1991.

[8]See Phil Williams, "Transnational Criminal Organisations and International Security," *Survival*, Vol. 36, No. 1 (Spring 1994), pp. 96–113, and "The Nature of Drug Trafficking Networks," *Current History*, Vol. 97, No. 618 (April 1998) pp. 154–159.

such as the Italian Mafia families allow their local branches considerable discretion.[9]

Perhaps even more significant, a major study of organized crime in Holland noted great variations in collaborative forms and concluded that

> the frameworks need not necessarily exhibit the hierarchic structure or meticulous division of labor often attributed to mafia syndicates. Intersections of social networks with a rudimentary division of labor have also been included as groups in the sub-report on the role of Dutch criminal groups, where they are referred to as cliques. As is demonstrated . . . there can be sizeable differences in the cooperation patterns within these cliques and between the cliques and larger networks of people they work with on an incidental basis.[10]

In other words, there is a growing recognition that organized crime often operates through fluid networks rather than through more formal hierarchies.

Important as they are, none of these studies systematically explores the advantages of network structures as an organizational form for criminal activities. This chapter identifies both the characteristics of networks that make them an ideal form for organizing criminal activities and the specific characteristics of criminal networks themselves. This conceptual analysis of criminal networks also draws insights from social network and business theories and moves beyond theory to draw implications for policymakers concerned with developing strategies to combat criminal networks.

BACKGROUND ON NETWORK ANALYSIS

Networks are one of the most common forms of social organization. They are simultaneously pervasive and intangible, ubiquitous and invisible, everywhere and nowhere. Networks are not an exclusive orga-

---

[9]FBIS, "BKA Registers Less Damage Caused by Gangs," *Frankfurter Rundschau*, July 13, 1998, p. 3.

[10]Cyrille Fijnaut, Frank Bovenkerk, Gerben Bruinsma, and Henk van de Bunt, *Organized Crime in the Netherlands* (The Hague: Kluwer Law International, 1998) p. 27.

nizational form and often exist within more traditional hierarchical structures, cutting through divisions based on specialization or rank. It is also possible to have networks in which hierarchical organizations are key participants. Networks are also an important complement to markets, making them more efficient, reducing transaction costs, and providing increased opportunities for both buyers and sellers.

These characteristics—their pervasiveness, their capacity to coexist both within and outside hierarchies, their ability to make markets more efficient by facilitating directed flows of information and commodities—give networks an elusive quality. In some respects, they appear little more than plastic organizations that can be molded in many different ways.

Networks vary in size, shape, membership, cohesion, and purpose. Networks can be large or small, local or global, domestic or transnational, cohesive or diffuse, centrally directed or highly decentralized, purposeful or directionless. A specific network can be narrowly and tightly focused on one goal or broadly oriented toward many goals, and it can be either exclusive or encompassing in its membership.

Networks facilitate flows of information, knowledge, and communication as well as more-tangible commodities. As communications have become cheaper and easier, networks have expanded enormously. Indeed, technological networks facilitate the operation of larger and more-dispersed social networks and can even act as a critical force multiplier for certain kinds of social networks. Against this background, the analysis here seeks to:

- Delineate very briefly the underlying concepts and ideas that encourage and facilitate analysis of criminal organizations in terms of network structures. These include social network analysis, a growing literature on network business organizations (a concept developed most fully in the idea of the virtual corporation), and previous studies of organized crime that have emphasized the importance of networks rather than the more traditional hierarchical structures.

- Identify the characteristics of networks that make them attractive to criminals and to elucidate further the major characteristics of criminal networks.

- Specify critical roles in criminal networks, bearing in mind that there are network roles that relate to the functioning of the network and substantive roles related specifically to the nature of the criminal enterprise. In some cases, these two roles might overlap; in others, however, they will be quite distinct.

- Highlight and assess the operations of criminal networks. In effect, the analysis will examine a case study of a criminal network that penetrated a legal institution—the Bank of New York.

- Outline ways in which governments and law enforcement agencies can attack networks more effectively. This requires an analysis of network vulnerabilities and how these can be exploited.

Some Underlying Analytic Concepts

The Network. A network can be understood very simply as a series of nodes that are connected. The nodes can be individuals, organizations, firms, or computers, so long as they are connected in significant ways. The focus here, of course, is on networks that originate and operate in order to obtain financial rewards through and from illicit activities. As such, this analysis draws on three separate strands of research: social network analysis, discussions of network business organizations, and previous work on criminal organizations (work that departs significantly from the emphasis on formal hierarchies that was long part of the dominant paradigm in the study of organized crime).

Social Network Analysis. Social network analysis originated in several fields, including anthropology, sociology, and social psychology. Perhaps the most important of the early pioneers was J. L. Moreno, who, in the 1930s, developed the notion of a "sociogram." This was

> a picture in which people (or more generally any social units) are represented as points in two-dimensional space, and relationships

among pairs of people are represented by lines linking the corresponding points.[11]

The essence of this type of approach is its focus on "the relationships or ties between the nodes or units in the network." These ties can be based on a variety of underpinning factors, such as "kinship, material transactions, flow of resources or support, behavioral interaction, group co-memberships, or the affective evaluation of one person by another."[12] In many cases there will be some kind of exchange between the nodes, whether of commodities or services (broadly defined to include information and favors). Whatever the basis for the relationship, however, the network concept emphasizes the linkages among actors.

Accordingly, social network analysis examines such issues as the importance or prominence of particular individuals in the network; the concept of centrality, i.e., the individual in the network with the most—or most important—ties to other actors; the notions of closeness and distance based on communication paths among the actors in the network; the notion of cohesive subgroups, that is, subsets of actors among whom there are relatively strong, direct, intense, frequent, or positive ties; the extent to which the relationships and transactions within it are regulated by explicit or tacit rules; and the number and diversity of actors within the network. Whatever the focus, however, there is a recognition of the flexibility and dynamism of social networks, qualities that stem from the ways in which ties are constantly formed and strengthened or weakened and broken.

Partly because of this dynamism, some sociologists conclude that network-based organizations are capable of superior performance than are more traditionally structured hierarchical organizations, especially in terms of adaptability to changes in their environment. This conclusion is reinforced by a growing literature on business networks—literature that has particular relevance to the discussion here since organized crime is perhaps best understood in quasi-

---

[11]Stanley Wasserman and Katherine Faust, *Social Network Analysis: Methods and Applications* (Cambridge: Cambridge University Press, 1994) pp. 11–12.

[12]Ibid. p. 8.

Clausewitzian terms as the continuation of business by criminal means.

Business Networks. The focus on networks in business has emerged in response to the limitations, rigidities, and inefficiencies associated with strict hierarchical structures, the need to exploit globalization through partnerships and strategic alliances, and a desire to emulate the Japanese success with the *keiretsu* (regularized networks of suppliers that enhance the efficiency of the production process). It has also emerged out of a recognition that understanding the opportunities provided by "structural holes" (see below) can be critically important for the success of a business in a competitive environment.[13]

The notion of business networks has been developed most explicitly, however, in the concept of the virtual corporation and its dependence on what are sometimes termed "agile networks."[14] Such notions place considerable emphasis on flexible internal communication networks; connections to other organizations; shared interests in obtaining certain outcomes; the need to respond rapidly to external opportunities and challenges; the capacity for environmental scanning, rapid information-processing, and quick decisionmaking; and the capacity of the organization to learn and adapt.

It is clear even from this abbreviated discussion of network organizations in business that a rich and varied research agenda has resulted in many important insights, some of which are as relevant to the functioning of criminal networks as they are to any other kind of enterprise network. Indeed, another important strand of research underlying the present analysis is the studies of networks that have been undertaken by analysts focusing directly on organized crime.

---

[13]Ronald S. Burt, *Structural Holes: The Social Structure of Competition* (Cambridge, Mass.: Harvard University Press, 1992).

[14]See Alf Steinar Saetre and David V. Gibson, *The Agile Network: A Model of Organizing for Optimal Responsiveness and Efficiency,* www.utexas.edu/depts/ic2/aamrc/saetr-ex.html.

Dimensions of Criminal Networks

Although networks are an important, and somewhat neglected, form of criminal organization, they are not the sole or exclusive form. The traditional hierarchical model long associated with Mafia families in the United States, for example, does not need to be jettisoned: After all, it is possible to have networks of hierarchies, hybrid organizational forms with some hierarchical components and a significant network dimension, and even a network of networks. If networks come in a great variety of shapes, however, they vary along several critical dimensions.

First, a network can be created and directed by a core of organizers who want to use it for specific purposes (a "directed network") or it can emerge spontaneously as a mechanism to add efficiency to the functioning of a market (a "transaction network"). The Colombian cocaine trade in the 1980s and early 1990s was very much a directed network—at least at the core—which came into existence to transport cocaine to the United States. The heroin trade from Southeast Asia, in contrast, is far more of a transaction network, in which brokers play a critical role at almost every stage of the process. Producers supply heroin to independent distributors, and it is then passed along a chain of brokers until it reaches the retail market. In practice, of course, a directed network can be part of a larger transaction network, and it appears that with the demise of the large, vertically integrated networks operating out of Medellín and Cali, the Colombian cocaine trade has increasingly taken on this hybrid quality.

Second, networks can range from small, very limited associations at the local level to transnational supplier networks that move a variety of goods, either licit or illicit—or even both—across national borders. Membership can be determined by a particular characteristic, such as ethnicity, or can be relatively open. Supplier networks are likely to be multiethnic when instrumental considerations outweigh the desire or need to maintain a high degree of exclusiveness.

Among the larger criminal networks, it is possible to identify both key individuals and key companies or firms through which they operate. One of the best examples of an extensive transnational criminal network is that revolving around Semeon Mogilevich. Based in Hungary,

Mogilevich is reputed to have close links with the Solntsevo criminal organization in Moscow, with prostitution activities in Frankfurt, with the Genovese family in New York, and with Russian criminals in Israel. For several years, Mogilevich operated in part through a company called Magnex YBM operating in the United States and Canada (where it was engaged in money laundering—the process of turning criminal proceeds into clean money by hiding origin and ownership—and stock fraud) and also had a network of companies in the Bahamas, the British Channel Islands, and the Caymans. Significantly, as a key figure in this transnational network, Mogilevich is far less vulnerable than a traditional Mafia don or family head, and, despite continued allegations about his role, he has never been convicted of any crime.

Third, networks can be highly structured and enduring in nature or they can be loose, fluid, or amorphous in character, with members coming and going according to particular needs, opportunities, and demands. Some individuals or even small organizations will drift in and out of networks when it is convenient for them to do so. Other networks will have a more enduring membership. In yet other cases, there will be some members who provide continuity (and direction) to the network, while others will play an occasional or ephemeral part. There will be both "embedded ties" and enduring relations based on high levels of trust, mutual respect, and mutual concern; but also more fleeting relations based on nothing more than a short-term coincidence of interests. A similar dynamism is evident in the way in which criminals develop and use front companies, creating them wherever opportunities exist and abandoning or closing them whenever they become the targets of law enforcement investigations.

Last, networks can be focused very narrowly on a single purpose or on the supply of a single product, or they can supply a broader range of illegal products or engage in more diverse criminal activities. Colombian and Mexican drug trafficking organizations, for example, engage in a very narrow range of activities. Although there has been a tendency to traffic in more than one kind of drug, essentially they are in the drug trafficking business and little else. Russian and Chinese criminal organizations, in contrast, have a very diverse portfolio of criminal activities, trafficking in drugs, stolen cars, arms, prostitution, antiquities, and endangered species, yet also engaging in various forms of extortion and financial fraud.

Whatever their precise characteristics, networks provide criminals with diversity, flexibility, low visibility, durability, and the like. Indeed, their attractions are very considerable:

- Networks can often operate clandestinely. The more visible a criminal enterprise the more likely it is to be attacked by law enforcement. One of the most significant points about networks, however, is that they are not immediately and obviously visible. Criminal networks can hide behind various licit activities, can operate with a lower degree of formality than other types of organization, and can maintain a profile that does not bring them to the attention of law enforcement. In some cases, of course, the network will be exposed. Significantly, though, when the FBI began to investigate the Mogilevich criminal network, there was considerable surprise at its extensiveness.

- Even when they are targeted by law enforcement, many criminal networks are inherently dispersed, with the result that they do not provide obvious centers of gravity or loci for law enforcement attacks. Lacking a physical infrastructure or a large investment of sunk costs that would add significantly to their vulnerability, networks can also migrate easily from areas where risks from law enforcement are high to areas where the risks are much lower.

- Criminal networks, especially when they are transnational in character, can exploit differences in national laws and regulations (Israel, for example, only criminalized money laundering in 2000) by engaging in what might be termed jurisdictional arbitrage. Throughout the 1990s, for example, criminals from the former Soviet Union flooded into Israel, exploiting both the law of return and the lack of anti–money laundering measures. In some cases, money from Russia was used in Israel to buy up virtually bankrupt businesses that would then start to make "profits" that flowed back to Russia. In some instances transnational criminal organizations also create jurisdictional confusion, making it difficult for any single nation's law enforcement agencies to act effectively against them. Laundering money through a series of firms and banks in multiple jurisdictions, for example, makes it arduous and costly for law enforcement to follow the money trail.

- Networks also offer opportunities for both redundancy and resilience. In network structures, it is easier to create redundancies than it is in more formal and rigid organizations—so that even if part of the network is destroyed it can still operate. Furthermore, degradation of a network does not necessarily lead to its demise: Networks are very resilient and can easily be rebuilt.

In view of these advantages, it is not surprising that network structures have become particularly prevalent in contemporary organized crime, whether in the United States, Europe, or states in transition such as Russia, Ukraine, other newly independent states of the former Soviet Union, South Africa, and Cambodia, or even China and Cuba. Accordingly, the analysis now looks at the main characteristics of criminal networks, characteristics that help make them extremely difficult to combat.

TYPICAL CHARACTERISTICS OF CRIMINAL NETWORKS

Network Cores

Networks of any substantial size will generally have both a core and a periphery, reflecting asymmetries of power, influence, and status within the network. The core is characterized by dense connections among individuals who, in the case of a directed network, provide the steering mechanism for the network as a whole. Usually the originators of the criminal enterprise, the core members initiate specific criminal activities, arbitrate disputes, and provide direction. Their relationship is often underpinned by bonding mechanisms that help to create high degrees of trust and cohesion.

In many cases, bonding will be directly related to family or kinship: Many Italian Mafia groups are still organized along family lines, while Turkish drug trafficking and criminal organizations are often clan based. Other bonding mechanisms include ethnicity and common experience in which the participants develop a strong sense of trust and mutual reliance.

Membership in youth gangs or time spent together in prison can also provide critical bonding mechanisms. In the United States, the Mexican Mafia (which is not actually Mexican) started as a prison gang in

Southern California but has developed much more extensively. Yet, it is the common experience that continues to give the core of the network a capacity to operate with confidence that disloyalty or defection are unlikely.[15]

If network cores exhibit strong collective identities, cohesion does not necessarily enhance—and can actually reduce—the capacity to obtain information and "mobilize resources from the environment." Indeed,

> recent trends in network analysis posit an inverse relationship, in general, between the density/intensity of the coupling of network ties on the one hand and their openness to the outside environment on the other.[16]

This explains the attraction of a two-tier structure in which the weaknesses of the core in carrying out the functions of information acquisition are more than offset by the periphery.

Network Peripheries

This zone features less dense patterns of interaction and looser relationships than the core. Yet, these characteristics play a critical role in networks, exhibiting and exploiting "the strength of weak ties."[17] In effect, the periphery allows the network to operate at a far greater distance—both geographically and socially—than would otherwise be the case, facilitating more-extensive operations, more-diverse activities, and the capacity to carry out effective intelligence collection.[18]

---

[15]The analysis here and the discussion of bonding mechanisms rests heavily on Ianni, 1974, pp. 282–293.

[16]See David Stark and Gernot Grabher, "Organizing Diversity: Evolutionary Theory, Network Analysis, and Postsocialist Transformations," in Stark and Grabher, eds., *Restructuring Networks: Legacies, Linkages, and Localities in Postsocialism* (New York and London: Oxford University Press, in press).

[17]Mark Granovetter, "The Strength of Weak Ties," *American Journal of Sociology,* Vol. 78 (1973) pp. 1360–1380.

[18]Ibid. and Burt, 1992.

The Cali cartel, for example, was generally thought of as a highly centralized and structured organization. In fact, it was a networked structure with a set of key figures at the core and a periphery that included not only those directly involved in the processing and transportation of cocaine, but also taxi drivers and street vendors who were an invaluable source of information at the grass-roots level.

For criminal networks, this capacity is critical, because it allows them to anticipate and thereby to neutralize many law enforcement initiatives. Indeed, sensitivity to both threats and opportunities is vital to the continued functioning of criminal networks, making them highly adaptable. In this sense, criminal networks resemble agile corporations: The ability to obtain advance warning is complemented by a capacity for rapid reconfiguration of internal structures and operational activities. Because they have limited fixed assets, networks not only have limited exposure to risks but also adapt in ways that further reduce this exposure and exploit the paths of least resistance.

Criminal Networks As Defensive Structures

If criminal networks usually have early warning mechanisms that provide the first line of defense against law enforcement, there are also additional defensive mechanisms that can be integrated very effectively into their network structures. As I have noted elsewhere:

> Two-tiered networks . . . with both core and periphery have formidable internal defense mechanisms. While it is possible for law enforcement to infiltrate the periphery of the network, getting in to the core is much more difficult partly because entry is dependent on a high level of trust that is based on bonding mechanisms rather than functional utility. Moreover, there will usually be several nodes in the network which act as built-in insulators between core and periphery, distance the core leaders from operations, and make it very difficult for law enforcement to strike at the center of gravity as opposed to nibbling around the edges. The concomitant of this, of course, is that the periphery is where the risks from law enforcement are greatest. Ultimately, however, this is not too serious a problem . . . if parts of the periphery are seriously infiltrated or compromised, they can

simply be discarded and new members recruited for the outer reaches of the network.[19]

Compartmented networks are good at protecting not only the core membership but also information (while also having effective information flows from periphery to core) that could compromise criminal operations or the integrity of the network. Criminal networks compartmentalize knowledge and information, making it difficult for law enforcement to have more than localized effects on their operations. This is not to deny that, on occasion, there will be defectors or informants whose testimony enables law enforcement to inflict considerable damage, as happened to the Sicilian Mafia in the 1980s.

For the most part, however, networks are very good at self-protection. This is even true when they operate outside the home state—where ethnicity and language become additional defensive mechanisms. U.S. law enforcement agencies, for example, find it difficult to infiltrate Nigerian and Chinese networks in the way they did La Cosa Nostra. Electronic surveillance is also highly problematic since many of the networks use unfamiliar languages or dialects. An additional problem is that criminal networks based on ethnicity are generally located in ethnic communities that offer cover, concealment, and a constant supply of recruits. This is an important factor, for example, in explaining the success of Turkish drug trafficking organizations, and more recently Albanian criminal networks in Western Europe.

Criminal Networks As Facilitators of Cooperation

Even distinctively ethnic networks are not exclusive in their collaboration. Part of the reason for this is that although such networks do not lack organizational identities, they are not overly preoccupied with organizational form. Criminal networks come together with one another when it is convenient or beneficial for them to do so without this being a threat to their identity or *raison d'etre*.

Connections among different criminal networks became a major feature of the organized crime world during the 1990s. Colombian-

---

[19]Phil Williams, "Drug Trafficking Networks," *Current History* (April 1998) pp. 154–159.

Sicilian networks brought together Colombian cocaine suppliers with Sicilian groups possessing local knowledge, well-established heroin distribution networks, extensive bribery and corruption networks, and a full-fledged capability for money laundering. Italian and Russian criminal networks have also forged cooperative relationships, while Colombian and Russian criminals have been meeting in various Caribbean islands to engage in guns-for-drugs deals. The importance of these network connections has been evident in increased seizures of cocaine imported to, or transshipped through, Russia. There have also been reports of Colombian money laundering activities taking place in Russia and Ukraine, something that would not be possible without some kind of network collaboration. The result of such collaboration, of course, is the creation of a network of networks. These super-networks or pan-networks come into existence for various reasons and operate at a variety of levels, varying in scope, duration, and intensity. As Clawson and Lee put it:

> At the lowest level are simple buyer-seller deals involving relatively small investments, little advance planning, and relatively little interaction between the parties. At the highest level is what might be called strategic cooperation, which encompasses the principles of long-term agreements, large volume shipments of both drugs and money, and the creation of specialized infrastructure to facilitate these flows.[20]

The latter can appropriately be characterized as strategic alliances.

Some criminal networks develop steady supplier relationships with one another along the model of the Japanese *keiretsu,* while others develop contract relationships for the provision of certain kinds of services, such as transportation, security, contract killing, and money laundering. Turkish drug traffickers in Belgium, for example, can buy services from Georgian car thieves to meet their transportation needs.

Even some of the more traditional criminal organizations, such as the Sicilians, are also reliant on networks of cooperation to ply their crim-

---

[20]Patrick L. Clawson and Rensselaer Lee III, *The Andean Cocaine Industry* (New York: St Martins, 1996) p. 84.

inal trade. Indeed, Turkish drug traffickers in Italy have links with the Sicilian Mafia, the Sacra Corona Unita, and the Calabrian 'Ndrangheta. In 1993, a narcotics trafficking network in Turin involved Calabrians, Turks, Pakistanis, and members of the Cali drug trafficking organization, forming what was clearly a highly cosmopolitan criminal network. In short, the internal flexibility of network organizations is mirrored in the capacity of criminals to form and operate flexible alliances and other cooperative ventures. Indeed, network structures facilitate cooperation among criminal enterprises in the same way they facilitate cooperation in the licit business world.

Criminal networks are also able to draw on a whole set of support structures, whether through acts of paternalism in the community or through more strictly financial considerations. Among the support structures are groups that provide false documentation, front companies, transportation, and a financial infrastructure that can be used to move the proceeds of crime. The creation of false documents facilitates the movement of various kinds of contraband and people and offers an extra layer of protection for those involved in criminal activities. A hint of the scale of the support structure was revealed in May 1998 when agents from the Los Angeles branch of the Immigration and Naturalization Service (INS) disrupted a counterfeit document operation and seized more than 24,000 counterfeit documents, sophisticated printing equipment, and 50,000 blank social security cards.[21]

Criminal Networks As Boundary Spanners[22]

Another closely related advantage of criminal networks is their capacity to flow around physical barriers and across legal or geographical boundaries. Networks transcend borders and are well-suited for business operations in a world where responding to the opportunities and challenges posed by globalization has become an imperative. It is no exaggeration to suggest a natural congruence between transnational

[21]INS, "INS Busts Major Counterfeit Document Ring," Press Release, May 21, 1998.

[22]I am grateful to my colleague Professor Kevin Kearns of the Graduate School of Public and International Affairs, University of Pittsburgh, for bringing this term to my attention.

or cross-border activities and network structures, irrespective of whether the networks operate exclusively in the legitimate sector or in supplying illicit (prohibited or stolen) goods and services.

The capacity to cross national borders creates several advantages for criminal networks. It enables them to supply markets where the profit margins are largest, operate from and in countries where risks are the least, complicate the tasks of law enforcement agencies that are trying to combat them, commit crimes that cross jurisdictions and therefore increase complexity, and adapt their behavior to counter or neutralize law enforcement initiatives. One important boundary crossing, of course, is that between the criminal world and the "upperworld." Criminal networks extend across this boundary in ways that are sufficiently important to require discussion in a separate category, which follows below.

Criminal Networks As Creators and Exploiters of Corruption

In a series of widely cited studies, Ronald Burt has provided considerable insight into the effective functioning of networks by means of his concept of structural holes. Defining structural holes as the separation between nonredundant contacts, Burt contends that the information benefits provided by large diverse networks are greater than those provided by small homogenous networks since size and diversity provide more nonredundant contacts.[23] Consequently, extending networks to cover structural holes provides important competitive advantages.

Closely related to this, he argues, relationships can be understood as social capital that can be exploited to benefit the enterprise. Networks provide access to people with specific resources, which create mutually advantageous information benefits and exchange relationships.[24] This is as relevant to organized crime as to business and helps to explain why criminal organizations extend their networks into the licit world. Further, extending a network into government provides access to both information and power.

---

[23]Burt, 1992.
[24]Ibid.

For criminal networks, spanning structural holes is particularly beneficial when it also involves crossing from one domain into another. By crossing from the underworld into the worlds of government, business, and finance, criminal networks not only identify and exploit new criminal opportunities, but also enhance their capacity to protect existing activities and opportunities.

The specific connections that facilitate criminal entry into the licit world can be understood as gateways or portals, while the relationships at the boundaries of the criminal network and the world of government and/or licit business can prove vital to a whole series of criminal operations and activities. For criminal organizations involved simply in theft, for example, the critical node is the person who can fence the goods, who in effect transfers them from the criminal network back into the world of legitimate business and commerce.

At a more sophisticated level are found the lawyers, accountants, bankers, and other financial professionals who help criminals both to conceal and to invest their profits. This facilitates the flow of criminal proceeds back into the legitimate financial system, where it rapidly becomes indistinguishable from money that has been obtained by legal means. In New South Wales, for example, lawyers, accountants, and financial managers have been categorized as "gatekeepers" for organized crime and targeted accordingly.

Perhaps most important of all, however, are the members of law enforcement agencies and government officials whose link to criminal networks involves exchange of information or protection for money. In the case of politicians, the exchange can be about personal gain but might also be about assistance in mobilizing the vote, support for electoral campaigns, criminal assistance in providing information about political opponents, or even in intimidating and, in extreme cases, eliminating political enemies. In the case of law enforcement personnel or members of the judiciary, the aim of the criminals is to minimize risks by undermining enforcement efforts, suborning the judicial process, and neutralizing the criminal justice system.

In short, criminal organizations extend their reach by coopting individuals and organizations in ways that facilitate, enhance, or protect their activities. The corruption networks they create are dynamic rather than static, increasing in significance as corrupted officials become more senior. In these circumstances, the exchange relationship between them becomes much more substantial in terms both of the favors done by the official and the payoffs provided by the criminal. While the official is not part of a criminal enterprise, he has become a vital node in a criminal network, providing important services including timely intelligence about law enforcement initiatives. In countries such as Turkey, Mexico, Colombia, Nigeria, and Russia, criminal networks have extended their reach into the domains of both commerce and government, thereby increasing their capacity to accrue large profits while simultaneously reducing the risks they have to confront.

Criminal Networks As Robust and Resilient Organizations

Networks are highly resilient, partly because of what might be termed loose coupling. Charles Perrow distinguishes between tightly coupled and loosely coupled systems. He contends that tightly coupled systems are the least stable because disturbances involve a chain reaction or, at the very least, serious knock-on effects. In contrast, "loose coupling gives time, resources, and alternative paths to cope with the disturbance and limits its impact."[25] Criminal networks—apart from the core—are based largely on loose coupling. Even if some parts of the network are destroyed, the effects are limited since other parts are left intact. In a loosely coupled network, knock-on or cascading effects are limited and damage to one part of the network does not undermine the network as a whole. Loose coupling also preserves more diversity, in response offering considerable latitude in the decision of which parts of the network should respond, in what manner and in what location.

Resilience, however, stems not only from the capacity to limit the damage that is inflicted but also from the ability to mitigate consequences. Criminal networks often develop certain forms of redundancy that facilitate recovery if part of the network is degraded or damaged. In legitimate business, redundant contacts in networks—

---

[25]See Charles Perrow, *Normal Accidents* (New York: Basic Books, 1984) for a fuller analysis of tight and loose coupling. The quote is from p. 332.

and indeed redundancy in general—are generally seen as wasteful and inefficient. For criminal networks, such costs are greatly outweighed by the benefits of redundancy in the face of attack and degradation by law enforcement.

The more redundancy in the network, the more options there are to compensate for law enforcement successes whether in finding new ways of moving illicit commodities to the market or alternative routes and methods of repatriating profits. In effect, redundancy enables members of the network to take over tasks and responsibilities from those who have been arrested, incarcerated, or killed by law enforcement. The diversity of different connections allows the network to function even if some connections are broken—not least because the nodes and connections that remain intact can be redirected. In effect, network redundancy makes it possible to maintain organizational integrity even in an extremely inhospitable environment.

Criminal Networks As Synergistic Organizations

Although social networks exist independently of technological networks, there are major synergies among the two distinct network forms. The ability of transnational criminal networks to exploit the information and communication networks that developed during the 1990s provides major multiplier benefits. While exploitation of information technologies is certainly not the sole prerogative of network-based organizations, networks are extremely well placed to exploit new technological opportunities. Indeed, many criminal organizations have been using technology as a force multiplier to carry out their entrepreneurial activities with greater efficiency at lower cost.

There is one set of technologies in particular that could give them an enormous advantage in their continued competition with law enforcement—strong forms of encryption. One of the most potent weapons of law enforcement in its struggle with organized crime has been the capacity to monitor communications among criminals—in effect, to identify and listen in on network connections. Encryption provides opportunities to neutralize this capacity and offers criminal networks a form of strategic superiority—the equivalent of an effective strategic defense initiative, at very low cost and based on off-the-shelf technologies.

In sum, criminal networks provide moving and elusive targets that operate across enemy lines, infiltrating law enforcement agencies and governments, avoiding confrontation in favor of cooption and corruption. They are resistant—although not impervious—to damage and have qualities that facilitate recuperation and regeneration. Before looking more fully at criminal networks in action, however, it is necessary to identify some of the critical roles that their members must play if the network is to maximize its potential.

ROLES IN CRIMINAL NETWORKS

Networks feature a considerable division of labor among members. Indeed, it is possible to identify a series of critical roles, some of which occur in all networks, and others that are found in specific types of "business" in which criminal networks are involved. In some networks, the tasks will be implicit and intuitive; in others, they are explicit and formal. In most criminal networks, the following roles are likely to be discernible:

Organizers. Those core individuals and groups that provide the steering mechanism for the network. These organizers will generally determine the scale and scope of activities and guidance and impetus for their execution.

Insulators. Individuals or groups whose role is essentially to insulate the core from the danger posed by infiltration and compromise. These individuals transmit directives and guidance from the core to the periphery of the network. They also ensure that communication flows from the periphery do nothing to compromise the core.

Communicators. Individuals who ensure that communication flows effectively from one node to another across the network as a whole. Their responsibility is to transmit directives from the core group and provide feedback. In some cases, insulators and communicators will be at odds because of competing impulses inherent in their differing responsibilities; in other cases the same individuals will combine both roles and make appropriate trade-offs.

Guardians. Enforcers concerned with the security of the network who take measures to minimize vulnerability to external attack or infiltration. Precautions about exactly who is recruited to the network combine with measures to ensure loyalty through a mix of ritual oaths and latent coercion directed against the new members or their families. Guardians act to prevent defections from the network, or in the event that such defections take place to ensure that the damage is minimized.

Extenders. Those whose role is to extend the network by recruiting new members, by negotiating with other networks regarding collaboration, and by encouraging defectors from the world of business, government, and law enforcement. Where the extenders are successful, the network will have access to the portals of the licit world discussed above. Among the tactics that extenders typically use are voluntary recruitment through bribery and corruption and involuntary recruitment through coercion, sometimes leavened by the addition of rewards or inducements. Their targets typically include important and powerful politicians who can provide a high degree of protection, bureaucrats in particularly sensitive or pivotal positions, and financial managers who provide access to legitimate financial institutions.

Monitors. Those who ensure the effectiveness of the network and whose responsibilities include reporting weaknesses and problems to the core organizers, who can then initiate remedial action. These network members are particularly crucial in ensuring implementation and providing guidance on appropriate corrective measures where necessary. They ensure that the network is able to adjust to new circumstances and maintain the high degree of flexibility that is critical to the capacity to circumvent law enforcement.

Crossovers. People who have been recruited into a criminal network but who continue to operate in legal institutions, whether governmental, financial, or commercial. Such people not only meet Burt's test of nonredundant contacts, but by operating in a different sphere from most of the network, they are able to provide invaluable information and protection.

These roles appear widely across criminal networks, regardless of their particular specialties. There are, however, other more specific roles that also have to be filled. Drug trafficking networks, for example, require chemists to oversee the processing of raw materials into finished products. Although such individuals are critical to the productive capacity of the network, their network role might be very limited. Detailed case studies of networks, of course, require the delineation of both general network roles and specific functional roles. The main purpose of the foregoing analysis here, though, is to illuminate the general functions and structures of networks.

CRIMINAL NETWORKS IN ACTION

Criminal networks are characterized by diversity in composition, density of connections, size, structure, shape, underlying bonding mechanisms, degree of sophistication, and scope of activities. This section briefly examines several criminal networks that were either transnational in scope or were the localized components of a transnational network.

The Spence Money Laundering Network

In one case, a money laundering network in New York that was not very sophisticated succeeded in laundering over $70 million for Colombian drug traffickers. The network was a fascinating mix. It included a taxi driver, an honorary consul-general for Bulgaria, a New York City police officer, two rabbis, a firefighter, and an attorney. The network was very amateurish in its methods, bringing large amounts of cash—which represented the proceeds of drug trafficking—to a Citibank branch on a regular basis and thus triggering a suspicious activity report. The deposits were transferred to a bank in Zurich where two employees forwarded the funds to the Caribbean account of a major Colombian drug trafficker. In spite of the diversity of those involved, the movement of the money across jurisdictions, the involvement of banking officials in Zurich, and the ultimate beneficiary, the network exhibited a surprising lack of sophistication.

The Cuntrera-Caruana Clan

A more sophisticated, network-based criminal group is the Cuntrera-Caruana family, which has played a critical role both as network extenders and as a network core for many other evolving criminal networks. For a long time, however, the family was overlooked, and its role in both drug trafficking and money laundering networks was only barely discerned. One reason is that "although structurally they were at the center of things, geographically they were at the outskirts. They did not come from Palermo, they did not move to New York."[26]

Exiled from Sicily in the 1960s, the Cuntrera-Caruanas initially went to Brazil before establishing themselves in Venezuela and Montreal. The clan has been described as

> a close wicker-work of blood-relations composed of family-nucleuses in different countries all over the world, joined with an equal wicker-work of economical and industrial connections, intended to improve their networks for international traffic in narcotics and money-laundering.[27]

As such, the clan provided important nodes in a whole series of drug trafficking and money laundering networks and was critical in linking Colombian drug suppliers with 'Ndrangheta families that distributed cocaine in Italy. In spite of its relatively low profile, the clan has suffered some setbacks. Three of the Cuntrera brothers—Pasquale, Paolo, and Gaspare—were deported from Venezuela in September 1992 and arrested on their arrival in Rome. In 1996, Pasquale Cuntrera was sentenced to 20 years, while his brothers both received 13 years. Nevertheless, the clan continues to operate and is involved in extensive criminal activities in Canada, partly through links with

---

[26]See Tom Blickman, "The Rothschilds of the Mafia on Aruba," *Transnational Organized Crime*, Vol. 3, No. 2 (Summer 1997) pp. 50–89. The analysis in this paragraph draws heavily on this article.

[27]Ibid.

outlaw motorcycle gangs, Asian-based criminal organizations, Colombian and South American groups, Eastern European–based organizations and Aboriginal-based organized crime groups.[28]

Outlaw Motorcycle Gangs

Another good example of criminal networks is provided by outlaw motorcycle gangs operating predominantly in the United States and Canada, but also in Britain and Scandinavian countries. The most famous are the Hell's Angels, which gradually evolved into criminal organizations, controlling prostitution and engaging in drug trafficking, often specializing in methamphetamine. Individual chapters of the Hell's Angels are organized along hierarchical lines that one close observer has compared to "little armies" with a president, vice president, secretary treasurer, sergeant at arms, and a road captain.[29] At the same time, these nodes are part of a much larger network that is held together by the same ethos, symbols, and sense of identity that often pits them against other outlaw motorcycle gangs.

Although the Angels have approximately 95 chapters in 16 different countries, their presence does not always go unchallenged. Part of this stems from a natural rivalry among different outlaw motorcycle gangs, and part of it stems from conflicts over drugs markets, something that has become increasingly intense as Mexican organizations have taken over much of the methamphetamine trade. In Canada, for example, there was continuing conflict throughout the 1990s between the Hell's Angels and Rock Machine, a gang based in Quebec and Ontario. Periodic outbreaks of open warfare among these groups resulted in several deaths. Even more intense was the conflict in Denmark between the Angels and the Texas Bandidos, a conflict that, on occasion, involved the use of rocket propelled grenades and antitank weapons.[30]

---

[28]Criminal Intelligence Service Canada, *Annual Report on Organized Crime in Canada*, 1998.

[29]Yves Lavigne, *Hell's Angels* (Secaucus N.J.: Lyle Stuart, 1996) p. 68.

[30]Dean E. Murphy, "Biker War Barrels Across Scandinavia: Swedish Legislator Campaigns to Evict Gangs After Rumbles Kill 6," *Los Angeles Times,* August 1, 1996, p. A19.

Immigrant Smuggling Networks

Criminal networks engage in a variety of enterprises, the most lucrative of which is alien smuggling. In 1998, the U.S. Immigration and Naturalization Service, in Operation Seek and Keep, dismantled a network of alien smugglers who for three years had smuggled up to 300 Indian nationals a month to the United States. Their business grossed an average of $70 to $80 million annually. The network had arranged air travel from India to Moscow to Cuba, boats to the Bahamas, and then either boats or planes to Miami. On occasion some of the illegal immigrants went from Cuba to Ecuador and were either brought into the United States via Miami or through Mexico and the southwest border.

The major investigative instrument used by the INS was wiretapping, which led to over 35,000 calls being intercepted. Most of the arrests were made between November 14 and November 19, 1998, and took place in the Bahamas, New York, New Jersey, Miami, Jacksonville, Tampa, Los Angeles, Fort Worth, Houston, Philadelphia, and San Juan, Puerto Rico—a diversity of locations that highlights the network structure of the people-trafficking organization.[31]

Crossover Figures

Most criminal networks extend into the licit world for support. Some of the larger and more powerful criminal networks, however, take this process to considerable lengths and in effect create crossover figures who have very high-level positions in government. Perhaps the most striking examples of this are Guilio Andreotti in Italy and Raul Salinas in Mexico. In the case of Andreotti, he was at the pinnacle of a pattern of collusive relationships between the Christian Democrats and the Mafia, relationships that until the 1980s offered protection and contract opportunities for the criminals and financial payoffs and political support for the Christian Democrat Party.

In Mexico, Salinas was able to amass a personal fortune as his reward for providing high-level protection and support for drug traffickers.

---

[31]INS Press Release, *U.S. Dismantles Largest Global Alien Smuggling Cartel Encountered to Date*, November 20, 1998.

Over $130 million was deposited in Swiss banks, much of it via Citibank in New York. General Guttierez, who was head of Mexico's antidrug unit yet in the pay of major drug traffickers, provides another example of the capacity of criminal networks to insinuate themselves into licit institutions in ways that are highly corrosive of the power, authority, and purpose of these institutions. It is this capacity that makes criminal networks so difficult to attack. Indeed, the next example highlights how a criminal network can embed or nest itself in a legitimate financial institution.

RUSSIAN INVOLVEMENT IN CRIMINAL NETWORKS

Money Laundering and Capital Flight Through the Bank of New York

In autumn 1999, reports appeared that about $15 billion from Russia had been laundered through the Bank of New York. Several officials at the bank were soon suspended. Further revelations suggested links between the Mabetex construction scandal in the Kremlin (in which a Swiss firm paid bribes for a very lucrative renovation contract), the Berezovsky Aeroflot scandal, an Italian criminal organization's money laundering activities, and the funds laundered through the Bank of New York. There were also allegations that one of the key figures in the money laundering was the aforementioned Semeon Mogilevich, a key organized crime figure based in Budapest.

Over the following 15 months, many aspects of the original story were either denied or qualified. Estimates of the amount of money involved were more than halved, and it was also suggested that most of this was capital flight and tax evasion money rather than the proceeds of crime. There was also a sense of frustration in U.S. law enforcement circles: Without the full cooperation of the Russian authorities, proving that prior crimes had been committed in Russia was virtually impossible.

In spite of all these qualifications, the Bank of New York money-laundering operation reveals very clearly the advantages that accrue through embedding a criminal network in a legitimate institution. In effect, what occurred was that a network of people wanting to move money out of Russia took advantage of a Bank of New York policy that had aggressively sought correspondent relationships with Russian banks without always exercising due diligence. It was this fundamentally sanguine approach—even though in 1994 there was congressional testimony indicating that 40 percent of Russian banks were controlled by organized crime—that made the bank and its officials vulnerable.

Central figures in the scandal were Lucy Edwards and her husband, Peter Berlin, both of whom pleaded guilty to a series of charges. According to court testimony, Lucy Edwards was approached by some Russians in Moscow she had met in her work at the East European division of the Bank of New York. The Russians controlled a bank called DKB and offered to pay Edwards and Berlin if they would assist in moving money from Russia through the Bank of New York. Berlin opened an account at the Bank of New York so that the Russians could obtain access to electronic banking software called micro/CASH-Register, which enabled them to wire-transfer money out of the account. Berlin created a front company, Benex International Company, Inc., (Benex). Lucy Edwards installed the software in a computer located in an office in Forest Hills, Queens, managed by individuals working for DKB. The Russians transferred funds into the Benex account almost daily, then the micro/CASH-Register software was used to transfer it to other accounts around the world.

In July 1996, Peter Berlin opened a second account at the Bank of New York in the name of BECS. In the fall of 1998, the Russians acquired control of Flamingo Bank and wanted a new bank account through which they could transmit funds on behalf of Flamingo. Berlin opened a third account in the name of Lowland and once again obtained micro/CASH-Register software. The Russians set up an office for Lowland in New Jersey.

In April 1999, Flamingo Bank began transferring large sums of money into the Lowland account using micro/CASH-Register software located in Russia to wire-transfer funds out. This facilitated contravention of Russian currency regulations and avoidance of custom duties and taxes. The scheme was also used to pay $300,000 in ransom on behalf of a Russian businessman who had been kidnapped in Russia. With around $7 billion passing through the accounts, Berlin and Edwards received a total of $1.8 million in commission payments.

In effect, this was a premier example of the way in which a criminal network, by coopting a critical and trusted bank official, is able to circumvent banking supervision and due diligence requirements, and embed or nest its activities within a legal and indeed highly reputable institution. In many ways, this is very typical of the style of Russian organized crime, the only difference being that, in this case, a Western financial institution and not simply a Russian bank was compromised.

Russian Organized Crime

Russian organized crime is a sprawling phenomenon that differs from city to city. It embraces ethnically based, non-Russian groups such as Chechens and Azeris and has developed symbiotic links with state and law enforcement apparatus in Russia. Russian criminal organizations also control a considerable portion of Russia's economic activity and have infiltrated key sectors of the Russian economy, such as banking, the aluminum industry, and the St. Petersburg oil and gas sectors. Although some of the major organizations have hierarchical structures, Russian organized crime can only be fully understood in terms of network connections between the underworld and the upperworld.

There are three manifestations of this phenomenon that are particularly important. First, there is cooperation among criminals, businessmen, and politicians who are part of the new iron triangle of network relationships that dominate Russian life. Using network analysis and software tools such as Analyst's Notebook, discussed more fully below, it is possible to trace some of these linkages. In some cases, the connections are made through common financial interests in one or more companies—interests that often make strange and surprising bedfellows.

Next, there is the use of violence to manage relationships that are anything but cooperative. For example, when criminal networks attempt to extend their influence into legitimate businesses and meet resistance, those resisting are often eliminated. Indeed, contract killings have become an important instrument of organized crime—and also a very visible indicator of organized crime infiltration of licit businesses or economic sectors.

Finally, there are figures who operate in both domains. One of the most notable of these figures is Yuri Shutov, a St. Petersburg Duma deputy who, until his arrest, ran a much-feared assassination squad. Shutov's team carried out a series of contract killings aimed at eliminating criminal rivals, removing obstacles to criminal takeover of the energy sector, and neutralizing threats from law enforcement authorities and reformist politicians.

CONFRONTING CRIMINAL NETWORKS

It is clear from the foregoing examples of criminal networks in action that they are formidable. This does not mean that they are invulnerable, however. Indeed, there are several ways in which governments and law enforcement agencies can respond more effectively to the challenges posed by criminal networks.

Although criminal networks are resistant to disruption and have high levels of redundancy and resilience, they are not impervious to attack by law enforcement. The nature of these networks, however, suggests that the attacks need to be carefully orchestrated, finely calibrated, and implemented in a comprehensive and systematic fashion. Indeed, there are several important prerequisites for initiating effective attacks on networks, especially clear delineation of objectives and enhanced intelligence assessments.

In attacking networks, it is vitally important to determine the major objectives: Are they to destroy the network, simply to degrade its capacity to carry out criminal actions, or to detach the network from its support apparatus in the licit world? The objectives can range from making operations more difficult for the network through creating instability in the environment to more direct attacks on the network itself that are aimed at disruption of its activities, dislocation or degradation of its capabilities, or even its compete destruction. While all are legitimate objectives, it is essential that there is clarity about precisely which of them is being chosen.

Even clear articulation of objectives is no guarantee that they will be achieved. One of the major problems in dealing with criminal networks is the absence of adequate models about precisely how these networks function. This is paralleled in the business world by a lack of understanding of why business networks succeed or fail. In neither domain has there been sufficient comparative work identifying patterns of success or failure. Specifically in relation to criminal networks, there has been little sustained empirical research on how these networks respond to different law enforcement initiatives. It is clear that networks react quickly and effectively to measures such as interdiction efforts and, for example, move their operations or find new modes of concealment and deception.

What is less clear, however, is the exact nature of their response when damage is inflicted upon them. If part of the network is compromised, for example, is it simply jettisoned or amputated and other components given increased responsibilities in an attempt to compensate, or are efforts made to regenerate the damaged portion of the network? Similarly, it is not always clear where the network starts and ends and whether an apparently successful attack has actually fulfilled its objective of significantly degrading or destroying the network. Damage assessment is always difficult; when it involves networks, it is even more problematic than usual.

To overcome these problems it is essential to develop more effective intelligence about criminal networks. In this connection, various software companies, working closely with the law enforcement community, have developed some important tools to assist with the intelligence analysis task. The three major packages are Analyst's Notebook produced by I2 (see www.i2inc.com), Orion Leads produced by Orion Scientific Systems (www.orionsci.com), and Watson Powercase (formerly owned by Harlequin) available from Xanalys (www.xanalys.com). Although these packages differ slightly in both power and usability (with some trade-offs between these two characteristics), they all have capabilities that assist in the identification of criminal networks. All of them, for example, have a component that facilitates telephone toll analysis of patterns of interaction among key individuals. The results of this analysis can be fed into what is generally referred to as association, network, or link analysis.

Such an approach helps to identify and assess the relationships or connections among people and organizations involved in crime, in effect helping to understand and visualize the network. Although it is often used for tactical purposes and for specific cases, link analysis could also be a valuable tool for strategic purposes. It could help, for example, in identifying some of the more important nodes in the network, in identifying key individuals who carry out the various network functions identified in the previous section, and in locating the portals or gateways through which the criminal network successfully crosses into the licit world. It could also be used strategically as an aid to damage assessment. In effect, the analytic intelligence process facilitates both identification and mapping of criminal networks.

Understanding network structures and operations makes it easier to identify vulnerabilities against which concentrated attacks should be directed. Particularly important in this connection is the identification of critical nodes.[32] A critical node in a network is one that generally has a high level of importance and a low level of redundancy. The importance can reflect the existence of certain specialized skills (which can be substantive in terms of the specifics of the criminal enterprise or related to the operation of the criminal network as a network) or the position of the node within the network. The low level of redundancy stems from the lack of adequate substitutes for those with these skills.

In terms of network functions, a critical node might be a person who is well connected and the focus for dense connections. If this person is removed and there are no readily available communication links, then the network could be severely degraded. On the other hand, even a few alternative communication links can provide the basis for reestablishing enough connectivity for the network to continue to function.

In addition to those nodes that are obviously critical, there are those that can become critical because of more general damage inflicted on

---

[32]This theme is developed in some very interesting ways by Sparrow, "Network Vulnerabilities and Strategic Intelligence in Law Enforcement," *International Journal of Intelligence and Counterintelligence,* Vol. 5, No. 3, Fall 1991. Sparrow's analysis provided both ideas and inspiration for this section.

the network. These nodes—the ones that are important but highly redundant—can become critical if they are attacked simultaneously or in close succession to one another. While this requires effective coordination, it is certainly an option that needs to be considered.

In attacking networks, it is also critical to target the boundaries, either from one network to another or from the criminal world to the upperworld. Particularly important in this connection are network extenders and crossover figures (defectors from the licit world), individuals who, in effect, straddle the boundary between the licit and illicit sectors and provide an important gateway for the criminals into licit financial, political, administrative, or business institutions.

Indeed, it is essential to disentangle the crime-corruption networks (and the nature of the exchanges between criminals and their clandestine supporters in the licit world) and thereby provide opportunities to detach the network from its various support structures. In part, the struggle between law enforcement and organized crime networks can be understood in terms of a competition in crossovers—informants and defectors from criminal networks on the one side and corrupted politicians, bureaucrats, law enforcement personnel, and members of the judiciary on the other.

Closely related to their ties to the upperworld, criminal networks can become deeply embedded in certain social, political, and economic structures that need to be attacked as a system. Perhaps the best example is the world of offshore financial centers and bank secrecy havens, which can be understood as a set of interlocking services provided to criminal networks that enables these networks to move, hide, and protect the proceeds of their criminal activities. Ironically, the providers of these services enjoy the protection of sovereignty. In effect, this puts the network crossovers out of reach and makes it necessary to attack the support system and not simply the network itself.

The other obvious target for attack is the network core. If the network is functioning effectively, however, and the insulation processes are working as intended, then this will prove extremely difficult. If the core figures are identified and removed, one of two results is possible. The first is that the network is so well-established—with the steering mechanisms so deeply embedded in operational procedures that operations have become more or less independent of the core group—that it can continue to function. A variation on this is that some of the figures who have been close to the core, but not necessarily part of it, can substitute for those removed from the steering group. The second possibility is that attacking the core group will significantly degrade the network and along with other measures, such as an attack on the gateways, will either force it to cease operating or, at the very least, significantly degrade its capacity and reach.

In effect, the options being discussed so far are all part of an external attack on the network. It is also possible to initiate internal attacks on criminal networks, however, where the objective is to create dysfunctional relations that seriously degrade the capacity of the network to function effectively. One option, for example, might be to destroy trust through misinformation and actions designed to create suspicion and acrimony. One way of doing this would be to identify some of the network crossovers and, rather than remove them, use them to feed misinformation into the network. Not only could this have a corrosive internal effect, but also it could encourage the criminals to move in directions that make them increasingly vulnerable to external attack.

One other important component of the response to defeating criminal networks is that governments and law enforcement agencies, in effect, need to mimic network structures. One of the advantages criminal networks enjoy is that they are smart, future-oriented organizations locked in combat with governments that, by contrast, are often hobbled by a variety of constraints. Governments still operate along hierarchical lines and are further hindered by bureaucratic rivalry and competition, interagency antipathies, and a reluctance to share information and coordinate operations. Working from John Arquilla's and David Ronfeldt's proposition that it takes a network to defeat a network, the most successful attacks on criminal networks are likely to be those carried out by innovative law enforcement structures that transcend the normal bureaucratic way of doing business.[33] Joint task forces, in which there are a pooling of resources and

---

[33]John Arquilla and David Ronfeldt, eds., *In Athena's Camp: Preparing for Conflict in the Information Age* (Santa Monica, Calif.: RAND, 1997).

information and a concerted attack on a particular criminal network, provide an important value-added approach to attacking criminal networks. They can be particularly useful where they involve transnational cooperation in response to a transnational criminal network.

Joint undercover operations have been particularly successful in this respect, largely because they provide access to crucial parts of the criminal network. Operation Green Ice in the early 1990s, for example, involved law enforcement agencies from eight countries and resulted in around 200 arrests in the United States, Spain, Italy, and Britain. In some cases, cooperation of this kind is even being institutionalized. One of the most forward looking agencies in responding to transnational criminal networks through the creation of its own networks has been the Financial Crimes Enforcement Network at the U.S. Treasury (FINCEN). Although FINCEN has its problems—and has been criticized, among other things, for its lack of performance indicators—its importance reflects the way in which the U.S. government has decided to attack not only criminal kingpins and criminal networks, but also the proceeds of crime. A key part of this has been an effort to combat money laundering by making it more difficult to introduce dirty money into the financial system without triggering either cash transaction reports (CTRs) or suspicious activity reports (SARs). In addition, the United States has developed laws for asset seizure and asset forfeiture that allow the confiscation of criminal profits. FINCEN has played a key role in this strategy and has acted as liaison with the financial community, encouraging banks to take on responsibilities such as "know your customer" requirements and the exercise of due diligence in all transactions. It is also the repository for the information provided by the banks through the CTRs and SARs.

FINCEN is one model of what has become known as a financial intelligence unit (FIU). Many other countries have developed their own variants of these units. Australia, for example, has its Transaction Reports and Analysis Center, known as AUSTRAC, while in Bermuda there is a Financial Investigation Unit. Generally FIUs have reporting and analytic functions; in some countries FIUs also have investigative responsibilities. The challenge, however, is that dirty money has become highly mobile, moving rapidly through multiple jurisdictions before being hidden in safe havens that place a high premium on financial anonymity and bank secrecy. The response to this has been to create a network of FIUs known as the Egmont Group. Established in 1997, the Egmont Group FIUs meet regularly for plenary sessions, while also exchanging information through a secure web site. As of May 2000, the Egmont Group had 53 operating units in as many countries. Although the national FIUs vary considerably in terms of skills, resources, and available technology, the network facilitates a multinational effort to combat money laundering. Given the speed, ease, and anonymity with which money can be moved around the global financial system, the Egmont Group, by itself, does not level the playing field, but it does make it less uneven.

Such developments are important, particularly when combined with what is a growing trend toward intelligence-led law enforcement. Yet there is still a gap between the prevalence and sophistication of criminal networks on the one side, and law enforcement networks on the other. Closing this gap and developing more-effective strategies to attack criminal networks has to be one of the priorities in government efforts to combat transnational organized crime in the 21st century. This requires changes in attitudes and ways of thinking, in organizational structures, and in the relationship between intelligence and action. Without these changes, criminal networks will continue to retain important advantages over those who are trying to combat them.

Chapter Four

GANGS, HOOLIGANS, AND ANARCHISTS—THE VANGUARD OF NETWAR IN THE STREETS[1]

John P. Sullivan

Editors' abstract. Street-level netwar is just around the corner. Sullivan (L.A. County Sheriff's Office) observes how gangs, hooligans, and anarchists are evolving so as to develop capacities for waging netwar. He concludes his analysis by affirming the need to build interagency networks to counter the networked adversaries now coming over the law enforcement horizon. This chapter updates and expands upon a series of his earlier papers, notably "Urban Gangs Evolving as Criminal Netwar Actors," Small Wars and Insurgencies, Vol. 11, No. 1, Spring 2000.

Contemporary society is undergoing significant changes that promise to alter—and in fact are altering—the nature of conflict and crime. Two significant factors are technological and organizational changes that enhance the power of relatively small groups. The information revolution allows small groups to exercise power and extend their influence in seconds, across vast distances. Accompanying this access and ability to move information via the Internet, cell phone, fax, and emerging digital technologies is a shift from hierarchies to network forms of organization. These two factors are ushering in an era of asymmetric threats, where nonstate actors can extend their influence

---

[1]This chapter's discussion of the capacity of street gangs to evolve into netwarriors originated in John P. Sullivan, "Third Generation Street Gangs: Turf, Cartels and Netwarriors," *Crime & Justice International*, Vol. 13, Number 10, November 1997. An expanded version of that paper appears in *Transnational Organized Crime*, Vol. 3, No. 2, Autumn 1997. The author thanks Robert J. Bunker for his insightful comments on earlier versions of this chapter.

and challenge states and their institutions to gain social, political, or economic influence.

This chapter examines the potential for certain small groups—in particular, urban criminal gangs—to embrace network forms of organization and doctrine, and to utilize advanced information technology to wage netwar. This potential lies on the blurry border between crime and warfare and threatens to change the faces of gangsterism and terrorism. It challenges state institutions such as the police, the military, courts, and political leadership to develop responses that protect the public and preserve civil order. The mutation of traditional urban gangs into a new generation that is organized and operates in the manner of many highly networked triads, cartels, and terrorists is highlighted. The chapter shows that politicization, internationalization, and sophistication are three factors that influence gang organization and help explain how a net-based threat can mature. Finally, the example of Black Bloc anarchists, who utilize sophisticated networked tactics and violence to stimulate political action, is reviewed to illustrate the difficulties police encounter when facing networked adversaries.

THREE GENERATIONS OF URBAN GANGS: THE EVOLUTION TO NETWAR

The information revolution is changing the nature of conflict and crime. Organizational changes benefit from these new technologies. As described by RAND analysts John Arquilla and David Ronfeldt, networks can prevail over hierarchies in this postmodern battlespace: "Power is migrating to small, nonstate actors who can organize into sprawling networks more readily than can traditionally hierarchical nation-state actors."[2] This conforms to the prediction made by eminent military historian Martin van Creveld,

> In the future war, war will not be waged by armies but by groups whom today we call terrorists, guerrillas, bandits and robbers, but

---

[2]John Arquilla and David Ronfeldt, "A New Epoch—and Spectrum—of Conflict," in John Arquilla and David Ronfeldt, eds., *In Athena's Camp: Preparing for Conflict in the Information Age*, Santa Monica, Calif.: RAND, 1997, p. 5.

who will undoubtedly hit on more formal titles to describe themselves.[3]

Thus, netwar may result, in some cases, in a distinct and perhaps re-
fined form of what we generally refer to as terrorism. In netwar, alliances between and among state and nonstate actors can fuel conflict.
Combinations of "internetted" transnational criminal organizations
(TCOs), triads, cartels, cults, terrorists, gangs, and other entities replace their more state-oriented predecessors. Again, Arquilla and
Ronfeldt are instructive with their view that

> "netwar" [is] an emerging mode of conflict (and crime) at societal
> levels, involving measures short of war, in which the protagonists
> use—indeed, depend upon using—network forms of organization,
> doctrine, strategy and communication.[4]

In the past, loosely organized violent criminal entities like gangs were
discounted as serious terrorist threats and excluded from discussions
of classic terrorism. But recognition of new technological and organizational factors shows that such entities are especially well-suited to
exploit new forms of conflict. Arquilla and Ronfeldt observe that

> netwar is about Hamas more than the PLO, Mexico's Zapatistas more
> than Cuba's Fidelistas, the Christian Identity Movement more than
> the Ku Klux Klan, the Asian Triads more than the Sicilian Mafia, and
> Chicago's Gangsta Disciples more than the Al Capone Gang.[5]

With a focus on gangs that especially concern urban police agencies,
this chapter explores the convergence of some of these groups on the
road to netwar. Gangs operating in urban areas have gone through
three generational changes—from traditional turf gangs, to market-
oriented drug gangs, to a new generation that may mix political and
mercenary elements. Particularly important for this evolution are

---

[3]Martin van Creveld, *The Transformation of War*, New York: The Free Press, 1991.

[4]John Arquilla and David Ronfeldt, "The Advent of Netwar," in John Arquilla and David
Ronfeldt, eds., *In Athena's Camp: Preparing for Conflict in the Information Age*, Santa
Monica, Calif.: RAND, 1997, p. 227.

[5]Ibid.

trends in the degree of politicization, internationalization, and so-phistication of a gang.

Politicization refers to the scope of political activity embraced by the gang. For example, most gangs are largely criminal enterprises, but some have begun to adopt varying degrees of political activity. At the low end, this activity may include dominating neighborhood life and creating virtual "lawless zones," application of street taxes, or taxes on other criminal actors. Gangs with more sophisticated political at-tributes typically co-opt police and government officials to limit in-terference with their activities. At the high end, some gangs have ac-tive political agendas, using the political process to further their ends and destabilize governments.

Internationalization refers to the spatial or geographic reach of the gang. Most gangs are extremely local in nature, generally spanning several blocks of turf. Other gangs operate as confederacies of smaller "cliques," working across entire metropolitan regions and nationally across state boundaries. At the high end, some gangs are cross-border, transnational, or even international in scope with outposts in foreign cities.

Sophistication addresses the nature of gang tactics and strategies, the use of weapons and technology, and organizational complexity of the gang. Gangs can demonstrate their sophistication in one or more of these areas. For example, a gang could use sophisticated infantry tac-tics to conduct ambushes, utilize networked organizational forms, or utilize information technology. The simplest indicator of sophistica-tion is the ability to seek and forge alliances or networks of smaller gangs. This potential is often underestimated, since networks (such as the umbrella "Crip," "Blood," "Folk," and "People" groupings) do not generally adopt the hierarchical forms of traditional organized crimi-nal entities familiar to police. It is important to recognize that net-works are not less sophisticated than hierarchies, just different. To a large extent, the diffusion of information technologies and/or net-worked organizational forms among urban gangs remains largely un-studied.

Each generation, as described below, has different characteristics, which affect the potential for what gangs may do on their own, when drawing inspiration from major criminal and terrorist organizations, or even when linking up with these organizations. The characteristics of these three gang generations are summarized in Figure 4.1. The attributes of gangs, hooligans, and anarchists in terms of the indicators of politicization, internationalization and sophistication are illustrated in Table 4.1.

The First Generation—From Traditional Street Gang to Soccer Hooligans

Traditional street gangs operate at the low end of extreme societal violence. Primarily turf-oriented, they are characterized by loose leadership with a focus on loyalty and turf protection in their immediate neighborhoods. Members of such gangs engage in opportunistic criminal activity, including rivalry for drug sales as lone individuals and as a gang competing with other gangs. These *first-generation* (turf) gangs are limited in political scope, are localized (often by city blocks), and are not highly sophisticated.

While traditional street gang activity does not constitute terrorism in the classic sense, it acts as a conflict generator that may provide the necessary foundation for some groups to evolve into complex internetted criminal entities, or even netwarriors. Examples of traditional

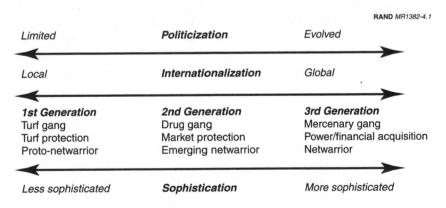

Figure 4.1—Characteristics of Street Gang Generations

Table 4.1

Netwar Potential of Gangs, Hooligans, and Anarchists by Indicator

Gang or Group	Politicization	Internationalization	Sophistication
Crips/Bloods	Low	Medium	Low-Medium
People/Folk	Low	Medium	Low-Medium
18th Street[a]	Low-Medium	High	Medium-High
Mara Salvatrucha[a]	Low	High	Medium
Football Firms[b]	Low	Medium-High	Medium-High
Gangster Disciples[a]	High	Medium	Medium
Black P Stone Nation	Medium	Medium	Medium
Vice Lords[a]	High	Medium	Medium
Calle Treinta[a]	Medium-High	High	Medium
Pagad[a]	High	Low-Medium	Medium
Hard Livings[a]	High	Low	Medium
Anarchists/Black Bloc[a]	High	High	Medium-High

[a]Displays emerging third-generation gang characteristics/netwarrior potential.
[b]For example, Inter City Firm, Service Crew, Ultra Sur, Tiger Commandos, etc.

gangs include the Chicago-based "Folk" and "People" Nations and Los Angeles–based "Crips," "Bloods," and their ubiquitous Hispanic counterparts including the notorious "18th Street" gang. Such street gangs, while violent, have been viewed as outside of the realm of true terrorism because of their lack of formal organization. In fact, one noted Los Angeles gang investigator, Wesley McBride, describes traditional gang activity as "disorganized crime in action."[6] Yet informal, network-like organization is one of the factors that may make netwar more difficult to combat than classic terrorism.

Analyst Robert J. Bunker sees this as one of the factors that may contribute to the rise of nonstate "soldiers." Other factors contributing to this emergence include failed communities with disintegrating social structure and increased sophistication. In Bunker's view, terrorism is an insidious form of warfare based on social rather than political con-

---

[6]This characterization is chronicled in Robert J. Bunker, "Street Gangs—Future Paramilitary Groups," *The Police Chief,* June 1996.

siderations.[7] He sees "armed nonstate actors intent on legitimizing forms of behavior that current societies consider to be criminal or morally corrupt" as a major security threat, with networked threats potentially organizing into "new warmaking entities, the smallest of which will be subnational groups, such as armed bands, private armies, and local crime organizations."[8] In that context, a web of dispersed, interconnected "nodes" gains relative advantage, since doctrine or common purpose is more important than hierarchical leadership. This view is similar to the observations of Arquilla and Ronfeldt regarding the emergence of netwar where societal-level conflicts are waged through internetted modes of communication, leading to the enhanced standing of nonstate actors.

While street gangs may be in a position to exploit changes in social organization, the majority of street gangs remain firmly in the first (turf focus) or second (nascent organized crime groups with a drug focus) generation. Gangs that span the first and second generations include some Crip and Blood sets, as well as a few Hispanic gangs such as 18th Street. These gangs tactically employ limited quasi-terrorism to support their activities. Like other organized crime groups, such street gangs use criminal terror, targeting each other, or interfering agencies, to sustain spatial or economic spheres of influence—"turf" or "markets." Finally, some gangs are becoming increasingly mobile, gaining in sophistication, developing links with organized crime entities, and maintaining caches of weapons financed by the large sums of money obtained from multistate or cross-border narcotics distribution networks.[9]

---

[7]See Robert J. Bunker, "The Terrorist: Soldier of the Future?" *Special Warfare,* Vol. 10, No. 1, Winter 1998, "Epochal Change: War over Social and Political Organization," *Parameters,* Vol. 27, No. 2, Summer 1997, and "Street Gangs—Future Paramilitary Groups," *The Police Chief,* June 1996, for a discussion of the potential emergence of the nonstate (criminal) soldier and related threats to security.

[8]Robert J. Bunker, "Epochal Change: War over Social and Political Organization," *Parameters,* Vol. 27, No. 2, Summer 1997.

[9]The 20,000 member, multiethnic Los Angeles drug gang 18th Street is a pertinent example. See Rich Connell and Robert Lopez, "An Inside Look at 18th St.'s Menace," *Los Angeles Times,* November 17, 1996, for an in depth review of the gang's operations. A discussion of 18th Street's cross-border activities and links with Eme (the Mexican Mafia) is found in Rich Connell and Robert Lopez, "Gang Finds Safe Haven and Base for Operations in Tijuana," *Los Angeles Times,* September 19, 1996.

An interesting variation on the traditional turf gang is the rise of soccer hooligans in Europe; for them, the important turf is not a neighborhood per se, but more a "virtual turf," as represented by their favorite local or national soccer team. Football hooligans are largely a European gang form, and soccer or football violence is a major social ill throughout Great Britain and the rest of Europe. Roving bands of thugs looking for street battles mar football games throughout the continent, injuring participants, police, and bystanders alike. These bands, known as "firms," differ from normal football supporters, since violence for violence's sake takes precedence over team success. Firms travel from stadium to stadium via ferry and train to engage their adversaries. They act like ethnonationalist marauders. British firms include the infamous "super-hooligan" armies, such as the Inter City Firm from West Ham and the Service Crew from Leeds. Other notable firms include Chelsea's Headhunters and Milwall's Bushwhackers. These firms generally have about 20 hardcore members, with total membership into the hundreds. British police divide football supporters into three categories. Category A involves typical fans; Category B consists of hooligans who occasionally embrace violence; and Category C is reserved for hardcore violent hooligans. All can engage in violence under certain circumstances.

British police estimate that there are approximately 600 hardcore (or Category C) hooligans in Great Britain. Up to 250–300 associates who participate in violence when under the influence of alcohol (Category B) supplement the violence orchestrated by the hardcore Category C nucleus. Category A members, the more typical football supporters, provide the outer fringe of the phenomena. Continental firms include Ultra Sur, which supports Real Madrid; Milan's Tiger Commandos and the Red & Black Brigade; as well as Barcelona's Boixos Nois (Crazy Boys).

What appears to have originated as spontaneous violence is now carefully coordinated and orchestrated. Hooligans are integrating network and hierarchical command and control to enhance their combat effectiveness with technology. Firms of opposing hooligans now use wireless technology and the Internet (email, etc.) for both marshalling their own combatants and challenging their opponents. For example, Milwall's Bushwackers are believed to have used Inter-

net tools (an interactive web site and real-time messaging) to organize and coordinate the violent activities of hooligans traveling to a
match in Cardiff, where 14 people were hurt and 6 arrested. Messages
had been posted both during and throughout the week prior to the
game, demonstrating the dangers of organized gangs of hooligans using the Internet. Police and the National Criminal Intelligence Service
(NCIS) in the U.K. believe the Cardiff melee confirms their fears that
organized soccer firms are turning to modern technology, including
computers and mobile phones, to avoid police intervention.[10] Wireless technology allows the firms to coordinate their battle plans and
issue tactical operations to precision-guide both pulsing and swarming attacks across Europe's borders.[11]

Second-Generation Street Gangs—The Influence of Organized Crime

Entrepreneurial, drug-centered street gangs occupy the *second generation*. Interested in market protection, these gangs view drugs as a
business. They use violence to control their competition and assume
a market rather than a turf orientation. With a broader, market-
focused political agenda, these gangs operate across a greater geographic (sometimes multistate) area, tend to have centralized leadership, and conduct more-sophisticated operations, particularly for
market protection. Second-generation gangs, like the first, are firmly
established criminal threats.[12] But while first-generation gangs operate much like clans, these second-generation types operate more like
professional businesses. Drug selling becomes a necessity rather than
an optional individual activity.

Second-generation street gangs utilize violence in a manner similar to
that of drug cartels, narcoterrorists, and other organized crime
groups. These gangs' activities—partly because of their focus on mar-

---

[10]"Football hooligans use Internet for rampage," *ITN Online,* August 8, 1999,
www.itn.co.uk/Britian/brit19990808/080801.htm.

[11]Tim Reid and Chris Boffey, "Behind the Thugs, 'Generals' Call Shots," *Sydney Morning
Herald,* June 22, 1998.

[12]See Malcolm W. Klein, *The American Street Gang,* New York: Oxford University Press,
1995. The distinction between street and drug gangs is found in Table 4-4, p. 132.

ket protection—fall into the center of the range of politicization, internationalization, and sophistication. Market protection may necessitate a step toward the use of violence as a political means, and these groups can be expected to exploit both violence and new forms of technology to extend their place in the criminal marketplace.

Crime and violence often blur when criminal enterprises seek to incapacitate enforcement efforts by the state, its police, and security forces. The use of violence as political interference has traditionally been a phenomenon of failed states, but a similar phenomenon occurs when street gangs exploit "failed communities" and dominate community life. As this exploitation occurs in the United States (particularly in neglected inner cities), it reflects what has already occurred in other countries where organized crime has taken hold in embattled "lawless zones." Particularly worrisome is narcoterrorism, which emerged as a result of collaboration between terrorist or guerrilla insurgents and drug cartels, and which involves the use of terror in support of cartel business. While the Andean cocaine trade and Colombian cartels are notorious for the use of narcoterrorism, recent instances of such terror are also evident in Mexico.

The extensive terror and disruption practiced by narcotics producing and trafficking cartels can incapacitate government enforcement efforts, paralyze the judiciary, fuel corruption, and destabilize local communities and nation states alike. Colombian experiences include bombings of public facilities, banks, commercial facilities, and aircraft and assassinations of supreme court justices, cabinet-level officials, magistrates, police officers, and journalists. Mexican cartels have targeted police officials in both Mexico and U.S. border states as well as journalists. Needless to say, cartels in both nations also target rival cartels.

Colombia's Medellín and Cali cartels and, more recently, the Arellano-Félix and other cartels in Mexico have employed extreme violence to further economic—and de facto emerging political—goals. Colombian *narcotraficantes* have recruited members from guerrilla insurgencies, such as FARC (Fuerzas Armadas Revolucionarias de Colombia), ELN (Ejército de Liberación Nacional) and M-19 (April 19th Movement); utilized guerrillas to conduct mercenary operations when their terrorist capability eroded; and solicited terrorist specialists, perhaps including the ETA (Basque Fatherland and Liberty), for technical assistance. Colombian cartels also on occasion target insurgent groups, such as M-19 and FARC themselves, while Peru's *Sendero Luminoso* (Shining Path) has used the drug trade to support its insurgency. Such examples illustrate the complex intertwining of politics, crime, and terrorism.

The interaction between cartels and terrorists poses a significant threat to legitimate governments. In this mix, drugs finance terrorists, while terrorists protect drug operations. Building upon relationships initially forged as marriages of convenience, the armed militias supporting contemporary cartels are equipped with modern weapons and seek to disrupt and preempt rather than merely react to government enforcement efforts. For example, FARC and ELN, with an estimated combined strength of 25,000 members, routinely disrupt oil pipelines and carry out wide-ranging acts of urban terrorism to undermine public support for the government.[13] The links between terrorist and organized crime (including drug traffickers) extends beyond the Andean drug states to form an integral node in the global network of organized criminal entities.

Drug cartels are not the only organized crime groups to engage in terrorist-like violence (or quasi-terrorism) to support their goals. Other nonstate actors—including millennialist cults and such transnational criminal organizations as the Asian Triads and other criminal enterprises—are also players. The Sicilian Mafia, the Russian mafiyas, and recently the Chinese 14K Triad have embraced bombings and high-order violence to influence government activity and achieve their goals. The Sicilian Mafia[14] has links with Colombian cartels, La Cosa Nostra (the American Mafia), and many Asian organized crime groups. Its tactics have included assassination, bombings, and kidnapping. Targets have included the police (including senior officers of the Carabinieri), magistrates, justices, and their families. Public

---

[13]John G. Roos, "The Enemy Next Door: Good Reasons to Hammer South America's Drug Cartels," *Armed Forces Journal International,* March 2000. p. 40.

[14]An early journalistic account of the evolution and power of the Sicilian Mafia's networking is found in Claire Sterling, *Octopus: The Long Reach of the International Sicilian Mafia,* New York: W.W. Norton & Company, 1990.

buildings, such as police stations, railway stations, restaurants, courts, and department stores, have also been attacked.

The Sicilian Mafia provides many useful insights into networking and politicization. The Sicilian Mafia (like the Colombian and Mexican cartels and at times the Russian mafiyas) is highly politicized, at times even capturing state institutions. As described by the Italian Parliamentary Anti-Mafia Commission, the Sicilian Mafia is a

> genuine criminal counter-power, capable of imposing its will on legitimate states. Of undermining institutions and forces of law and order, of upsetting delicate economic and financial equilibrium and destroying democratic life.[15]

Indeed, Manuel Castells sees the Sicilian Mafia as part of a "global, diversified network, that permeates boundaries and links up ventures of all sorts."[16] This network, which allows traditional criminal entities to escape the controls of a given state at a given time, includes the American Mafia, the constellation of Russian *mafiyas*, Colombian and Mexican cartels, Nigerian criminal networks, Chinese triads, and the Japanese *Yakuza* (itself a network of constituent entities including the *Yamagachi-gumi*, with 26,000 members in 944 networked gangs; the *Inagawa-kai*; and the *Sumiyoshi-kai*).

Castells observes that the Sicilian Mafia, by linking with this diffuse global network, remains one of the most powerful criminal networks in the world. Drawing from its deep penetration of the Italian State, particularly its political influence over the Christian Democratic Party, the Mafia was able to exert its presence throughout Italy's political, business, and financial infrastructure. When the state sought to reassert its autonomy, the Mafia reacted with unprecedented violence and brutality. When the state prevailed in its efforts to limit the Mafia's control of government institutions, the Mafia strengthened its international linkages and enhanced its networking, providing flexibility

---

[15]"Report of the Anti-Mafia Commission of the Italian Parliament to the United Nations Assembly," March 20, 1990, as cited in Claire Sterling, *Thieves World: The Threat of the New Global Network of Organized Crime,* New York: Simon & Schuster, 1994, p. 66.

[16]Manuel Castells, "The Perverse Connection: the Global Criminal Economy" (Chapter 3) in Manuel Castells, ed., *End of Millennium,* Oxford, U.K.: Blackwell Publishers, 1998.

and versatility to the organization. As a result, the Sicilian Mafia and its allied organizations embrace networking both internally in their own organizations and externally with other criminal organizations as a way to counter state interference. Thus flexible networking among local turfs and strategic alliances with other traditional criminal enterprises allow the Sicilian Mafia, like its partners, to achieve global reach.[17]

Russian organized crime groups, loosely known as the mafiyas, have utilized similar approaches, adding trafficking in weapons of mass destruction and local political dominance in parts of the former Soviet Union to their repertoire. Russian mafiyas operate in 60–65 countries and include about 6,000 separate groups, 200 of which operate in the United States. Within this series of loose, dynamic networks of semiautonomous entrepreneurs are a collection of gangs, crews, and individuals operating in a dozen major networks in up to 20 American cities. With a flexible and diffuse structure, these mafiya nodes are responsive to new threats and opportunities and are virtually impossible to decapitate.[18] These groups are flexible and often seek alliances with a variety of entities, including former KGB operatives and other criminal enterprises, including partnerships or service provision to relatively low-level gangs. A good example of a contemporary TCO, these mafiyas are often better organized and financed than the police. Their activities cross so many borders that intergovernmental coordination is awkward. In brief, they are less constrained by bureaucracy, have more money, and have access to the best technology.[19]

Triads are also players in the quasi-terrorist game. The Hong Kong–based 14K Triad has conducted an extended terror campaign in Macao.[20] Utilizing petrol bombs, attacks on police and journalists

---

[17]Ibid.

[18]Mark Galeotti, "Russia's Criminals Go Global, *Jane's Intelligence Review*, March 2, 2000, http://jir.janes.com/sample/jir0659.html.

[19]Richard C. Paddock, "A New Breed of Gangster Is Globalizing Russian Crime," *Los Angeles Times*, September 23, 1998.

[20]Niall Fraser and Adam Lee, "Bombers 'Targeting' Casino Watchdogs," *South China Morning Post*, October 24, 1997.

with secondary devices,[21] and attacks against infrastructure and public places, this campaign illustrates the use of terror as a political tool by a criminal enterprise. The 14K campaign exists within the vacuum of power resulting from the transition from Portuguese to Chinese rule. This transition has yielded an intertriad war, where organized crime groups including Hong Kong triads, mainland Chinese gangs from Guangdong Province, and corrupt security officials interact in an ever shifting web of alliances, business deals, and violence. Within this framework, the 14K campaign seeks to influence the police, government agencies, casino watchdogs, and other criminal groups.

These trends in transnational crime—and criminal netwar—abroad have, to some extent, worked their way into the United States. Further penetration must be guarded against, especially to prevent their linking up with American gangs. A danger is that gangs like the Bloods, Crips, or Gangster Disciples will be drawn increasingly into serving as wholesale trafficking and distribution networks for foreign cartels.

Third-Generation Gangs: Social, Political, and Mercenary Objectives

The overwhelming majority of street gangs are first or second generation. However, a small number of gangs in the United States and South Africa are also moving into the next generation. A *third-generation* gang is a mercenary-type entity focused on power and financial acquisition. These criminal netwar actors have significant political aims, aspire to operate at the global end of the spectrum, and are more sophisticated than first- and second-generation gangs. Currently, no gang is fully within the third generation, but some are moving swiftly along the spectra of politicization, internationalization, and sophistication. These gangs can be expected to utilize terrorism (including for hire) to advance their influence and can be expected to challenge the institutions of the nation state.

---

[21]In September 1998, 14K bombers conducted one bombing to attract police and the media and shortly afterward set off a second device injuring four police and ten journalists. See Steve Macko, "Macau Mobsters Sending a Bloody Warning to Police?" *ERRI Emergency Services Report*, Vol. 2-251, September 8, 1998.

For the most part, these gangs are mercenary in orientation; however, in some cases they share some political and social objectives. Examples of third-generation evolution among gangs in Chicago, San Diego, Los Angeles, and Cape Town, South Africa are provided in the following discussion.

The potential for third-generation street gangs first emerged in Chicago. It is home to approximately 125 gangs, including some of the most complex, well-disciplined gangs in the nation. A few demonstrate the sophistication and political savvy of third-generation evolution. These gangs not only create and sustain alliances, but act as paramilitary groups embracing terrorism or terrorism for hire.

The "El Rukn" gang (currently known as the "Black P Stone Nation"), which sought to carry out terrorism for hire on behalf of Libya is an early example. Members of the gang once traveled to Libya, trained for terrorist operations in abandoned buildings, and sought to arm themselves with rocket-propelled grenades. Under the leadership of Jeff Fort, the gang diverted economic development funds into criminal activity and later endeavored to conduct terrorist operations for hire. If successful, the gang might have blown up airplanes and buildings on behalf of Moammar Qaddafi, and obtained payments in the millions of dollars. Fort is currently in prison for a 1986 mercenary conspiracy.

An example of high-level violence among Chicago gangs is demonstrated by the Four Corner Hustlers faction of the Vice Lords, who plotted to blow up the Chicago Police Department Area 4 headquarters using an M72A2 light antiarmor weapon system rocket.[22]

Politicization, which signals a shift from a second- to a third-generation gang, is also found among Chicago gangs. Gang "empowerment" and political evolution are illustrated in the example of the 30,000-member Gangster Disciples (GDs) and their political arm 21st Century VOTE. The GDs conduct drug trafficking activities in 35 states, in some cases controlling entire housing projects, schools, and streets. The gang has evolved from a turf-based street gang utilizing murder

---

[22]Chicago Crime Commission, *Gangs Public Enemy Number One*, Chicago, Ill., 1995.

and drive-by shootings to settle disputes into a complex organization that conducts overt political activity.

Police intelligence gathering can be hampered as gangs evolve into third-generation types with political agendas—in Chicago, a 1982 injunction barred surveillance, infiltration, and dossier compilation on groups, including gangs, that played political roles. Such well-intentioned but naive restraint can have particularly alarming consequences.[23]

In addition to maintaining a political action committee funded through a "street tax" on gang members, GD members infiltrate police and private security agencies, sponsor political candidates, register voters, and sponsor protest marches. A rival gang, the Unknown Conservative Vice Lords, has embraced similar activities, including establishing a nonprofit political association known as the United Concerned Voters' League. This political dimension signals a shift from a second- to a third-generation gang, where power acquisition rather than simple market protection is a motivating concern.

Internationalization is another indicator of gang evolution. Gangs in Los Angeles and San Diego are notable in this respect. Los Angeles gangs currently have outposts in Tijuana, Mexico; Managua, Nicaragua; San Salvador, El Salvador; and Belize. Police in these jurisdictions are concerned about the evolution of these gangs into drug mafias. In El Salvador, home to an estimated 10,000 L.A.-style gangs, some elements are merging with local *maras,* yielding the potential for lethal criminal coalitions. Traditional L.A. rivals Mara Salvatrucha (MS) and Eighteenth Street (18th Street) have outpost affiliates that have clashed in San Salvador, utilizing AK-47s and hand grenades. In Tijuana, 18th Street has expanded its scope to include trafficking in stolen vehicles from the United States and moving illicit small arms into the United States.

---

[23]Consider Japan's intelligence failure involving the Aum Shinrikyo. Japanese police avoided close scrutiny of the group because of its protected status as a "religious" group, thus missing critical warning signs and precursors that ultimately resulted in a sustained campaign of chemical and biological attacks.

San Diego's Calle Treinta (30th Street, Barrio Logan, or Logan Heights) gang has ventured into the third generation in conjunction with the Arellano-Félix (Tijuana) cartel. Members of Calle Treinta have been recruited in true mercenary fashion, with as many as 30 members abandoning their turf orientation and turning to acting as proxies for the Arellano organization. This is not the only recent example of gangs acting as mercenaries for drug mafias. Brazilian street gangs active in Rio de Janeiro's *favelas*[24] provide an interesting parallel to the Logan Street gang experience, selling their services to drug traffickers and embracing combat training to strengthen their hold on their *favelas* and to ward off competition from others.[25]

During their heyday, Calle Treinta's international gang combat left a trail of over 200 bodies from San Francisco to Tijuana to Venezuela.[26] Notorious activities include a 1993 San Diego shootout killing 26 and the 1993 slaying of Cardinal Juan Jesús Posadas Ocampo in Guadalajara. Calle Treinta members have been linked to assassinations, drive-by shootings, and torture slayings on both sides of the border. In November 1997, a Mexican judge issued an arrest warrant for seven "Logan 30s" members, believed to be in San Diego, for the Tijuana assassination attempt of J. Jesús Blancornelas, publisher of the Tijuana paper *Zeta*.[27] Blancornelas observes that the cross-fertilization of street gangs and drug traffickers results in "a binational army of killers working in both Mexico and America."[28] Barrio Logan provided the cartel with imported *pistoleros*, who had the advantage of being able

---

[24]The parallels between Brazilian and American street gangs bear watching. For a glimpse into the contemporary similarities between the Brazilian and North American urban condition, especially "failed communities" see "Like Teheran and São Paulo" in Robert D. Kaplan, *An Empire Wilderness: Travels into America's Future*, New York: Vintage Departures, 1999.

[25]"Drug Traffickers Hire Mercenaries," *News from Brazil from SEJUP (Servico Brasileiro de Justica e Paz)*, Number 281, July 24, 1997, www.mapinc.org/drugnews/v97/n259/a01/html.

[26]Sebastian Rotella, *Twilight on the Line: Underworlds and Politics at the U.S.-Mexico Border*, New York: W.W. Norton & Company, 1997, p. 152.

[27]"7 San Diego Gang Members Sought in Tijuana Ambush," *Los Angeles Times*, August 9, 1998.

[28]J. Jesus Blancornelas, "The Drug Crisis Isn't Just in Mexico," *Los Angeles Times*, September 27, 1998.

to float between nations, thus complicating apprehension and prosecution.[29]

An emergence of third-generation street gangs has also occurred in South Africa, where both street gangs and countergang vigilante groups are politically active and conduct quasi-terrorist or terrorist campaigns.[30] Pagad (People Against Gangsterism and Drugs) is notable in this regard. Primarily Muslim in orientation, Pagad is a vigilante group battling gangs in Cape Town and the Western Cape. Its activities have resulted in a near war, reaching the level of civil conflict with terrorist attacks (bombings and armed assault). Pagad maintains a paramilitary, G-Force, organized into small cells at the neighborhood level. These cells have the capacity to operate undetected and independent from central control. Some members are believed to be veterans of Islamist campaigns in Bosnia, Lebanon, and Afghanistan. Pagad also has links, including shared leadership, with the Islamist-oriented Qibla Mass Movement.[31] Pagad currently is believed to be nurturing an alliance with the Pan Africanist Congress (PAC), an outgrowth of the relationship between PAC's former military wing Apla and Qibla, forged during the armed struggle against apartheid.[32]

Cape area gangs, with an estimated 400,000 members, have evolved a political dimension as well. The Hard Livings gang is in conflict with Pagad for both political and criminal transgressions. Hard Livings developed links with the African National Congress during Hard Livings' opposition to South Africa's apartheid regime. These links were formed when Umkhonto weSizewe guerrillas relied upon gang safe houses and trafficking routes. Now, both Pagad and the Hard Livings gang participate in conventional politics—and South African police observe that these gangs have better organizational networks in Cape Flats and other Western Cape towns than do traditional political actors. For example, Pagad spent considerable effort arranging political

---

[29]Rotella, 1997, p. 143.

[30]Stefaans Brümmer, "Pagad, Gangs Mix It up with Politics," Johannesburg *Mail & Guardian,* May 9, 1997.

[31]Damian Daniels, "Pagad's True Colours Revealed," Johannesburg *Mail & Guardian,* September 2, 1998.

[32]Stefaans Brümmer, "Mixing Gangsters with Politics," *Electronic Mail & Guardian,* May 12, 1997, www.mg.co.za/mg/news/97may1/12may-pagad.html.

alliances for the 1999 elections; and Hard Livings sponsors the Die Suid-Afrikaan party and Community Outreach Forum (CORE). Hard Livings member Rashied Staggie, an influential CORE member and frequent spokesman for its political positions, including support of the South African Olympic Committee's bid to host the Olympics, publicly voiced an interest in standing for office in that election. Staggie is the twin brother of Rashaad Staggie, who was assassinated in a Pagad attack in 1996.[33]

VIOLENT NETWAR—THE ADVERSARIES AFTER NEXT?

Few criminal gangs, even in the third generation, have become full-fledged terrorist organizations as well. But there is some cross-fertilization occurring in terms of organizational designs, strategies, and tactics. Thus, it behooves an analyst concerned about the future evolution of gangs to be cognizant of trends in the world of terrorism that could lead to imitation by criminal gangs or, worse yet, to links and alliances with criminal gangs, either as proxies or partners.

Trends Linking Terrorists and Criminal Gangs

Today's terrorists, characterized by religious and social motivations, stand at the threshold of netwar. The lighting rods of postmodern conflict, these terrorists eschew terrorism's classic political motivations and instead seek broader religious or social change. This resurgence of religious terrorism involves Shi'a and Sunni Islamists seeking to replace the secular state, such Christian fundamentalists as the Army of God in the United States, the racist Christian Identity movement, and some fundamentalist Jews in Israel. Millenarian cults have also sprung up, perhaps best exemplified by the Aum Shinrikyo's international conspiracy to wield weapons of mass destruction, which culminated in a campaign of chemical and biological assaults, in-

---

[33]Gustav Thiel, "Cape Drug War Heads for the Polls," Johannesburg *Mail & Guardian,* April 4, 1997, "Drug Lords Aim at Political Power," *Electronic Mail & Guardian,* April 4, 1997, www.mg.co.za/mg/news/97pr1/4apr-drug.html; and Alex Duval Smith, "Who Is Rashied Staggie?" *Electronic Mail & Guardian,* January 20, 1998, www.mg.co.za/mg/news/98jan2/20jan-staggie.html.

cluding the sarin attack on the Tokyo subway. Aum continues to exist and is once again building membership under the name Aleph.

These postmodern religious zealots rely in large measure on common philosophy rather than rigid individual command and control to reach their ends. Exemplified by the Aryan Nation's ambassador-at-large Louis Beam's "Leaderless Resistance," calls to action through "fictional" manifestos, web pages, Usenet news groups, and radio shows allow like-minded extremists, in a decentralized way, to select their own course of action under a common philosophical umbrella. The Army of God, for example, provides a call to action in "Rescue Platoon." This "fictional" web site account gives operational and tactical advice and recommends targets (both venues and thinly veiled individuals)—thereby, in actuality, laying the blueprint for a terrorist antiabortion campaign.[34]

An internetted coordinating structure allows what Bruce Hoffman has termed the "amateurization of terrorism," where an ad hoc amalgamation of like-minded individuals can come together for a specific, perhaps one-time operation[35] with or without coordination from a nonstate network actor. A good recent example of such a nonstate network is found in the interaction between Osama bin Laden's *al-Qaeda* (The Base) and the International Islamic Front Against Jews and Crusaders.

Bin Laden not only provided philosophical guidance to his own network (al-Qaeda), but was able to bring together an amalgam of separate but like-minded groups, including Gama's al-Islamiya (Egypt), the Ulema Society and Harkat-ul-Ansar (Pakistan), and the Jihad Movement (Bangladesh), into a networked alliance.[36] This alliance, sharing the goals of the Islamic Army for the Liberation of the Holy Places, was able to come together to conduct specific operations in the near-simultaneous bombing of the U.S. embassies in Kenya and

---

[34]David Maccabee (a pseudonym), "Rescue Platoon-AOG Rescue Platoon," (no date), www.geocities.com/CapitolHill/Lobby/8735/rescue.htm.

[35]Bruce Hoffman, "Responding to Terrorism Across the Technological Spectrum," *Terrorism and Political Violence,* Vol. 6, No. 3, Autumn 1994.

[36]"Islamic Group Threatens the United States," *ERRI Daily Intelligence Report,* Vol. 4-231, August 19, 1998.

Tanzania. A similar "one-time" coalition, guided by Bin Laden, is thought to be responsible for the 1993 bombing of New York's World Trade Center. Such groups are not susceptible to preemption through pressure on a state sponsor, since they are not guided by a single "opposing state." Rather, they occupy an amorphous space, cloaked in anonymity and deniability.

The groups discussed in this chapter all share a common evolution toward becoming violent, pernicious network-based threats to domestic and international security. Those at the lower stages (turf gangs and soccer hooligans) are similar to the more advanced groups (cartels and terrorist networks), in that all exhibit the adoption of networked organizational designs and strategies and an ability to integrate new technological tools to enhance their virulence and lethality. All types are difficult to counter precisely because of their multicephalic, networked nature. For example, the killing of Pablo Escobar weakened the Medellín cartel but did little to permanently inhibit the Colombian cartels, and the imprisonment of Larry "King" Hoover and other key GD leadership has had little effect on the gang's continued success. While all use violence to varying degrees to pursue their ends, some of these groups are particularly dangerous because of their ability to learn to exploit the benefits of the network form. Like some militant net-activists, violent networked groups benefit from the obscurity and near simultaneous transmission of information inherent in new network-enhancing technology.

It appears that contact with other sophisticated networked groups may help to "revolutionize" groups at the lower stages of progress, enabling them to become full-fledged criminal and terrorist netwar actors. Such cross-fertilization—whether it occurs in prison, where less-sophisticated gangs learn from more-seasoned organized crime groups, or when groups come together to conduct criminal activity—has clearly sped movement toward enhanced sophistication. As these groups become more sophisticated and obtain more skills and technology through the vast amount of illicit funds at their disposal, it may only be a matter of time until they learn to attack cyber as well as physical infrastructures, and to use information warfare to their advantage.

Groups with a natural network structure are able to quickly learn to exploit new technologies and organizational ideas to make the group more successful. Such was the case when insurgent guerrillas conducted joint operations with Colombian cartels; when ANC guerrillas converged with South African gangs, and Islamist guerrillas came in contact with Pagad; when Mexican cartels recruited San Diego gangs; and when Los Angeles street gangs merged with local *maras* in El Salvador. Such cross-fertilization has particularly ominous overtones as these networked gangs come into continuing contact. Such contact allows these criminal partners to exploit existing smuggling and distribution channels, obtain funds for purchasing weapons and technology, co-opt local police and government officials, and learn from each other's experience. When more-evolved TCOs, terrorists, and gangs converge, the potential for truly more terrible netwar actors arises.

To combat groups evolving along this path requires that police, security agencies, and other government offices quickly recognize the threats that can be posed by internetted adversaries, and thus craft networks of their own. Police, military, and security forces must learn to integrate network forms into their hierarchical structures to enable rapid, robust, and flexible response across organizational and political boundaries. Otherwise, networked adversaries may exploit the gaps and seams between governmental organizations. These must be filled through innovative, multilateral, interagency collaboration. An important strategic intelligence task for police and law enforcement agencies combating these groups is identification of evidence of the diffusion of information technologies and/or networked forms among urban gangs, hooligans, and violent anarchist cells.

Lessons from Observing Social Activists and Anarchists

The disturbances associated with the World Trade Organization (WTO) meeting in Seattle, Washington, between November 29 and December 3, 1999, are illustrative for police and law enforcement officials seeking to cope with networked adversaries. The WTO meeting brought 8,000 delegates from 135 nations, as well as a number of nongovernmental organizations for the WTO ministerial meeting. A number of dignitaries, including the U.S. president and the secretary of state, were present in Seattle for the sessions. Prior to the meeting, the Federal Bureau of Investigation warned Seattle authorities of the potential for disturbances, cyber activity, and demonstrations from a wide variety of social activist organizations.

When the WTO meeting sought to initiate its sessions, a confluence of loosely organized social activists converged on the sessions in both physical and cyber space. What ultimately transpired involved a unique combination of social netwar and violent demonstrations as anarchists, legitimate demonstrators, and opportunist criminals converged and met a police force unprepared for postmodern net conflict. The Seattle police encountered a diverse, yet cohesive coalition of demonstrators with the objective of shutting down the WTO. This loosely organized coalition used and exploited intelligence, the principles of mass and maneuver, good real-time communications, and well-practiced techniques to meet its objectives. Ultimately, the police were overrun by this unique combination, demonstrating their failure to discern between lawful demonstrators, anarchists, opportunists, and bystanders. As a result, the WTO sessions and ceremonies were disrupted, and police credibility was challenged, particularly as the demonstrations were seen on global television and simultaneous demonstrations spilled over in cities across the globe.

What occurred in the Battle of Seattle, which resulted in 601 arrests and $3 million in property damage,[37] was more than a tactical and intelligence failure of a single police force. Rather it became an example of the challenges that hierarchical organizations face when confronting networked adversaries with faster reaction cycles. The opposing force (demonstrators and anarchists) in Seattle were systematic, well-organized, and well-funded. Using extremely good modern communications, including live Internet feeds, they were able to execute simultaneous actions by means of pulsing and swarming tactics coordinated by networked and leaderless "affinity groups." The social netactivists were able to combine simultaneous physical protests and

---

[37]"Seattle Police Department After Action Report," November 29–December 3, 1999, WTO Ministerial Conference, www.cityofseattle.net/spd/SPDmainsite/wto/spdwtoaar.htm.

cyber activity (including virtual sit-ins and denial-of-service attacks) to communicate their antiglobalist, anti-WTO message.

While over 160 social activist groups (including the Direct Action Network, Rainforest Action Network, The Ruckus Society, Free Mumia supporters, Earth First!, Green Peace, and others, including the Zapatista National Liberation Army, which was among the first groups to embrace "social netwar") physically converged on Seattle, a group of approximately 180 black-clad anarchists fomented violent action,[38] while other unknown actors in Seattle and elsewhere struck in cyberspace.

> Hackers probed the Seattle Host Organization's web site nearly 700 times looking for weaknesses and tried to hack into the web site 54 times in a continuing attempt to bring down the site.[39]

While the email attacks (directed denial-of-service attacks) were repelled, the WTO home page at www.wto.org was brought down briefly. The WTO also faced simultaneous infrastructure attacks as saboteurs claiming to be the Green Rennet Brigade cut power to the WTO's Geneva headquarters.[40]

The legitimate social activists are not the concern in this discussion.

> While the vast majority of the protestors were non-violent, there was a small band of people dressed in all black, with masks covering their faces, who caused destruction and tried to incite the crowd to violence.[41]

---

[38]Jonathan Slyk, "Smashing Seattle: How Anarchists Stole the Show at the WTO," *Anarchy: A Journal of Desire Armed*, Spring/Summer 2000, p. 53–56.

[39]Kery Murakami, "WTO Web Site Was Target of Assault from Hackers, *Seattle Post-Intelligencer*, seattle P-I.com, January 19, 2000, http://seattlep-i.nwsource.cm/local/comp19.shtml.

[40]*Battle in Seattle: Strategy and Tactics Behind the WTO Protests*, Washington, D.C.: The Maldon Institute, December 16, 1999, p. 25.

[41]Lawrence Jarach, "Dueling Diatribes—ACME and Albert," *Anarchy: A Journal of Desire Armed*, Spring/Summer 2000, p. 40.

These black-clad actors were the Black Bloc, a "loosely organized cluster of affinity groups and individuals"[42] who roamed downtown Seattle, remaining constantly in motion, pulsing and swarming to avoid engagement with the police as they targeted corporate organizations. The Black Bloc was most active on November 30 (N30), causing property damage to sites owned by Occidental Petroleum, Bank of America, US Bancorp, Washington Mutual, Old Navy, Banana Republic, the GAP, NikeTown, Levi's, McDonald's, Starbucks, Warner Bros., and Planet Hollywood. Slingshots, sledgehammers, paint balls, and eggs filled with glass etching solution were employed.[43]

According to a communiqué by the N30 Black Bloc,[44] anarchists planned to physically strike out against capital, technology, and civilization, exploiting the inability of police forces to shift from their planned method of addressing illegal demonstrations through crowd dispersal. The Black Bloc refers more to a tactic employed by loosely organized, fluid, and dynamic groups that come together in short-term affinity groupings, than to any defined, regular group. This loosely organized coalition, embracing network organization and tactics, frustrated police efforts to gain the situational awareness needed to combat the seemingly chaotic Seattle disturbances. The groups facing the Seattle police were not "disorganized." They were organized as a network, something hierarchies have difficulty addressing.

CONCLUSION: BUILDING A RESPONSE TO NETWORKED THREATS

Responding to networked crime groups will require new and innovative responses from states and their law enforcement agencies. Such traditional responses as "counterleadership targeting" (e.g., the takedown of Pablo Escobar or attempts to capture Osama bin Laden) have dominated strategy to date, with little long-term effect. Networked groups are resilient. Their organizational flexibility allows them to absorb such shocks and rechannel their efforts to remain intact. This or-

---

[42]ACME Collective, "N30 Black Bloc Communiqué," *Anarchy: A Journal of Desire Armed,* p. 47–51.

[43]Ibid.

[44]Ibid.

ganizational ambiguity frustrates hierarchical organizations, since their efforts to contain the opposing criminal forces are frequently too rigid or too slow to stop these adversaries.

Networked threats require a response from either a network or a hybrid (that is, a blend of a hierarchy and a network). The capabilities to address these types of threats may be contained in another agency or another discipline altogether. The traditional organizational "stovepipes" of law enforcement and state agencies give networked adversaries the ability to exploit the seams and gaps of traditional responses. Recent interagency coordination among federal entities to build flexible, tailored response capabilities shows that some progress is being made at the national level. A concluding point of this chapter is that much more progress must be made at the local and metropolitan levels, particularly to coordinate activities among a range of local, state, and federal offices, in areas that may range from intelligence gathering to emergency response.

An example of such an effort is found in the Los Angeles County Operational Area's Terrorism Early Warning Group (TEW).[45] TEW was designed to address emerging threats and to deny networked adversaries the ability to maneuver in the gaps between traditional response organizations. TEW is a hybrid organization that blends networked organizational features with traditional government structures. Bringing together law enforcement, fire service, and health authorities at all levels of government, TEW serves as a mechanism for monitoring terrorist trends and potentials, assessing threats, sharing information, and rapidly disseminating alerts and warnings. TEW is organized as a plenary open forum with three subcommittees: a Playbook Committee for developing response information (target) folders and playbooks to guide response, an Emerging Threat Committee to address upcoming threats in the five–ten year range, and a Net Assessment Group for providing decision support in actual incidents (threats or attacks).

---

[45]Additional discussion of the role of networked responses to networked threats, including a discussion of TEW and its emerging counterparts, can be found in Robert J. Bunker, "Defending Against the Non-State (Criminal) Soldier: Toward a Domestic Response Network," *The Police Chief,* November 1998.

Prior to an incident or specific threat, TEW serves as an indications and warning apparatus. When a specific threat warrants a potential response, the Net Assessment Group is convened from the multidisciplinary, interagency response community to assess the effects of the event, monitor situation status, and provide the Unified Command with potential courses of action to mitigate or respond to the situation. The Net Assessment Group has a networked architecture and is designed to plug into existing response structures and command pathways. Figure 4.2 portrays TEW's Net Assessment Group organization.

This organization includes an officer in charge (OIC) or command group, and an analysis/synthesis element for coordinating the group's tasking and decision support products. The consequence management element assesses the effects of an incident and the necessary elements of response. The crisis management/investigative liaison (law-intel) element provides investigative support and obtains infor-

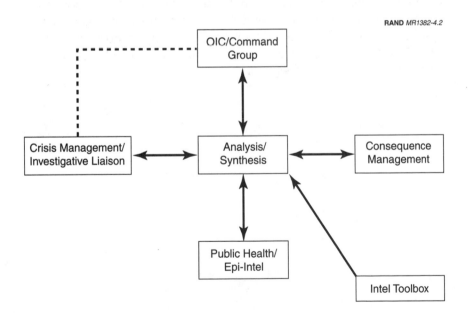

RAND *MR1382-4.2*

Figure 4.2—Terrorism Early Warning Group/Net Assessment Organization

mation about the threat derived from investigative sources. The public health/epidemiological (epi-intel) element links disease surveillance (human, veterinary, and food surety) with incident response. Finally, TEW utilizes an "intelligence toolbox" consisting of virtual reachback to technical specialists or advanced technology tools (forensic intelligence support such as plume modeling, field analytical and laboratory capabilities, decision models, and gaming) to help project incident consequences and the event horizon.

Since its inception, TEW has employed open source intelligence (OSINT), including links to networked Internet-based OSINT tools and relationships with external academic specialists to scan the horizon for emerging threats and to act as a forum for monitoring specific threats, building response capabilities, and supporting decisionmaking efforts.

As such, TEW forms a network to speed an integrated and effective response to a range of threats—networked and otherwise. Recognition of these threats and use of the benefits that may stem from networked forms of organization are essential to limiting the growing, evolving specters of big-city gangs, transnational organized criminal enterprises, and terrorism, as well as the possible blending of these adversaries into compound netwarriors.

Part II

SOCIAL NETWARS

NETWORKING DISSENT: CYBER ACTIVISTS USE THE INTERNET TO PROMOTE DEMOCRACY IN BURMA

Tiffany Danitz and Warren P. Strobel

Editors' abstract. If only social netwar could bring down dictators! This case reflects some of the limitations that activist NGOs face when confronting resolute authoritarians. Danitz (Pew Center on the States) and Strobel (Knight Ridder News Service) study the network of civil society actors who tried, during the 1990s, to free Burma from the dictatorial rule imposed by the State Law and Order Restoration Council (SLORC). So far, that hierarchy has prevailed. Moreover, the United States Supreme Court recently overturned the selective purchasing legislation passed by some states and cities in response to the networked social activism that Danitz and Strobel describe. Despite this setback, the SLORC's transnational opponents have adopted a protracted strategy, one imbued with and enlivened by the belief that the information revolution sharply raises the costs of repression. We thank the United States Institute of Peace for permission to publish this condensation of the January 1999 report they sponsored and published under the same title (USIP, Virtual Diplomacy Series, No. 3, February 2000, www.usip.org/oc/vd/vdr/vburma/vburma_intro.html).

INTRODUCTION

On Monday, January 27, 1997, the huge U.S. conglomerate PepsiCo announced to the world that it was terminating its last business operations in Burma. News of the decision, one that the company had long resisted, raced across financial and political newswires.

But to denizens of the Internet who monitored events regarding Burma, it was already old news. A copy of Pepsi's statement, for which they had long labored and hoped, had crisscrossed the Internet the day before, a Sunday.[1] A battle by global, electronically savvy activists was finally over. With computer modems, keyboards, electronic mail, web sites, long hours, and organization, they had forced the soda-and-snack-food giant to leave a land ruled by a regime that the activists considered illegitimate and repressive. Stealing PepsiCo's thunder by spreading word of the decision before corporate officials had a chance to spin the news was a major coup for the activists.

The case of Burma raises intriguing questions about the effect of modern computer communications on the balance of power between citizens and elected officials, and among local, national, and international power structures and, ultimately, their effect on the conduct of diplomacy in the 21st century. Geographically dispersed but knitted together by the Internet, Burmese and non-Burmese activists from the United States as well as from Europe and Australia joined a long-standing effort to bring democracy to Burma (a small, and to many, obscure Southeast Asian nation). Their global campaign raised constitutional and national policy questions in the United States, as a state government and local councils passed foreign policy legislation without consulting Washington. Many of these decisions may violate international trade agreements between the U.S. federal government and foreign entities. Then, in April 1997, President Clinton signed federal legislation banning any new investment by U.S. companies in Burma. As this chapter will show, these legislative decisions were made because of a global grassroots campaign run to a considerable degree on the Internet, and despite the presence of only a negligible Burmese constituency in the United States.

We offer evidence that the Internet was crucially influential in enabling civil-society actors to force the passage of a series of laws regarding business and political dealings with Burma. The Internet was also used to sway international public opinion and pique the interest of more-traditional news media.

---

[1]See Appendix C [in original full paper by Danitz and Strobel].

In particular, we find that, among its many and still unfolding uses, the Internet—by its very nature—lends itself as a potent tool for advocates organizing for action on international issues.

Following the example of the Chinese student dissidents in Tiananmen Square in 1989, the Burmese activists and their allies added the advances of technology to their struggle. Burma remains in the grip of a powerful military junta, known until recently as the State Law and Order Restoration Council. The representatives elected democratically in 1990 remain unseated. China's rulers found it difficult to stop the outside world from seeing what happened in Tiananmen Square or prevent student dissidents from communicating with the outside world in 1989. SLORC has found it even harder to clamp down on the surreptitious flow of information across borders in the Internet age. Burma's rulers have tried to stanch the flow by passing harsh laws fixing criminal penalties for unlicensed fax machines and computer modems and insisting on state control over international Internet connections. Inside the country, the technological structure is woefully underdeveloped, which is one of the reasons the prodemocracy campaign has been led from the United States instead of Burma.

While the role of the Internet is important, it is not a replacement for other forms of interaction and communication. But it is a powerful supplement. Traditional face-to-face lobbying is still more effective than computers. In addition, using the Internet has inherent limitations for grassroots activists. Its use is limited to those who have access to the technology, and its openness allows information to be manipulated by those holding opposing points of view.

TECHNOLOGICAL REVOLUTION AND INTERNET ACTIVISM

There is much in recent history to suggest that the appearance of new information technologies has aided grassroots, or "citizen," activists in challenging nondemocratic regimes by widely exposing the offending issue, by facilitating public education about the issue, and by promoting and mobilizing "netizens" in actions against the regimes. In doing so, the activists have augmented the effects of their activities on international relations, challenging the management of diplomatic affairs traditionally carried out by states and their diplomatic representatives. Nevertheless, is the promise greater than the reality? This study seeks to examine the use of the Internet by the Burma prodemocracy activists as a case study with that question in mind.

It is also reasonably well-established that new communication technologies, including the 15-year-old revolution in real-time television, have given new powers to nonstate actors, challenging officials' primacy in international and internal affairs.[2] Ordinary citizens have used the handheld videocamera, the telephone, the fax machine, and other communication technologies to make their causes known, from the "people power" revolution in the Philippines to the antiapartheid movement in South Africa and the Zapatista rebellion in Mexico.[3]

The past decade is replete with examples of how advanced-information flows have played a central role in helping grassroots activists, who seek democratic rule, to counter dictatorial regimes. The 1989 revolutions throughout Eastern Europe were fueled by both personal media, such as hand-passed videocassettes and newsletters, and mass media beamed in from abroad, allowing citizens in one place to learn of, and then mimic, political dissent elsewhere.[4] While the peaceful demonstrations in Tiananmen Square were in progress, information was the crucial umbilical cord between the Chinese students, their cohorts around the world, and an international audience. One technology often blended with and fed into another, in a sort of "feedback loop," as news sent out of China by foreign reporters was "smuggled" back in via hundreds of fax machines. The dissemination of information and news facilitated by the new technology helped delegitimize the regime significantly in the eyes of the international community and the Chinese people.

---

[2]See, for example, Warren P. Strobel, *Late-Breaking Foreign Policy: The News Media's Impact on Peace Operations* (Washington, D.C.: U.S. Institute of Peace Press, 1997).

[3]For example, see Brook Larmer, *Revolutions Without Guns: Nonviolent Resistance in the "Global Village,"* unpublished work-in-progress presentation, U.S. Institute of Peace, April 27, 1995.

[4]Ibid.; Ted Koppel, "The Global Information Revolution and TV News," address to the United States Institute of Peace conference, *Managing Chaos,* Washington, D.C., December 1, 1994; Johanna Neuman, *The Media: Partners in the Revolutions of 1989,* Atlantic Council Occasional Paper (Washington, D.C.: Atlantic Council Publications, June 1991).

Nevertheless, because information and communications increasingly form the base of international transactions, the dictator finds himself in a dilemma. Modern states require citizens—whether doctors, businessmen, or inventors—to have access to the latest sources and forms of information in order to compete in the global marketplace. "But the more they [i.e., dictators] permit these new technologies, the more they risk their monopoly of control over information and communication."[5]

Another view is that new information and communication technologies do not give an inherent advantage—either to governments or other centralized authorities, on the one hand, or citizens, on the other. In this analysis, new forms of information distribution cause temporary changes in the societal structure, but these soon dissipate. "When the political system absorbs a new technology, the public may know a temporary high of influence before the balance of power returns to a shared custody over policy."[6] Whereas McLuhan declared "the medium is the message,"[7] in this view, the intrinsic characteristics of the medium are less important than who uses it and how. The fundamental nature of technology is "its irrepressible ambivalence."[8] Put another way, "Cyberocracy, far from favoring democracy or totalitarianism, may make possible still more advanced, more opposite, and farther apart forms of both."[9]

A third point of view concentrates on what might be called the darker side of the destabilizing changes hailed by the technological optimists—that technology advances social disintegration, increases the divide between the information "haves" and "have-nots" and hastens

---

[5]Secretary of State George Shultz, quoted in David Ronfeldt, "Cyberocracy Is Coming," *The Information Society Journal*, Vol. 8, No. 4, 1992, pp. 243–296. Also at www.livelinks.com/sumeria/politics/cyber.html. For a recent discussion of the "dictator's dilemma," see Christopher R. Kedzie, *Communication and Democracy: Coincident Revolutions and the Emergent Dictator's Dilemma*, Dissertation (Santa Monica, Calif.: RAND, 1997).

[6]Johanna Neuman, *Lights, Camera, War* (New York: St. Martin's Press, 1996), p. 255.

[7]See, for example, Marshall McLuhan and Quentin Fiore, *The Medium Is the Message* (New York: Random House, 1967).

[8]Jacques Ellul, *The Technological Bluff* (Grand Rapids, Mich.: William B. Eerdmans Publishing Company, 1990), p. 76, quoted in Ronfeldt, "Cyberocracy Is Coming," 1992.

[9]Ronfeldt, "Cyberocracy Is Coming," 1992.

the spread of racist, pornographic, or other undesirable materials.[10] More to the point of this chapter, new technology is said to create a ruling "knowledge elite" and aid the powers of centralization—to the point where governments can threaten and intrude on the privacy of their citizens.[11] Critics of the Clinton administration's policies with regard to electronic privacy and government databases have raised these concerns in a more than theoretical way.

This chapter strives to cast fresh light on these issues by tracing the effects of the rapidly growing and changing global computer network known as the Internet. The Internet has characteristics in common with other technological innovations throughout history—the ability to more rapidly replicate information and transmit it in large quantities over great distances. But the Internet also has distinct advantages and disadvantages that flow from its particular characteristics. More than any other technology, it permits its users to create and sustain far-flung networks based on common interests or concerns of the members, where none existed before.

A SHORT HISTORY OF HOW THE INTERNET CAME TO PLAY A ROLE IN THE BURMA CRISIS

In the early 1990s, a few Burmese exiles opposed to the regime in Rangoon began communicating on the Internet via electronic mail. Among the first was Coban Tun, an exile living in California who re-distributed newspaper reports from Bangkok, Thailand, and other information about Burma on the Usenet system, using an electronic mailing list called *seasia-l*.[12] The first regular and consistent source of information on Burma available on the Internet was BurmaNet. It took shape in Thailand in late 1993, the brainchild of student Douglas Steele. In October 1993, at the Internet Center at Bangkok's Chula-longkorn University, he perused an online Usenet newsgroup called *soc.culture.thai* and Thai newspapers that carried the only in-depth English-language accounts of events in neighboring Burma. Steele re-

---

[10]See, for example, Stephen Bates, *The Potential Downside of the National Information Infrastructure* (Washington, D.C.: The Annenberg Washington Program, 1995).

[11]Ronfeldt explores this concern in depth in "Cyberocracy Is Coming," 1992.

[12]Various interviews and electronic correspondence with Coban Tun.

alized that the Internet could be used to provide information about human rights abuses and the usurpation of democracy in Burma.[13] Steele began keying in, verbatim, reports on Burma from *The Bangkok Post, The Nation,* and other sources and sending them out on the Internet without comment. Unadulterated news remains BurmaNet's editorial hallmark today. The effort got a vital boost before the year's end. Steele received a $3,000 grant from the Soros Foundation's Open Society Institute to purchase modems and electronic mail accounts, testing whether it was feasible to train the large Burmese exile community in Thailand to be active online.[14]

Far more important than the news that was transmitted was the new network itself, which provided information, and in so doing empowered members of the Burmese diaspora. This educated elite, scattered around the world in the 30 years since the events of 1962 and cut off from their homeland, for the first time had access to the same up-to-date information and a means to communicate. "Once it was so obvious that people were using it, that it was useful to them, more and more came on. Pretty soon you had, if not the entire Burmese exile community in the world, but all the ones who have $20 a month and a modem," Steele recalled. "There's a lot of Burmese in exile, but they weren't together and the Net allows them, in one way, to be together." The Internet's power to connect and organize geographically disparate individuals and groups would be dramatically displayed in the activist campaigns behind the Massachusetts selective purchasing legislation and the Pepsi boycott.

BurmaNet—maintained on a computer server run by the Institute for Global Communications (IGC), a computer network serving peace and human rights activists[15]—grew rapidly. The number of electronic subscribers went from a handful, to 30, to 100, to 400 in its second year, until it was impossible to keep track of the real "readership," because BurmaNet's reports were posted on the Usenet system and re-

---

[13]Interview with Douglas Steele, Washington, D.C., February 2, 1997; A. Lin Neumann, "The Resistance Network," *Wired,* January 1996, Vol. 4.01, p. 108.

[14]Ibid.

[15]Neumann, "The Resistance Network," 1996; Martha FitzSimmon, ed., *Communicators of Conscience: Humanitarian and Human Rights Organizations' Use of the Internet* (New York: The Freedom Forum Media Studies Center, 1994), p. 25.

printed in paper newsletters.[16] As of January 1997, BurmaNet had 750 known subscribers worldwide.[17]

A difficult decision faced the activists in 1994: whether to allow the Burmese regime's embassy in Washington and other known SLORC representatives to subscribe to BurmaNet and "post" messages giving Rangoon's viewpoint. The decision was made to allow SLORC to join, in the interests of free speech and full debate—which is, after all, a strong part of the Internet's culture. According to Steele, "it's actually sort of beneficial to have this on the Net," because the regime, by its very nature, is able to communicate little beyond its standard propaganda. Activist Michael Beer of Nonviolence International agrees. "Very often they come across as looking ridiculous," said Beer, a veteran among those using the Internet and working for political change in Burma. But by seeing SLORC's viewpoint, like a Kremlinologist of old, "you can then get in their heads. . . . we can sit in their shoes."[18]

At about the same time BurmaNet was ending the international drought on news about Burma and helping both form and inform an international network whose members were dedicated to ending SLORC's rule, related efforts got under way to challenge the regime's choke-hold on information within Burma. This effort was and continues to be hampered by the regime's intelligence apparatus and the lack of any significant private Internet connections inside Burma itself. In September 1996, SLORC passed the "Computer Science Development Law," which metes out a prison sentence of 7 to 15 years and fines of up to $5,000 for anyone who owns an unregistered modem or fax machine.[19]

---

[16]Steele interview, 1997.

[17]*The BurmaNet News,* No. 603, January 3, 1997. The full text of BurmaNet's daily news reports are archived at ftp://Sunsite.unc.edu/pub/academic/political-science/freeburma/bnn/.

[18]Steele interview, 1997. Interview with Michael Beer, Washington, D.C., December 19, 1996. See also, The Associated Press, "Asian Rebels Use Internet," *The* (Annapolis) *Sunday Capital,* April 23, 1995, p. A12.

[19]BurmaNet Editor, *The Free Burma Movement and the Internet,* unpublished manuscript. The writer, while known to the authors, requested anonymity.

Still, information seeped in and out. Despite SLORC's stiff controls, exile groups along Burma's borders with Thailand and India began feeding news—which had first been transmitted on the Internet—back into Burma on computer diskettes or simple, two-sided newsletters. (Rank-and-file SLORC soldiers have been among the customers.) The BBC and the Democratic Voice of Burma, a Burmese-language radio station operating in Norway, broadcast news picked up via the Internet into Burma.[20] Burmese prodemocracy activists use the Internet to publicize news from within Burma that is taken out of the country in other ways and for safe (encrypted) communications between various prodemocracy groups or between them and supporters in the United States and elsewhere. In terms of cost, rapidity, and ease of use, the Internet is a significant advantage over previous technologies for this purpose.[21] These efforts and their effects inside Burma will be discussed in more detail later.

In 1994 and 1995, a new front was opened in the struggle for political change in Burma, as students and expatriates in the United States began to organize the Free Burma campaign, whose central goals included pressuring American and European companies to cease doing business with SLORC. The Internet was again the most frequent communication medium of choice for organizing and exchanging information. By this time, powerful new Internet tools were available, especially the web and associated technologies that make it possible to view and share audio, video, and graphics. With the necessary computer hardware and software and a click of a mouse, interested parties and, more particularly, activists anywhere in the world could listen to a speech by Aung San Suu Kyi; transmit Free Burma campaign materials, such as posters and flyers; or look through a virtual keyhole into Burma itself. Within days of the December 1996 student demonstrations—the largest in Rangoon since 1988—images of them, taken from a private videocamera that surreptitiously recorded the events, were available on the Internet.[22] Dozens of web pages now exist covering every imaginable facet of Burma.

---

[20]Ibid. Beer interview, 1996. See also Barbara Crossette, "Burmese Opposition Gets Oslo Radio Service," *The New York Times,* July 19, 1992, p. 11.

[21]Information provided by Mike Mitchell, International Republican Institute.

[22]December 15, 1996, email message posted on BurmaNet.

SLORC has responded by paying an American company to set up its own web site, www.myanmar.com. The site, which was registered in Laurel, Maryland,[23] features pictures of the country and information about tourism, business, and development—no politics whatsoever. SLORC almost certainly monitors the public Internet discussion dominated by prodemocracy activists. A known SLORC representative, who uses the electronic mail address <OKKAR66127@aol.com>, regularly transmits the regime's official statements on BurmaNet and the *soc.culture.burma* newsgroup. Others who are believed to be representatives of, or at least sympathetic to, the regime also participate in the debate.[24]

In the summer of 1997, SLORC and its representatives appeared to have begun a more aggressive attempt to use the Internet. While the timing may have been coincidental, it should be noted that this took place shortly after the United States instituted federal sanctions against doing business in Burma. In May 1997, the regime began its own electronic mailing list, MyanmarNet, to compete with Burma-Net. It was moderated—i.e., articles are selected or rejected for electronic distribution to the list's subscribers—by the individual known as Okkar. Okkar stated that his policy would be (a) to accept most of the submitted postings, omitting "only the junk mails and very rude usages,"[25] and (b) to welcome submissions of news, information, and comments about political, social, and economic affairs in Burma that have "not been posted elsewhere such as soc.culture.burma and other mailing lists."[26] This ensures that BurmaNet cannot electronically "flood" MyanmarNet with its own content. In practice, MyanmarNet appears chiefly to echo the regime's point of view: Postings include text of the government-controlled *New Light of Myanmar* newspaper; other government statements; reprints of articles favorable, or at least neutral, to the regime; and information on business opportunities for foreign investors. However, in MyanmarNet's first weeks, Okkar did

---

[23]June 28, 1997, email message posted on BurmaNet.

[24]Interview with Beer, 1996; authors' monitoring of BurmaNet and related electronic mail lists.

[25]Posting in *The BurmaNet News*, No. 762, July 1, 1997.

[26]This was contained in an email welcome message after one of the authors electronically subscribed to MyanmarNet.

accept several reprints of articles critical of SLORC's handling of the economy and its reputed drug ties, which had been posted on Burma-Net by members of prodemocracy groups.

SLORC's ability to fight back outside its borders when the Internet is used against it appears to be limited to monitoring public Internet discussions and trying to publicize its own point of view. "The delete key can't do very much to you," Steele said. "The only currency that works on the Internet is the ability to persuade, entertain, whatever."[27]

Nevertheless, the year 1997 saw modest, but potentially significant, changes in SLORC's attitude toward the Internet within Burma. In mid-April, the government-controlled Myanmar Poste Telegraph and Telephone signed an agreement with a Singaporean firm for Burma's first digital communications link with the rest of the world. This modest-sized link, which uses Singapore as a gateway, will be available for businessmen with interests in Burma, as well as foreign businesses operating in Burma. The All Nippon Airways office in Rangoon, as well as several universities in the capital, reportedly now have Internet access.[28] It remains to be seen whether SLORC can keep its citizens' use of the Internet limited to business and academic matters.

THE MASSACHUSETTS SELECTIVE PURCHASING LEGISLATION

> Supporters of this and other bills to impose sanctions on Burma have been particularly successful in their use of electronic mail to keep their movement going, leading one activist to describe this as the first "cybercampaign."
> —Massachusetts Governor William Weld[29]

On June 25, 1996, with a group of "cyberactivists" and Burmese exiles looking on, Massachusetts Gov. William Weld signed into law a bill

---

[27]Steele interview, 1997.

[28]Various email messages.

[29]Gov. William Weld, remarks at Burma bill signing, June 25, 1996. Provided by the governor's office.

that bans corporations that do business in Burma from getting new contracts with the Commonwealth of Massachusetts. The "selective purchasing" legislation, as it is known, is one of more than a dozen such laws and ordinances directed against the SLORC regime that have been passed in cities and counties across the United States since early 1995. Forced to choose between lucrative local government contracts and the often-mediocre business opportunities in Burma, a host of American firms have chosen the former.

Such major brand names as Pepsi, Disney, Eddie Bauer, and Liz Claiborne have withdrawn from Burma, compelled by a combination of negative publicity, shareholder pressure, and selective purchasing legislation. The Massachusetts law alone was cited by Motorola, Hewlett-Packard, Apple Corp., and other major companies as the reason they pulled up stakes in Burma.[30] It has bitten hard enough that both the European Union and Japan have complained to the U.S. State Department and intend to challenge the law in the World Trade Organization (WTO).[31]

According to participants on all sides, the Internet—particularly electronic mail—played a defining role in the campaign to draft and pass the legislation. Activists had already organized on the Internet and used this ready-made network. The campaign itself was conceived through communications on the Internet. Information on conditions in Burma was fed to sympathetic legislators on the Internet. Email alerts were sent out at key points in the legislative process, generating letters to state legislators and Gov. Weld's office.

Although different, "older" technologies such as telephones or fax machines could have carried out these functions; at least one role the Internet played here would have been virtually impossible before its existence. The Burma selective purchasing bill was consciously modeled on almost identical legislation passed in the 1980s, in Massachusetts and elsewhere, that sought to prevent American businesses from operating in South Africa under the apartheid system. (See Table 5.1.)

---

[30]See, for example, Theo Emery, "Motorola, HP to Cut Ties to Burma: Cite Massachusetts Law Barring Business in Nation," *The Boston Globe,* November 29, 1996, p. B11.

[31]See, for example, "A State's Foreign Policy: The Mass That Roared," *The Economist,* February 8, 1997, pp. 32–33.

However, unlike with African-Americans concerned about South Africa—or Irish-Americans about Northern Ireland, for that matter—there was no existing constituency in the United States, outside of a few progressive groups, in the case of Burma. The Internet, because of its ability to create geographically dispersed networked communities, created the constituency necessary for action. It drew together activists as close to the state capitol building as Harvard, and as far away as Burmese exiles living in Europe and Australia. "This was truly the first time that this legislature had gotten involved with foreign policy on the face of the issue, without any hyphenated constituency to drive it," said State Rep. Byron Rushing (D), the selective purchasing legislation's sponsor and leading proponent in the state legislature. "The first thing the 'Net did in this campaign was to connect the Burmese wherever they are."[32]

The Internet also provided the advantages of stealth early in the Burma campaign, as long as the activists wanted it. Much of the networking took place outside the public eye. Once the drive to pass the selective purchasing legislation emerged with full force, it was a surprise to those who might have opposed it, including corporations and the of-

Table 5.1

States and Localities That Have Passed Selective Purchasing Legislation on Burma

Commonwealth of Massachusetts	Alameda County, Calif.
New York City, N.Y.	Berkeley, Calif.
Madison, Wisc.	Santa Monica, Calif.
Ann Arbor, Mich.	San Francisco, Calif.
Oakland, Calif.	Carrboro, N.C.
Takoma Park, Md.	Boulder, Colo.
Chapel Hill, N.C.	Los Angeles, Calif.
Portland, Oreg.	

NOTES: As of January 1999. List is meant to be suggestive; other localities also may have passed such legislation.

---

[32]Interview with Rep. Byron Rushing, Boston, January 23, 1997.

fice of Gov. Weld, who had a reputation as a pro-business Republican. When Weld's press secretary, Jose Juves, first heard about the legislation and checked into it—to do so, he used the web for the first time—"I was kind of shocked that the whole sort of ready-made organization . . . was out there."[33] Of all the companies with business in Burma, only the oil and gas concern UNOCAL Corp. took the effort to hire a local lobbyist. For many other companies, the first time they heard about the issue was after the selective purchasing bill had become law, and they were notified that they were on an official state list of affected companies. "They definitely came late to the dance," Juves said.[34]

As the bill slowly made its way through the state legislature in 1995 and 1996, activists used the Internet to push it along. Rushing, working with Simon Billenness of the Massachusetts Burma Roundtable and other activists, sent emails from home and office to keep supporters apprised of developments and to urge them to make their voices heard when the bill was at a key legislative juncture or in trouble. The electronic missives generated phone calls and letters to state senators and representatives from their constituents inside Massachusetts and activists outside the state, explaining the need for the legislation and pressing for passage.

The legislation very nearly died several times. Activists using the Internet rallied to overcome each obstacle. An amendment that would have added virtually every totalitarian regime in the world to the legislation—and thus buried it under its own weight—was killed, and a March 1996 Senate motion to table the bill and postpone it to the next legislative session was reversed. Billenness, through the Burma Roundtable, was central in using electronic communication to keep the issue alive in the legislature, repelling obstacles to passage and maintaining an electronic community behind the bill. The Internet, he said, "is very good at getting one message sent to a lot of people, with minimal cost and minimal time."[35] It was vital in keeping sub-

---

[33]Interview with Jose Juves, Boston, January 23, 1997.

[34]Juves and Rushing interviews, 1997.

[35]Interview with Simon Billenness, January 23, 1997.

scribers up-to-date on the status of the Burma bill and eliciting their support.[36]

The Internet, and the electronic network that stretched from Burma's borders around the world and back again, also meant that timely information was a key ally for activists in the campaign. Culling the news from BurmaNet and many other sources, the cybercampaigners were able to get accurate information on the conditions endured by the Burmese people. Without the Internet, "it's hard to imagine that we would have had as much information," Rushing said. "The thing that makes these things work is that you can go up to a rep[resentative] and say, 'Look, this is what's happening there.'" The information flow allowed proponents to meet and counter the objections of opponents or skeptics. And, vitally, it allowed them to be confident that they were in tune with the positions of Aung San Suu Kyi herself. A campaign "can blow up" if it does not have the support of the pro-democracy groups within the affected country, Rushing said. With the Internet, "we always knew how Suu Kyi was on this issue." The Massachusetts lawmaker acknowledges the problem of misinformation on the Internet. But he believes there are enough "voices" out there that the communication network quickly self-corrects inaccurate information.[37]

Weld had old-fashioned political reasons for signing the Burma bill. His opponent in the 1996 race for the U.S. Senate, incumbent Democratic Senator John Kerry, had wavered on the issue of federal sanctions against Burma and had supported continued U.S. antinarcotics aid to the SLORC regime. Weld saw an opening that would embarrass Kerry and help him pick up support among the state's progressive voters.[38]

But the Internet campaign helped bring the otherwise-arcane issue of Burma to Weld's attention and kept the pressure on. "CONTINUE TO FAX AND CALL GOVERNOR WELD . . . CALL DAILY!" Billenness urged

---

[36]See Appendix C [in original full paper by Danitz and Strobel].

[37]Rushing interview, 1997.

[38]See Wayne Woodlief, "Burma Bill May Gain Votes for Weld," *The Boston Herald*, June 13, 1996, p. 35; Michael Kranish, "Proposed Sanctions on Burma a Hot Issue for Weld, Kerry," *The Boston Globe*, June 14, 1996.

in an update sent to supporters on June 12, the day before the bill landed on the governor's desk. For good measure, Weld's newest fax number was included. During mid-June, Weld received a flood of letters imploring him to sign the bill. They came not just from Massachusetts, but from around the United States, as well as Japan, the United Kingdom, France, and Canada. One came from within Burma's western border, sent via a supporter in India. Sam Bernstein of Braintree, Massachusetts, was not alone when he told Weld: "If you do sign, your action will go a long way in helping me make up my mind about the upcoming U.S. senate race."[39] According to Juves, Weld received roughly 100 letters and 40 electronic mail messages regarding the legislation, which he described as a huge number for an issue that had nothing to do with bread-and-butter issues like street repairs, crime, or taxes. Weld saw samples of the letters. "I don't think it had an impact on his decision to sign the bill. . . . [But] it made him think about it more than he otherwise would have," said Juves, who disputed the widespread feeling that Weld had signed the bill merely for political advantage.[40]

At first, the activists assumed that Weld would veto the legislation and that they would have to try to convince the legislature to override the veto. They approached the governor's office aggressively. But once Weld's office signaled that he might sign the legislation, the relationship changed dramatically. The governor's office wanted to stage a media event to highlight his position on the bill. The Internet activists put their network into action, using email once again to encourage a large turnout and to make sure the governor's office had the background information it needed and quotes from activists such as Zar Ni of the Free Burma Coalition in Wisconsin. "For me, it was a big logistical help," said Juves, who was in charge of setting up the event.[41]

---

[39]Copies of letters provided by Gov. Weld's office.

[40]Juves interview, 1997. This would be consistent with others' findings about the effects of media and communication technology on decisionmakers, namely, that these technologies chiefly push issues to the top of the agenda and accelerate decisionmaking. See Strobel, *Late-Breaking Foreign Policy*, 1997, and Martin Linsky, *Impact: How the Press Affects Federal Policymaking* (New York: W.W. Norton, 1986).

[41]Juves interview, 1997.

Finally, the Internet campaign in Massachusetts, because of its very success, had another, derivative effect. The more-traditional news media, fascinated by the idea that a state could craft its own foreign policy and that the Internet could be used as a grassroots tool of political power, began to give significant coverage to the prodemocracy groups and what had happened in Massachusetts. These stories, of course, also highlighted the struggle in Burma. Juves' phone rang with inquiries from BBC Radio, Australian Broadcasting, Cable News Network, Bloomberg Business News, the Voice of America, Newsweek, and many other media outlets.

Juves said that the legislation might not have come to fruition without the Internet, or at least would have taken much longer to do so. Significantly, many of the people he dealt with were geographically dispersed, but they had the Internet. "People were really focused in on Massachusetts," he said. "Everybody's connected to one place."[42]

In the aftermath of the legislative victory, Rushing predicted that the issue of localities playing a role in foreign policy—something once unthinkable—will come more and more to the fore. Many cities and states are taking up the issue of human rights and whether and how to do business with nations that have a bad human rights record. Indeed, the issue of who controls foreign policy, and where economic sovereignty begins and ends, has become a more-than-theoretical concern.[43] Activists went back to the "cyberbarricades" after Japan and the European Union argued that the Massachusetts selective purchasing law violates world trade rules and urged Washington to "get its provinces back into line." Then, on November 4, 1998, a U.S. district judge declared the selective purchasing law unconstitutional, ruling in favor of the National Foreign Trade Council, an industry group, and stating that the law "impermissibly infringes on the feder-

---

[42]Ibid.

[43]One count found 27 state, county, and city sanctions dealing with Burma, Nigeria, Cuba, and Tibet. See Michael S. Lelyveld, "Massachusetts Sanctions Struck Down: Judge's Ruling May Set Precedent for State Bans," *Journal of Commerce*, November 6, 1998. For skeptical views of these developments, see "The Mass That Roared," 1997, and David R. Schmahmann and James S. Finch, *State and Local Sanctions Fail Constitutional Test*, Trade Policy Briefing Paper No. 3 (Washington, D.C.: The Cato Institute, August 6, 1998).

al government's power to regulate foreign affairs."[44] The issue is likely to be decided by the U.S. Supreme Court.[45] Both sides continue to make their arguments, and the Internet remains a vital tool for those who gave birth to the Massachusetts selective purchasing law.

THE FREE BURMA COALITION AND THE PEPSI BOYCOTT CAMPAIGN

> "This is the information backbone of a larger movement that aims to mobilize public opinion against the military leaders of Burma," Strider says. "It was the 'Net,'" he explains, "that helped mobilize activists on college campuses and elsewhere in their opposition to investment in Burma by Eddie Bauer."[46]

The advent of computers on university campuses linking student groups into national and international networks seems to have invigorated social activism and has transformed the character of student protests. It has also opened up the world to these students, shrinking the globe into a local community that provides a great number of issues on which to campaign. "We are beginning to see the formation of a generic human rights lobby at the grassroots level (on the Internet). People care even though they don't have a personal connection to the country," explained cyberactivist Simon Billenness.[47]

Computers have become so integrated into university life that they are a virtual appendage of scholars at study. Every freshman entering Harvard University is supplied with an email address and account. Students and student groups have united online, initially to converse—the outgrowth has been a heightened awareness on a number of issues, including human rights. Once on the Net, students meet campaigners and advocates. These people and the information they

---

[44]Lelyveld, "Massachusetts Sanctions Struck Down," 1998. See also Appendix C, Email No. 3 [in original full paper by Danitz and Strobel].

[45]"L.A. Bans Trade Ties to Burma Despite Federal Ruling," Inter Press Service, Washington, D.C., December 17, 1998.

[46]Neumann, "The Resistance Network," 1996.

[47]Interview with Billenness, 1997.

provide have taught students techniques for organizing electronically, and in return the students have joined in the action.

At the press conference held to announce Gov. Weld's decision to make Massachusetts the first state to slap sanctions on SLORC, many students turned up. Juves, Weld's press secretary, says that the Internet was responsible for helping students turn out, and that there were more students there than at any other bill signing.[48]

Plugging in has transformed the meaning of "tune in," at least on Harvard's campus, where "students cannot get by without using email. Most college organizations can't conceive of how this would be done without the Internet."[49]

As the selective purchasing campaign to deprive Burma of American investment and ultimately all foreign investment gained strength, student groups caught on. The selective purchasing resolutions were presented to city legislatures, and with each small success, the sanctions campaign widened and bolstered the prodemocracy campaign. Students and other activists organized shareholders that had been writing resolutions for consideration regarding their corporation's business ties in Burma. Shareholder resolutions were presented at each annual shareholder conference to educate investors on their companies' dealings in Burma and to call for corporate withdrawals from the country.[50] These efforts grew out of the more traditional forms of activism, roundtable discussion groups, and letter-writing campaigns.

When PepsiCo became a target of the campaign, student activists were able to connect with a tangible product, process, and outcome. They could start small, on campus, by educating their friends about Pepsi's operations and its cooperation with the military junta ruling Burma. From there, they could pass student resolutions, instigate student boycotts of all of Pepsi's subsidiaries and possibly cause changes in university food service contracts. Then they could move on to the

---

[48]Juves interview, 1997.

[49]Interview with student activist Marco Simons, Boston, January 24, 1997.

[50]For a fascinating example of how individuals using the Internet can affect investment policy, see Appendix C [in original full paper by Danitz and Strobel].

town meeting where their university sits, to the city council, and eventually to the state legislature.

In 1990, Pepsi entered Burma through a joint venture with Myanmar Golden Star Co., which is run by Thein Tun, once a small-time exporter of beans. Most Burmese who were working for Pepsi were connected in one way or another to the SLORC regime, said Reed Cooper, of the Burmese Action Group in Canada.[51] Pepsi ran a bottling operation in Rangoon that grew "from 800,000 bottles a day to 5 million" and added a new plant in Mandalay.[52]

In a Seattle resolution on Burma, which urged an "international economic boycott of Burma until the human rights violations cease and control of the government has been transferred to the winners of the 1990 democratic election," Pepsi was mentioned as one of the companies that supports the military regime and its "cruel measures against the Burmese people."[53] The resolution passed unanimously just after a similar boycott resolution successfully passed in Berkeley, California.[54]

Cyberactivist Billenness was building a campaign with a solid foundation at the local levels. His office was delivering ribbons of circular stickers proclaiming "Boycott Pepsi" across the country to various groups of activists. He had solicited and developed the support of the Nobel Peace laureates who attended the pivotal 1993 fact-finding mission to the border regions of Burma (they were not permitted into the country). The Nobel laureates joined in a call for an international boycott of products exported from Burma. The 1993 trip sparked a campaign that the grassroots organizers, like Billenness, Cooper, Larry Dohrs, and others, had slowly been orchestrating. The necessary definitive moment that legitimized their efforts had arrived.

---

[51]Telephone interview with Reed Cooper, Washington, D.C., March 1995.

[52]Michael Hirsh and Ron Moreau, "Making It in Mandalay," *Newsweek*, June 19, 1995, p. 24.

[53]Press release by The Seattle Campaign for a Free Burma, April 24, 1995. Also in Seattle City Council Resolution 29077.

[54]Ibid.

"This is how South Africa started," Billenness said. The strategy: to get selective purchasing legislation passed in town councils, then cities, then the states. Congress would be sure to follow, he believed.[55] Most of the roads and Internet lines connecting this network of Burma activists lead back to Billenness. So it is not surprising that he wanted to encourage a university campaign among American colleges to support the growing Burmese student movement.

The Pepsi Campaign at Harvard University

> There are few Burmese in the States, and relatively few people who even know where Burma is. But those who care are organized and effective, and it's because of the Internet. —Douglas Steele[56]

Students at Harvard tapped into the Burma Internet network, and soon after, they were successful in preventing a contract between PepsiCo and Harvard's dining services. Their activism also had an influence on the Harvard student body, by raising awareness as well as passing resolutions in the student government that affected the university's investments in Burma.

One of the students who became a ring leader for the Burma campaign on campus was Marco Simons.[57] The summer before his junior year at Harvard, Simons, who had written a paper on the human rights situation in Burma while still in high school, tapped into the Net via the newsgroup *soc.culture.burma*. Soon after, Billenness, who worked at the Franklin Research Institute for Socially Responsible Investing, contacted Simons. Billenness was trying to initiate a Burma group at Harvard. At this same time, autumn 1995, the Free Burma Coalition (FBC) was first appearing online. The FBC's web site was able to attract numerous students across the United States, and it became a hub for the network that would follow.

---

[55]Billenness interview, 1997.

[56]Quoted in Neumann, "The Resistance Network," 1996.

[57]This section is based on interviews with Marco Simons in Boston, January 1997, and Zar Ni in Washington, D.C., February 1997.

There were no Burmese undergraduate students at Harvard. There was one native Burmese graduate student and a few students who had either visited Burma or lived there as foreigners. For this reason, the three Harvard students who initiated the Burma group felt their first order of business should be to raise awareness. They set up a table at the political action fair at the start of the fall semester. They tested students who came by on their geographical prowess by asking them where Burma was on a map and which countries bordered it. Those who stopped to play the game were asked to leave their email addresses. Between 40 and 50 addresses were collected that day.

Simons describes the culture on campus as one that is virtually interactive. The only "real mail" (i.e., postal mail) he gets is from the university administration, he says. "Our internal organizing was done through email meetings," Simons said. The group communicated almost exclusively by email. As the campaign developed to include lobbying the student government on resolutions regarding Burma, Simons said, the activists communicated with the student government via email also. Thus, they combined the traditional avenues for social activism with the technology that the university setting made available.

Once they had the student email addresses, members of the fledgling group began encouraging students to join them in letter-writing campaigns calling for university divestment from various companies. They also tried to organize an honorary degree for Aung San Suu Kyi. Harvard became the first student government to pass resolutions supporting the Burmese prodemocracy movement. Since then, many campuses have passed similar resolutions, and many used the Internet to seek advice from Simons on how to engage in this campaign.[58]

Some of the resolutions passed by Harvard's student government required that the university send letters to companies operating in Burma, calling for corporate withdrawal. Simons says the students believed Harvard's name carried a lot of clout in corporate circles. These resolutions passed in January and February 1996.

---

[58]Simons and Billenness interviews, 1997.

Harvard University is itself a large investor, with a $7 billion endowment. The students decided to campaign for resolutions requiring Harvard Corp. to write to the companies it owns stock in that deal with Burma and register its desire for divestment.

The Burma activists at Harvard also attempted to localize their campaign whenever possible. Then they stumbled onto a link with Pepsi that allowed them to expand their campaign into a story that would later become a splash with the media.

"At first we didn't think we would have a Pepsi campaign at Harvard because Harvard contracted with Coke for a long time," Simons said.[59] Simons had been aware of the national campaign that Burma activists were waging against PepsiCo from his involvement with Billenness.[60] Billenness held a regular Burma Roundtable that was advocating for a "Boycott Pepsi" campaign, in conjunction with a national group of activists.

It was then that the *Harvard Crimson* ran a story stating that Harvard's dining services were planning on contracting with Pepsi instead of Coca-Cola. "Pepsi was trying to get the beverage contract on campus the whole time," Simons explained. "Coke's dining contract was up for renewal, and they were so dissatisfied with Coke's service, the dining services were thinking of going with Pepsi."[61]

The Burma activists decided to protest this contract on two fronts: first with the student legislature and then with Harvard dining services. As part of the contract, Pepsi would be giving $25,000 to student organizations at Harvard and $15,000 directly to the student government. The activists' strategy with the student legislature would be to attack the Pepsi donations with resolutions. These resolutions called for Harvard to explore investing options for the Pepsi contributions. They could outright refuse the money or, ironically, donate it to Burma-friendly groups like the Boycott Pepsi campaign. When they

---

[59]Simons interview, 1997.

[60]Selective purchasing laws were being considered across the country by local city councils.

[61]Simons interview, 1997.

began investigating these options, the students discovered that dining services had not signed the contract yet.

The students met with dining services in a lobbying effort. They also stayed in contact over the Internet with Michael Berry, then director of dining services, and Purchasing Director John Allegretto. Rand Kaiser, a PepsiCo representative, met with dining services and the activists to explain Pepsi's position. Simons says that Kaiser argued for constructive engagement with the Burmese military junta.[62] Kaiser was successful in casting Pepsi's investments in Burma in a positive light. After the meeting, Simons and Berry contacted one another over the email system. This allowed students to voice opposing arguments to those presented by Pepsi. Simons made a deal: He told dining services that the students would feel that they had adequate information if PepsiCo released a list of their suppliers for countertrade in Burma.[63] Dining services agreed to this request. The Harvard students asked PepsiCo to fax its list of suppliers. Dining services also made a separate request for the information. Simons says neither the students nor dining services ever received a list.

Meanwhile, the 1996 Pepsi shareholders meeting had commenced and a resolution was introduced to withdraw from Burma. PepsiCo's management effectively blocked the filing of the resolution on the basis that Burma did not represent a significant portion of its business. In reaction, Billenness wrote a letter to Pepsi and the shareholders explaining the effects that the Boycott Pepsi campaign had had on the company. He included the clippings from events at Harvard. This proved to be a boost to the students, who felt their efforts were extending beyond their campus.

---

[62]Constructive engagement is the free-market argument for investment in troubled regions. The argument goes that with investment, the standard of living is raised for the average person. This in turn raises the expectations for rights and freedoms from the government. At the same time, the heightened economy requires a free flow of information, which boils down to technology and freedoms of press and speech. These then open up a previously closed society.

[63]In Burma, the currency is virtually worthless, so foreign investors have to repatriate their profits before taking them out of the country. The human rights community firmly believed that PepsiCo was buying agricultural goods to sell to recoup its profits and that those goods were harvested with state-enforced slave labor. Macy's department stores had published a similar list, and the resulting pressure proved destructive to its investment.

The Burma student activists requested that their student government pass another resolution that specifically asked dining services to sign a contract with Coca-Cola and not Pepsi. This passed through the student legislature, and dining services renewed their contract with Coca-Cola. Dining services then went on the record explaining that Burma was a factor in its decision.

A media campaign ensued, and the Harvard students were courted by mainstream news organizations. Students downloaded press releases, conferred over the Internet with other student leaders in the FBC, and then sent their statements to the press. Stories appeared in the *Washington Post, USA Today, Boston Globe,* on the Associated Press wire, and in local newspapers.[64] In addition, Simons said he received overseas calls from the BBC and from a Belgian news outlet.

Other students who subscribed to the FBC web site and email list were able to follow what was happening at Harvard and use information generated there for campaigns on their own campuses. They also emailed and conversed with other students to discuss techniques and strategy, while learning from past mistakes.

Even with the help of the Internet, not every student campaign on Burma was a success. An effort at Georgetown University in Washington, D.C., did not go very far.[65] Another Boston university, Tufts, also saw a spark of student activism on the Burma issue. The Tufts activists were hooked up to the Net, which they used to communicate with the Harvard group. But the Tufts students were unable to convince their student government to pass a resolution that would end their dining services' contract with Pepsi. Kaiser had been to the student government to lobby in favor of PepsiCo in the wake of the Harvard campaign. Tufts students also admit that they did not have as good a relationship with dining services as the Harvard students did. The director of dining services deferred the Pepsi decision to the university president, who renewed Pepsi's contract.

---

[64]The many such articles include a front-page report by Joe Urschel, "College Cry: 'Free Burma' Activists Make Inroads with U.S. Companies," *USA Today,* April 29, 1996, p. 1A.

[65]Steele interview, 1997.

The Network

The Harvard group worked closely with several activists who have come to define a core for the Burma prodemocracy campaign. They are Billenness, Father Joe Lamar, Zar Ni of the FBC, and Larry Dohrs.

As noted, the bulk of the Harvard campaign was conducted over the Internet. Simons would post condensed versions of the *Free Burma* daily digest (a news-like account of events in Burma and developments in the Free Burma campaign, similar to BurmaNet) for the Harvard students. Previously written press releases, with quotes chosen through collaboration, were used throughout the campaign. Furthermore, most of the Harvard group's meetings were held over the Internet via email. "This would not have happened without the Internet," Simons said of the Pepsi campaign. "The Free Burma Coalition and possibly the whole movement would not have been nearly as successful this far and would look completely different," he added.[66]

The FBC, which is a network of student organizations, organized three international days of coordinated protest, one in October 1995, another in March 1996, and a fast in October 1996. These were coordinated almost exclusively on the Internet. The Harvard students joined in these events.

The Boston student network grew from contact with FBC and through outreach between local groups. Harvard University, Tufts University, Boston College, Brandeis, the Massachusetts Institute of Technology, and Boston University are all in contact with the Burma Roundtable set up by Billenness. Now the Boston network is reaching into the high schools. The Madison, Wisconsin–area groups are the only others reaching into the high school level. Boston and Madison are using the Internet to coordinate an organized effort to bring activism on Burma to the public schools.

The FBC provides information, advice, and an organized framework for these students to plug in and perform very simple tasks that help push the campaign along. The Internet is appealing to the students because (1) they have easy access through their university; (2) infor-

_____

[66]Simons interview, 1997.

mation is available quickly; (3) it is affordable; (4) things that make a difference can be done quickly and with few individuals; (5) the network has the ability to coordinate internationally; and (6) it is social. The three students who initiated the Burma campaign at Harvard remained the core group, producing most of the work on Burma. Because of the group's small size, they say there is no way they could have been so successful and effective without the larger outside network.

"Phone trees and snail mail are suboptimal because they are labor intensive and expensive," explained Simons, who spends an average of a couple of hours a day on the Internet. He thinks another advantage to the web of "spiders," as the activists call themselves, is the up-to-the-minute information that comes from people in Rangoon or the surrounding areas. Zar Ni's updates hit the web immediately.[67]

There was a preexisting network of activists that the FBC has drawn on. For instance, Simons was in his high school Amnesty International group. The actions were planned to raise awareness and strengthen the growing network of people. Very few activists are working on Burma exclusively. They may begin with the Southeast Asian nation but then expand to work on East Timor, Sri Lanka, environmental issues, and the like. Many of these smaller networks are relying on the FBC and the Burma campaign as a model for their own actions.

ASSESSING THE RESULTS

The prodemocracy activists are engaging in an information confrontation against SLORC. Both sides are producing information about events inside the country. Both are trying to paint a portrait of the other for the international community. But the prodemocracy advocates have used the Internet effectively in the Massachusetts campaign, U.S. citywide selective purchasing campaigns, and the Boycott Pepsi campaign. They have also used the Internet to contact journalists and inform them about their actions and about other issues in Burma, such as slave labor, student protests, and government crackdowns. For its part, SLORC has produced a web page and has relied

---

[67]Ibid.

heavily on UNOCAL, the California-based oil company, and ASEAN, a trade association of Asian nations, to promote a good image of their rule abroad. But the regime has not to date taken full advantage of the technology available to it. The audience for both sides is the world, particularly possible investors in Burma; the means is attempting to gain information superiority over the other side.

SLORC has regularly retaliated on the Internet. In late June 1997, SLORC waged a misinformation campaign using both traditional and modern techniques. To begin, representatives of the regime held a press conference exposing several American nongovernmental organization workers as conspirators working on behalf of the U.S. government to bring down the SLORC regime. They named several names and published biographies with pictures of those individuals on the Internet. The individuals involved dispute the claims made by SLORC, and the U.S. State Department refuted the charge that it was engaging in the support of terrorism on Burmese soil.[68]

The advantage of the Internet is diplomatic. It promotes dialogue between those in closed societies and the outside world. It can be argued that those in the Burma campaign are presenting their version of events to the world and SLORC via their Internet campaigns. It is a classic attempt at unraveling misinformation. If SLORC responds by matching the activists with its version of events, diplomatic resolution may be achievable. The opposing sides can utilize the forum provided by the Internet to develop their resolution within a global context. Currently, the Burma campaign has been trying to generate a cyberdebate with SLORC. "We keep asking the council to engage in a debate. This might not be much of a debate, but it's a start," a Burmese exile in Bangkok said in reaction to SLORC's posting of a web page and assigning Okkar to the Internet.[69]

Using the Internet as a forum for this purpose places the debate in the context of the global community. The activists may be able to use international sympathy to sever SLORC's connection with the global

---

[68]Tiffany Danitz, "Burmese Junta Says U.S. a Partner in Terrorism," *The Washington Times,* July 4, 1997, p. A9.

[69]William Barnes, "Generals Fight Back Through Internet," *South China Morning Post,* March 19, 1997.

community. On a limited scale, they have already been successful in doing so with sanctions legislation.

Although the Burmese activists on the Internet are not physically opposing SLORC's use of armed force, their "soft power" use of information may have some of the same effects. The information is being fed into the country through vehicles such as the previously mentioned Democratic Voice of Burma and *New Era Journal*, a newspaper written in Bangkok and distributed via refugees who make trips from the border regions into the country (this program receives partial funding from U.S. grants). The newly established Radio Free Asia is broadcasting into Burma in native languages. Information is also traveling into the country via computer diskettes that often are marked with a Disney logo or some other video game logo to deceive censors and customs officials.

In Burma, the Internet may just be a forum for the voice of the dissident, a place for the Burmese and the world to go to hear alternative information to what SLORC distributes in-country or internationally. If Burma's activists are successful in overcoming the SLORC regime and are able to institute democratic institutions, they may become an example of how the dissident voice in closed societies is capable of providing a rallying point for opponents to the government. However, it would only be one of many factors that created an atmosphere for success.

In theory, when a voice of opposition exists in a conflict where misinformation is used and the propagating party is in control of the state media and the opposition in an alternative media vehicle and its point of view is disseminated on a grassroots level, the opposition voice may eventually win over the constituency. A constant presence that provides alternative information, especially more-credible information, can have the effect of reversing the success of a misinformation campaign. Furthermore, outside the closed society or nation, the voice of dissent—which can flow more freely—will have the effect of countering the misinformation to the international community. If the international community chooses to believe the dissenters, it erodes the legitimacy of those propagating untruths.

The information-rich cyberhighway has inspired a number of people to engage in campaigns they may not have otherwise, because these campaigns are cheap and take little time and also little effort. Why else should an American Jane or Joe in Idaho take action on behalf of natives of Burma? These campaigns have educated a Burmese citizenship-in-exile in consensus building and in grassroots cyber-strategy.

THE INTERNET'S EFFECT ON ACTIVISM

Advantages of Using the Internet

Case studies and an Internet activist survey[70] indicate that the Internet, including electronic mail, the web, and its other facets, gives grassroots groups an important new tool for attempting to foster political change. Some of these advantages appear to be merely evolutionary improvements on "older" technologies such as the telephone and fax machine in terms of speed and cost. Other advantages appear to be truly revolutionary, reflections of the Internet's unique nature. Of course, no technology by itself guarantees a successful campaign, but the Internet gives its users more power when other forces come into play.

The Internet is inexpensive and convenient. Sending messages via electronic mail is far less expensive than using the telephone, fax machine, or other technologies, especially when activists must communicate over long distances and reach members of the network who live in remote areas (as in the case of Burma's borders). Moreover, as we saw in the PepsiCo case, organizers can distribute campaign materials (posters, photographs, recordings, and the like) far more cheaply—and, of course, more rapidly and easily—than would be the case if they used the postal mail or other means to distribute physical copies of the materials.

While some start-up costs are necessary (a computer, a modem, an Internet account), these are not beyond most individuals' means. Our

---

[70]See Appendix A [in original full paper by Danitz and Strobel.]

survey revealed that many activists make use of freely provided university email accounts.

Cost and the labor of the core organizers are, of course, vital considerations to grassroots groups and nongovernmental organizations that rely on grants and donations that make up their shoestring budgets.

The Internet is an organizational tool "par excellence." Without the Internet, it would have been virtually impossible in the case of Massachusetts—or many other cases not cited here—for activists to coordinate and bring the pressure to bear that they did. Burma activists were dispersed around the United States and around the world; but, because of the Internet, they might as well have been around the block. Neither did the fact that Massachusetts has a minuscule Burmese population matter. A "virtual community" for action was created and acted in concert once its members saw a target of opportunity in the selective purchasing legislation. Coordinating such a campaign via traditional telephone trees or fax machines would have been all but impossible because of the need to act quickly and the sheer physical distances involved.

Moreover, because the Internet permits them to rapidly exchange messages or send the same information to hundreds of recipients around the world, activists are better able to coordinate with a greater number of individuals and refine ideas. "Listservs" like BurmaNet are particularly suited for rapid brainstorming, because a single individual can send an idea in an email and can rapidly receive feedback from many different sources.[71] A handful of organizers can rapidly generate dozens of letters and emails to decisionmakers, the "cyber" equivalent of lobbying, with a few well-timed online appeals. The number of people involved in a campaign doesn't matter as much—it can be quite small—as it does in other activities, such as demonstrations and protests.

This seems to be a revolutionary state of affairs. Perhaps for the first time, the Internet allowed members of the international community to comment on and affect domestic, local legislation, a privilege once

---

[71]See BurmaNet Editor, *The Free Burma Movement and the Internet,* unpublished.

reserved for lobbyists or, at the very least, registered U.S. voters. This might be called "cyberdiplomacy."

The Internet puts information in the hands of organizers fast. In the Massachusetts and PepsiCo cases, proponents of measures against SLORC used the Internet to gather and transmit up-to-date information about conditions within Burma and the policies toward that country of various governments around the world. This helped make their arguments particularly effective and ensured there was no "disconnect" between them and the prodemocracy movement inside Burma. Knowledge, as they say, is power.

The Internet allows rapid replication of a successful effort. Organizers of a successful Internet campaign can immediately share their winning (or failed) strategies with cohorts anywhere on the globe.

A success in one locale does not automatically translate into success in another, because of local conditions and factors. But in the selective purchasing campaign, activists in New England emailed the text of proposed legislation, press releases, and other material to colleagues who wanted to wage a similar effort elsewhere.[72] They could then tailor the materials to their own local conditions. This, of course, is similar to the use fax machines have been put to for years. But with the Internet, many more sources can be reached at once. And with the web, for example, the materials can be posted permanently for downloading, anytime, anywhere.

The Internet also helped other related campaigns coordinate and "compare notes." These included activists trying to foster change in Nigeria or Tibet, or those who are primarily interested in environmental issues, such as the destruction of teakwood forests in Southeast Asia.

The Internet allows users to select their level of activity. Using the same type of computer and communications equipment, different activists can choose how active they want to be in a given campaign. They may elect to simply keep up on the news, by subscribing to BurmaNet, reading *soc.culture.burma,* and browsing the various Burma web pag-

---

[72]Ibid.; Billenness and Beer interviews, 1997 and 1996.

es. At a higher level, they may post articles and comments on the various newsgroups, add their names to electronic petitions, fill out surveys, and download campaign materials for use. At the highest level, they may use the Internet to organize and carry out a specific campaign for political change.[73]

The Internet helps publicize the cause and the campaign. Obviously, this is especially true when a campaign scores successes. There seems little doubt that the Internet—as the pamphlet, telephone, and fax machine did for previous generations of dissidents—helped activists broadcast news around the world about their campaign and about the situation of the people in Burma, prompting a wider public debate. This, of course, is the first goal of any global grassroots campaign.

In the Massachusetts selective purchasing and Pepsi cases, the campaign led to dozens of articles in the Boston newspapers, as well as articles in such national publications as *USA Today* and *The New York Times*. Once it became clear that Gov. Weld would sign the selective purchasing legislation, traditional media from around the world descended on Massachusetts. Radio and television outlets from Europe, Asia, and Australia were suddenly—and probably for the first time—focused on a local bill in a U.S. state legislature.

Wielding political power via the Internet is sufficiently new that many of the traditional media seemed to be drawn by the novelty of *how* the cyberactivists were doing what they were doing as much as what they were doing. Whether this novelty wears off as the Internet becomes a more widespread tool of political activism remains to be seen.

Either way, it has been noted elsewhere that grassroots political campaigns, which do not use force or violent coercion, depend heavily on words and images, as well as reason.[74] The Internet helps spread these words and images to what the activists hope will be a sympathetic public.

---

[73]BurmaNet Editor, *The Free Burma Movement and the Internet,* unpublished.

[74]Larmer, *Revolutions Without Guns,* 1995, p. 9.

The Internet-based activists have a leg up on non-Internet-based groups. Grassroots organizers, whether involved in the Burma campaign or other efforts, were among the first to understand the political powers of the Internet. While SLORC and international corporations doing business in Burma have begun to realize the power that the tool gives their adversaries and have tried to emulate it, the prodemocracy movement has been consistently ahead in its use of the Internet. This raises the question of whether the Internet is by its very nature more suited to decentralized groups and inimical to hierarchical organizations.

SLORC, because of the relatively impoverished nature of the country it rules, does not have the full infrastructure needed to make maximum use of the Internet. Even if it did, it is far from clear that it, corporations, or governments sympathetic to it could use the Internet in the same way. It is far easier for activists using a worldwide network to play "offense" by exposing SLORC and campaigning for change, as was done in these cases, than it is for their opponents to play "defense." It is unclear what SLORC would use the Internet for. Answering the activists' charges directly only gives them wider currency. The alternative is advertising and image making, such as that represented by www.myanmar.com. But many, if not most, Internet users are instinctively wary of authority and organization and are unlikely to warm to the enticements of a government or corporation.

Disadvantages of Using the Internet

There are several disadvantages, or potential disadvantages, to using the Internet that can limit its usefulness to grassroots groups engaged in political action. Many of these "downsides" depend on how the Internet is used. Like the advantages of the Internet noted above, some have to do with the medium's unique characteristics.

It is dangerous to rely solely on a single source of communication. Although the Internet was designed for robustness during an emergency, disruptions can and have occurred. In July 1997, Internet traffic "ground to a halt" across much of the United States because of a freak combination of technical and human errors, presaging what some Internet experts believe could someday be a more catastrophic meltdown.[75] On April 20, 1997, the Institute for Global Communications' computer server, which hosts BurmaNet and many other listservs related to peace and human rights, "crashed." Two days later, President Clinton announced that he was imposing federal economic sanctions on the SLORC regime. The IGC server was not restored until April 24, which meant that activists were seriously impaired in getting news and discussing this watershed development for several days.[76] The Free Burma Coalition "would probably fall apart if the Internet connection were all of a sudden turned off," Marco Simons said. "Maybe we rely too heavily on it."[77]

Other technologies, such as the telephone and the fax machine, still have advantages in particular situations, particularly if the sender needs immediate acknowledgment that the information has been received.

Communications over the Internet can be easily monitored. Without a doubt, SLORC and its sympathizers monitor the public discussion on BurmaNet and other channels of discussion. Such monitoring allows the Burmese regime and perhaps even corporations targeted by the campaign to electronically eavesdrop on prodemocracy groups' activities. However, several respondents to our Internet activist survey, as well as several interviewees, did not see this monitoring as necessarily a bad thing. As one BurmaNet subscriber put it: "I hope they read some of our stuff. They must learn in some manner."[78]

Private, one-to-one electronic mail messages are slightly more secure, but these can be "hacked" by anyone with sufficient technical knowledge.

A more potent option is strong encryption, which, in theory, allows only sender and receiver to read the decoded message. The encryption system known as Pretty Good Privacy (PGP) has allowed U.S. or-

---

[75]Rajiv Chandrasekaran and Elizabeth Corcoran, "Human Errors Block E-Mail, Web Sites in Internet Failure: Garbled Address Files from Va. Firm Blamed," *The Washington Post*, July 18, 1997, p. A1.

[76]See *The BurmaNet News*, No. 701, April 23, 1997.

[77]Simons interview, 1997.

[78]Electronic mail message on BurmaNet, dated May 27, 1997.

ganizations backing the National League for Democracy and other prodemocracy groups to maintain regular contact with groups on the Thai and Indian borders. It has been used to set up meetings inside Burma and to transmit, almost in real-time, debriefings of activists who come out of Rangoon or other cities to the border areas. A more recent development, in reaction to increased pressure by the Thai government against democracy activists, is the construction of secure web pages that require passwords for users to enter secure "chat rooms" where real-time conversations take place.[79] These and other technologies, however, remain out of reach of many Internet users.

Opponents may try to use the Internet for sabotage. This is related to the concerns noted above but represents a more active use of the Internet by the target of a political campaign, in this case SLORC, to trick, disrupt, or otherwise sow dissension in activists' ranks.

The available materials, including interviews, discussions on Burma-Net and other online forums, and our Internet activist survey indicate that this can be a problem at times. But it is not a debilitating one, nor one that erases other advantages of the Internet for global activism. Most of those who responded to the survey said they had not experienced incidents of attempted sabotage by SLORC and expressed little or no concern about damage to the campaign from such activity.

Nevertheless, because the Internet allows for anonymity, it is possible for provocateurs posing as someone or something else to try to cause dissension or sidetrack the campaign by posting messages for that purpose.[80] Okkar, who is obviously in sympathy with the regime, has from time to time posted messages on BurmaNet designed to confuse or undercut the anti-SLORC campaign. One such message, posted in February 1997, was purportedly a letter sent by a "Dr. Myron Segal" and relayed how the National League for Democracy had urged Japan not to help build schools and supply polio vaccine in Burma, in order to increase the people's suffering and dissatisfaction with SLORC rule.[81] Just a month earlier, movement leader Zar Ni had posted an

---

[79]Information provided by Mike Mitchell, International Republican Institute.

[80]BurmaNet Editor, *The Free Burma Movement and the Internet,* unpublished.

[81]See Appendix C [in original full paper by Danitz and Strobel].

email warning of SLORC attempts to cause dissension in the ranks. "Not just what we read as news but how we read it is going to help shape the course of action many of us take. So let's be careful in 'consuming' Burma news and reports," Zar Ni advised. He quoted from Rudyard Kipling: "Things are not quite what they seem. This is the Orient, young man."[82]

At other times, the Internet discussion has degenerated into rounds of finger pointing over real or imagined SLORC provocateurs, discussions that are often heavily tinged with Burmese history or ethnic politics. But U Ne Oo, a Burmese exile in Australia and long-time Internet user, argues that more recently, the Internet "seems to reach its maturity: there are less instances of SLORC being able to instigate the users [into] getting into squabbles."[83]

Information transmitted on the Internet is "unmediated" and can sometimes be of questionable accuracy. One of the advantages of the Internet for activists and many other users, of course, is the fact that it allows them to dispense with the traditional "filters" for news, including reporters and government officials. It allows users to self-select information they are interested in and retrieve data in far more detail than available in a newspaper or, certainly, a television program.

This same lack of structure, however, can present dangers, allowing for wide and rapid dissemination of information that is factually incorrect or propagandistic, including material that is racist, sexist, or otherwise hateful and incendiary.[84]

In the case of Burma, the problem of false or malicious information from SLORC was discussed above. Our research came up with no instances in which the prodemocracy movement in Burma and its international supporters took a major action or made a major an-

---

[82]Zar Ni, "How to Read Burma and Burma Reports," email posted on free-burma listserv, January 10, 1997.

[83]U Ne Oo, "The Grassroots Activism and Internet," article posted on BurmaNet, May 16, 1997.

[84]For example, see Graeme Browning, *Electronic Democracy: Using the Internet to Influence American Politics*, Wilton, Conn.: Pemberton Press, 1996, pp. 79–81.

nouncement of policy based on information that later turned out to be false.

Much appears to depend on the level of sophistication of the Internet user. As Rep. Rushing said of information, "Early on, you have to get it through your head, the fact that it is coming through a computer [does not] make it real, true." But, he added, "People pretty quickly tell you [if] something's not true. . . . I'm comfortable [that] the system's self-correcting."[85]

Access to the Internet is not equal and may highlight divisions between information "haves" and "have-nots." Not all who wish to play a role in the campaign for change in Burma, or in Burma's future generally, have access to the most modern tools of communication, including computers, modems, and the necessary telephone lines or other means to connect to the Internet. As already noted, access to encryption methods that allow for more-secure communication may be limited.

Our Internet activist survey found that English is far and away the language of choice for Burma activists. While few respondents said that language was a barrier to their participation, it may be that those for whom it is a problem simply are not online at all. There has been growing use of special fonts that permit the use of Burmese-language scripts on the Internet, but English still dominates the Internet discussion.

> Thus, the discussions are dominated by non-Burmese activists and those Burmese who can communicate effectively in English. As organizing and development of leadership revolve more and more around effective use of the Internet, those who cannot write fluently or persuasively in English risk becoming marginalized.[86]

Like language, funding is also a major issue here. Those exile groups that are better financed (usually by Western nongovernmental groups or charities) and are located in urban centers, as opposed to jungle border areas, may have more access to the Internet and more chances

---

[85]Rushing interview, 1997.

[86]BurmaNet Editor, *The Free Burma Movement and the Internet,* unpublished.

to shape the opposition to SLORC. The concerns of Burma's many minor ethnic groups may go uncommunicated and unaddressed.[87]

The Internet cannot replace human contact in lobbying and other campaign activities. This warning was made virtually unanimously by those we interviewed. The Internet and other communications media cannot replace human interaction. Rather, the Internet has its own distinct advantages and disadvantages, and is only one of the "arrows" in an activist's "quiver."

Even in the Massachusetts selective purchasing campaign, tools other than the Internet, including phone calls and face-to-face meetings, "were more important," said Michael Beer of Nonviolence International. "At some point, local, physical interaction is going to predominate."[88] Rushing, in describing the campaign, talked about the Internet's crucial role in electronic lobbying and in rapidly delivering information to those who needed it. But he also returned repeatedly to how he arm-twisted his colleagues in the state legislature. The Internet "supplements" that kind of lobbying, he said. "It can fill a big void if you can't do [it] face to face."[89]

In terms of a campaign's internal organization, the Internet can also bring changes in personal interaction. Because the Internet has become such a powerful tool of communication for campaigns, especially global campaigns,

> face-to-face group meetings are necessary less often. The function of group meetings, particularly of geographically dispersed people, is now less to work out detailed strategies and more to strengthen bonds of friendship and bring in outside speakers.[90]

Finally, a campaign that focuses on little else but external communication and publicity—rather than human contact and internal organization—may be in danger.

---

[87]Ibid. For more on this problem in U.S. society, see Browning, *Electronic Democracy*, 1996, pp. 76–79.

[88]Beer interview, 1996.

[89]Rushing interview, 1997.

[90]BurmaNet Editor, *The Free Burma Movement and the Internet*, unpublished.

> There is . . . a troubling tendency among modern nonviolent move-
> ments to fixate on the media to the exclusion of other important fac-
> tors. . . . [The media] have a notoriously short attention span, and
> there are always other conflicts that step up to take center stage.
> Once the media leave, what happens to the movement? If there is lit-
> tle in the way of sustaining organizations—or if the mobilization was
> media-driven—then it may crash and burn.[91]

The Internet may contribute to a lack of historical memory and archives for a full-scale political campaign. This is a general concern
with the growing use of computers and media that work without ever
putting documents into printed form. However, as noted elsewhere,
archives of the BurmaNet News and related materials are kept electronically.[92] For the sake of their successors, activists should pay careful attention to storing records of past debates, decisions, and actions.

Internet campaigns, because of their decentralized nature, may be
unstable. It is at the very least worth pondering whether—because of
its fast-changing, organic, and decentralized characteristics—the Internet gives rise to campaigns that grow, take action and then disappear more rapidly than in the past. Centralization and hierarchy have
many disadvantages, especially in the modern world, but they do
tend to lend themselves to stable structures.

CONCLUSION

In these cases of networked social activism aimed at the military regime in Burma, relatively insignificant constituencies in the United
States were able to influence American foreign policy using the Internet. The constituency's members—backed by a loose coalition of activists around the globe, with the modem as their common thread—
were so influential that they thrust the United States into negotiations
with the Europeans and Japanese at the World Trade Organization.
There, a complaint has been filed over the Massachusetts selective-
purchasing law aimed at Burma. Resolution of that case could have a
profound impact on local sovereignty issues.

---

[91]Larmer, *Revolution Without Guns*, 1995, pp. 18–20.

[92]See Emery, 1996.

However, the Internet does not guarantee the success of international grassroots campaigns aimed at social or political change. It is a powerful tool when used to organize far-flung activists; to rapidly share news or replicate successful strategies from one location to another; or to focus activists on a single, well-defined goal. Traditional approaches, such as face-to-face lobbying and "retail politics," remain vital to success in many political campaigns. In addition, reliance on the Internet brings risks of electronic sabotage, monitoring, or disruption by opponents.

Still, in the cases we studied, the Internet's capabilities provided a new tool for grassroots activists to counter powerful forces of multinational corporations and the regime in Rangoon. Since the Burma campaign raged across phone lines and fiber-optic cables, the use of the Internet to advance work on human rights and democracy has spread to Indonesia, Nigeria, Tibet, and East Timor, and has taken up such subjects as global warming and East Asian teak forests.

These and other campaigns are prime ground for further study of when and how the Internet can be best used, its limitations, and its still-to-be-felt effects on political power and sovereignty.

Chapter Six

EMERGENCE AND INFLUENCE OF THE ZAPATISTA SOCIAL NETWAR

David Ronfeldt and John Arquilla

Editors' abstract. Social netwar is more effective the more democratic the setting. We condense this chapter from our earlier RAND book, The Zapatista Social Netwar in Mexico *(1998). The case shows how the Zapatista movement put the Mexican government on the defensive during 1994–1998, a time when Mexico was evolving from an authoritarian to a more open, democratic system. NGO activism even impelled the government to call a halt to military operations on three occasions— yet the air of crisis also prompted the Mexican army to adopt organizational innovations that meant it too became a more networked actor. Until the "Battle of Seattle," this case, more than any other, inspired social activists to realize that networks—and netwar—were the way to go in the information age.*

The Zapatista National Liberation Army (EZLN) is composed of rural insurgents. But they are not ordinary, and they were quickly perceived by intellectuals (e.g., Mexico's Carlos Fuentes and Pablo Gonzalez Casanova) as representing the world's first postcommunist, "postmodern" insurgency:

> Many people with cloudy minds in Mexico responded to what happened in Chiapas by saying, "Here we go again, these rebels are part of the old Sandinista-Castroite-Marxist-Leninist legacy. Is this what we want for Mexico?" The rebels proved exactly the contrary: Rather than the last rebellion of that type, this was the first post-communist rebellion in Latin America (Fuentes, 1994, p. 56).

This marvelous argument makes an important point: The EZLN insurgency was novel. In addition, the features that make it so novel—notably the links to transnational and local NGOs that claim to represent civil society—move the topic largely out of an "insurgency" and into a "netwar" framework. Without the influx of NGO-based social activists, starting hours after the insurrection began, the situation in Chiapas would probably have deteriorated into a conventional insurgency and counterinsurgency, in which the small, poorly equipped EZLN may not have done well, and its efforts at "armed propaganda" would not have seemed out of the ordinary.

Transnational NGO activism attuned to the information age, not the nature of the EZLN insurgency per se, is what changed the framework. The EZLN was not a "wired" indigenous army. In the leader known as Subcomandante Marcos, it had a superb media spokesman, but the guerrillas did not have their own laptop computers, Internet connections, fax machines, and cellular telephones. These information-age capabilities were in the hands of most transnational and some Mexican NGOs—and they used them to great effect for conveying the EZLN's and their own views, for communicating and coordinating with each other, and for creating an extraordinary mobilization of support.

THREE LAYERS TO THE ZAPATISTA MOVEMENT

In retrospect, Mexico and Chiapas were ripe for social netwar in the early 1990s. Mexico as a whole—its state, economy, and society—was (and still is) in a deep, difficult transition. Traditional clannish and hierarchical patterns of behavior continued to rule the political system. But that system was beginning to open up. Presidents Miguel de la Madrid (1982–1988) and Carlos Salinas de Gortari (1988–1994) had started to liberalize the economy and, to a much lesser degree, the polity. Mexico was beginning to adapt to modern market principles. And independent civil-society actors, including a range of NGOs, were beginning to gain strength and to challenge the government for lagging at democratization and for neglecting social welfare issues.[1]

---

[1] On civil society and the NGOs, see Fox (1994) and Fox and Hernandez (1992).

Meanwhile, Chiapas, once an isolated backwater on Mexico's southern border, was becoming awash with outside forces. It was still characterized by tremendous, age-old gaps between the wealthy and impoverished—kept wide by privileged landowners who ran feudal fiefdoms with private armies, by dictatorial *caciques* (local bosses), and by the plight of poor *indigenas* (indigenous peoples) who wanted their lives improved and their cultures respected. Mexico's neoliberal economic reforms, especially those instituted by the Salinas administration, made matters much worse for many *indigenas*, and that set the stage for the organization and rise of the EZLN.[2]

Local economic and social conditions are important, but more to the point for this chapter is that Chiapas was increasingly subject to a plethora of transnational influences. During the 1980s, it became a crossroads for NGO activists, Roman Catholic liberation-theology priests, Protestant evangelists, Guatemalan refugees, guerrillas from Central America, and criminals trafficking in narcotics and weapons. These transnational forces were stronger and more distinctive in Chiapas than in two other nearby states—Oaxaca and Guerrero—that have been likely locales for guerrilla insurgencies. Transnational NGOs, notably those concerned with human-rights issues, were showing far more interest in conditions in Chiapas, and they had better connections there (mainly through the diocese and related Mexican NGOs in San Cristóbal de las Casas) than they did in Guerrero or Oaxaca.[3] This helps explain why Chiapas and not another state gave rise to an insurgency that became a netwar in 1994.

How, then, did network designs come to define the Zapatista movement? They evolved out of the movement's three layers, each of which is discussed below:

---

[2]Sources consulted include Collier (1994a, 1994b), Gossen (1994), Harvey (1994), Hernandez (1994a, 1994b), Nash (1995), and Ross (1995). Chiapas has a long history of rebelliousness over land issues and was viewed in Mexico City as being filled with truculent *indios*, according to a century-old but still interesting report by Stephens [1841] (1988).

[3]In Guerrero and Oaxaca, the *indigena* cultures and structures were also not quite as strong, distinctive, and alienated from the Mexican government as they were in Chiapas.

- At the social base of the EZLN are the *indigenas* from several Mayan language and ethnic groups. This layer, the most "tribal," engages ideals and objectives that are very egalitarian, communitarian, and consultative.

- The next layer is found in the EZLN's leadership—those top leaders, mostly from educated middle-class *Ladino* backgrounds, who have little or no Indian ancestry and who infiltrated into Chiapas to create a guerrilla army. This was the most hierarchical layer—at least initially—in that the leadership aspired to organize hierarchical command structures for waging guerrilla warfare in and beyond Chiapas.

- The top layer—top from a netwar perspective—consists of the myriad local (Mexican) and transnational (mostly American and Canadian) NGOs who rallied to the Zapatista cause. This is the most networked layer from an information-age perspective.[4]

These are very diverse layers, involving actors from disparate cultures who have different values, goals, and strategic priorities. This is far from a monolithic or uniform set of actors. No single, formalized organizational design or doctrine characterizes it—or could be imposed on it for long. The shape and dynamics of the Zapatista movement unfolded in quite an ad hoc manner.

The social netwar qualities of the Zapatista movement depend mainly on the top layer, that of the NGOs. Without it, the EZLN would probably have settled into a mode of organization and behavior more like a classic insurgency or ethnic conflict. Indeed, the capacity of the EZLN and of the overall Zapatista movement to mount information operations, an essential feature of social netwar, depended heavily on the attraction of the NGOs to the EZLN's cause, and on the NGOs' ability to impress the media and use faxes, email, and other telecommunications systems for spreading the word. But the nature of the base layer, the *indigenas*, also drove the EZLN in network directions, as discussed below. These distinctions about the layers are significant for

---

[4]Not much is done in this chapter with the point that tribal, hierarchical, and networked forms of organization have coexisted within the Zapatista movement. But for an explanation as to why this point may be significant, and a hint that more might be done with the point, see Ronfeldt (1996).

sorting out which aspects of the Zapatista movement correspond to netwar, and which do not.

To understand why a social netwar emerged in Mexico—and why an insurgency mutated into a social netwar—the analyst must look at trends outside Mexico involving activist NGOs.[5] Such NGOs, most of which play both service and advocacy roles, are not a new phenomenon. But their numbers, diversity, and strength have increased dramatically around the world since the 1970s. And mainly since the 1980s, they have developed information-age organizational and technological networks for connecting and coordinating with each other.[6] Thus, the NGOs' ability to swarm into Mexico in response to the EZLN's insurrection was no anomaly; it stemmed from a confluence of network-building efforts spread over a decade or two at global, regional, and local levels.[7]

Some of the activist NGOs were more radical and militant than others, and some were more affected by old ideologies than others. But, altogether, most were in basic agreement that they were not interested in seeking political power or in helping other actors seek power. Rather, they wanted to foster a form of democracy in which civil-society actors would be strong enough to counterbalance state and market actors and could play central roles in making public-policy decisions that affect civil society (see Frederick, 1993a). This relatively new ideological stance, a by-product of the information revolution, was barely emerging on the eve of the EZLN insurrection, but we surmise that it had enough momentum among activists to help give coherence to the swarm that would rush into Mexico, seeking to help pacify as well as protect the EZLN.

---

[5]Here, the term NGO includes many nonprofit organizations (NPOs), private voluntary organizations (PVOs), and grassroots organizations (GROs). It does not include international governmental organizations (IGOs), and what are sometimes referred to as government-organized NGOs (GONGOs), government-inspired NGOs (GINGOs), and quasi-NGOs (QUANGOs).

[6]Ronfeldt (1996) cites documentation for this general phenomenon. Mathews (1997) and Slaughter (1997) are significant additions to the literature.

[7]Our background comes in part from Frederick (1993b) and other chapters in Ronfeldt et al. (1993).

Thus, by the time of the EZLN's insurrection, the transnational NGOs that had been building global and regional networks, notably those concerned with human rights, indigenous rights, and ecumenical and prodemocracy issues, had counterparts to link with in Mexico City, San Cristóbal de las Casas, and other locales. Then, as NGO representatives swarmed into Chiapas in early 1994, new Mexican NGOs were created to assist with communication and coordination among the NGOs—most importantly, the Coalition of Non-Governmental Organizations for Peace (CONPAZ), based at the diocese in San Cristóbal.[8] (An NGO named the National Commission for Democracy in Mexico was established in the United States, but it was basically a public-relations arm for the EZLN.)

Were the EZLN's leaders aware of this potential? Did they foresee that numerous NGOs would swarm to support them? We have no evidence of this. Yet conditions in Chiapas were well-known to activists. Amnesty International and Americas Watch had each published a similar report of human-rights violations in the area, the former in 1986, the latter in 1991. Minnesota Advocates for Human Rights and the World Policy Institute published a joint report in August 1993 about soldiers beating and torturing a group of *indigenas* in May 1993. And the Jesuit Refugee Service, long active in the area to deal with Guatemalan refugee issues, became alarmed about the treatment of the *indigenas* in Chiapas and issued an "Urgent Call to the International Community" in August 1993. The Jesuits' demands are nearly identical to those voiced a few months later by many Mexican and transnational NGOs in January 1994.

What we see, then, is the emergence of a movement composed of several layers. The *indigenas* and the NGOs preferred nonhierarchical, network forms of organization and action, while the EZLN was drawn in this direction despite tendencies, as in any traditional Marxist armed movement, to want a hierarchy at its core. This overall bias in

---

[8]CONPAZ was formed by 14 Mexican human-rights groups that were active in Chiapas before January. They came together because they were troubled by the outbreak of war, wanted to promote peace, knew they would be more influential if they united, and lacked funding to operate well independently. CONPAZ's aims included coordinating the delivery of emergency supplies and services in the conflict zone, monitoring and denouncing human-rights violations, keeping up communication with affected communities, and generating international visibility for NGO activities.

favor of nonhierarchical designs made for affinities—and uneasy alliances—that would facilitate the mobilization of the NGOs on behalf of the EZLN and the *indigenas* and contribute to the solidarity of the movement once mobilized. Moreover, by the end of 1993, strong organizational and technological networks were in place to sustain a multilayered mobilization.

MOBILIZATION FOR CONFLICT

The insurrection did not begin as a social netwar. It began as a rather traditional, Maoist insurgency. But that changed within a matter of a few days as, first, the EZLN's military strategy for waging a "war of the flea" ran into trouble, and second, an alarmed mass of Mexican and transnational NGO activists mobilized and descended on Chiapas and Mexico City in "swarm networks" (the term is from Kelly, 1994). Meanwhile, no matter how small a territory the EZLN held in Chiapas, it quickly occupied more space in the media than had any other insurgent group in Mexico's if not the world's history.[9]

The EZLN in Combat—A "War of the Flea"

The EZLN's leaders may be credited with intelligence, flexibility, and innovation for working with indigenous ideas and institutions. Marcos in particular succeeded at adapting the EZLN's world views to those of the Maya. Even so, the EZLN—as a small guerrilla force confronting a far stronger state opponent—evidently intended, at least initially, to pursue a very traditional strategy of armed struggle: a "war of the flea" (a term popularized by Taber (1970) and repeated in regard to Chiapas by Ross (1995)).

This is often an optimal design for small, lightly armed, irregular forces. It allows insurgents to keep the initiative through surprise attacks by small units, following Mao's dictum of combining central strategic control with tactical decentralization (see Griffith, 1961, p. 114). Acts of sabotage against Mexico's economic infrastructure were to be features of the EZLN's campaign plan. Victory in such a war would hinge

---

[9]Point adapted from writings by Mexican commentator Carlos Montemayor (e.g., "La Rebellion Indigena," *La Jornada Semanal*, February 9, 1997).

on the ability of dispersed operational units (like the *focos* of Ernesto "Che" Guevara's theory of guerrilla warfare—see Guevara [1960], 1985) to pursue a common strategic goal, strike at multiple targets in a coordinated manner, and share scarce resources with each other through strategic and logistical alliances.

This strategic approach has antecedents throughout the history of Mexican wars and struggles for independence (Asprey, 1994, pp. 159–171). Emiliano Zapata, to whom the EZLN owes its name, waged a flea-like guerrilla war that played an important role in determining the outcome of the Mexican Revolution. Guerrilla operations were key forms of resistance in earlier periods as well, against both the Spanish drive (1815–1825) to maintain control over this part of its overseas empire in the wake of the Napoleonic Wars and the French effort to rule Mexico in the 1860s. Each time, guerrilla warfare succeeded against powerful opponents. The EZLN's leadership was cognizant of these historic episodes and of earlier uses of guerrilla warfare techniques by the Indians who had resisted the Spanish conquest in the 16th century.[10]

When the EZLN commenced hostilities on January 1, 1994, it thus continued in the long guerrilla-warfare traditions of Mexican insurgency and resistance. And, like so many previous movements, it quickly found itself in trouble—perhaps by adhering too closely to the basic tenets of the "war of the flea."

Two major problems emerged, one at the organizational level, the other at the tactical. First, at the outset of its campaign, the EZLN or-

---

[10]For the *Mexica*, as the Aztecs referred to themselves, guerrilla tactics emerged naturally, as a way to counter the Spanish invaders' advantages in firepower, cavalry, and body armor. As Prescott ([1843] 1949, p. 428) put it, "In the open field, they were no match for the Spaniards." Yet this deficiency drove the Indians to innovate, even to diverge from their own military traditions. Thus, "The *Mexica* themselves were fighting a different kind of war . . . all just fought as best they could, without many orders, but with instinctive discipline" (Thomas, 1993, p. 400). This drove the conquistadors to make doctrinal adjustments of their own, the most prominent being a shift from their traditional close-packed formations to what Díaz ([1568], 1963, pp. 353, 364) recalled, in his memoir, as a more loosely knit "skirmishing" approach. This was made necessary by the firepower of the guerrilla bands: "The enemy discharged so many stones from their slings, so many darts, and so many arrows, that although all our soldiers wore armor, they were wounded." Despite the hard fighting, the Spanish doctrinal innovation paid off with complete victory.

ganized itself into just a few units of almost battalion size (500–700 fighters), which was the optimal battle formation according to Mao (Griffith, 1961, p. 80). While separate detachments were formed out of these larger units, they remained under central command and control, which left them with little initiative to pursue further action in the wake of their occupations of the small towns in Chiapas. Consequently, much of the Zapatista force simply sat in place until orders were given to retreat into the Lacandón jungle. Also, these dispersed detachments were simply too far away to come to the aid of the main forces in a timely manner when the latter came under attack by the Mexican army.

This dispersion of the fighting forces turned out to be a serious problem for the main EZLN components, because it bled off fighters at a time when the EZLN fully expected to be engaging in pitched battles. Indeed, their tactical doctrine was also much influenced by Mao, whose dictum was that "Guerrillas concentrate when the enemy is advancing upon them" (from Griffith, 1961, p. 103). For example, in the firefight in the Ocosingo market, the EZLN units stood their ground, with most of one operating field unit, comprising hundreds of fighters, engaging the Mexican army openly. The results were disastrous: The insurgents were quickly pinned down and exposed to heavy fire from artillery and helicopters. There is some evidence that the EZLN military leadership tried to avoid this engagement by calling for a prompt retreat, but the Zapatista commander on the ground in Ocosingo continued to follow what he took to be his standing orders, and EZLN casualties were very high (scores dead, over one hundred wounded).[11]

EZLN leaders quickly became aware of the flaws in their traditional guerrilla strategy, and they promptly began adapting. They retreated from their exposed positions in the cities and towns and dissolved their large combat units, replacing them with much smaller fighting bands of roughly squad size (12–16 men). Their doctrine of open confrontation, which they expected would spark a national uprising

---

[11]Tello (1995) is a useful source on the first days of the fighting. The authors thank an anonymous Mexican military intelligence official for his comments on the EZLN's organization and doctrine.

(which showed no signs of emerging),[12] was replaced with a series of ambushes and minor skirmishes. Combat operations were thus dying out, and when the public, the media, and human-rights NGOs, both domestic and transnational, got involved, the EZLN was ready to shift gears to a very different sort of conflict in which the principal maneuvers would take place off the battlefield.

The most apparent organizational shift in the EZLN was its decentralization and downsizing of maneuver forces.[13] This took place within weeks of the initial attacks on the cities and towns of Chiapas. The other significant development was the EZLN's campaign to attract NGOs and other members of "global civil society" to their cause. As discussed below, these nonstate actors mobilized quickly, and they helped to constrain the Mexican government's military response to the uprising, even during a period when the United States may have been tacitly interested in seeing a forceful crackdown on the rebels. While reaching out to these nonstate political allies, the EZLN altered its own declaratory political goals, calling explicitly for reform instead of the overthrow of the government. As these changes occurred, the EZLN's "war of the flea" gave way to the Zapatista movement's "war of the swarm."

Transnational NGO Mobilization—A "War of the Swarm"

As word of the insurrection spread, U.S. and Canadian activist NGOs that had earlier participated in the networks opposing the North American Free Trade Agreement (NAFTA) and U.S. policy in Central America were among the first to mobilize to express support and sympathy for the EZLN's cause and to criticize the Mexican government's response. Also quick to mobilize were NGOs that belonged to the growing, highly networked human-rights and indigenous-rights movements. Soon a broad array of peace, ecumenical, trade, and other issue-oriented NGOs joined the mobilization.

---

[12]The EZLN proved to lack a strong nationwide structure. Moreover, despite exhortations by Marcos and other EZLN leaders, no other armed indigenous groups rose up elsewhere in Mexico in this period.

[13]Later, we shall see that the Mexican army decentralized in response. Thus, one type of decentralization was countered by another.

Within days, delegations were flowing into Mexico City and San Cristóbal de las Casas, where links were established with local NGOs and EZLN representatives. Demonstrations, marches, and peace caravans were organized, not only in Mexico but even in front of Mexican consulates in the United States. The NGOs made good use of computerized conferencing, email, fax, and telephone systems, as well as face-to-face meetings, to communicate and coordinate with each other. They focused on improving their ability to work together (as in the creation of CONPAZ) and began to struggle ceaselessly through fax-writing campaigns, public assemblies, press conferences and interviews, and other measures to make Mexican officials aware of their presence and put them on notice to attend to selected issues. The fax numbers of Mexican and U.S. officials were often posted in Internet newsgroups and mailing lists; if a number became inoperable, a new one was sometimes discovered and posted. In addition, the activists worked to ensure that the insurrection became, and remained, an international media event—swollen by the "CNN effect"—so that the EZLN and its views were portrayed favorably. Indeed, all sides waged public-relations battles to legitimize, delegitimize, and otherwise affect perceptions of each other.

Meanwhile, Marcos and other EZLN leaders kept urging NGO representatives to come to Mexico. Likewise, the NGOs already there began calling for other NGOs to join the mobilization. A kind of "bandwagon effect" took hold. A dynamic swarm grew that aimed to put the Mexican government and army on the defensive. NGO coalitions arose that were characterized by "flexible, conjunctural [*coyuntural*], and horizontal relations" held together by shared goals and demands (Castro, 1994, p. 123).[14]

What did the NGOs demand? The list included the achievement of democracy through nonviolent means; respect for human rights; a cease-fire and withdrawal by the army; peace negotiations, with the local bishop in Chiapas as mediator; freedom of information; and respect for the NGOs' roles, including access to monitor conditions in the conflict zone. Except for the commitment to nonviolence, the NGOs' collective agenda closely resembled the EZLN's. To some ex-

---

[14]Also see Reygadas (1994).

tent, this was a compromise agenda. At first, there were tensions (notably in meetings at CONPAZ) between those NGOs that wanted to voice solidarity with the EZLN and those that preferred neutrality. Some activists also had other agendas, notably to achieve the erosion if not the downfall of Mexico's ruling party, the PRI, since it was viewed as the linchpin of all that was authoritarian and wrong in Mexico's political system.[15]

Many NGO activists sensed they were molding a new model of organization and strategy based on networking that was different from Leninist and other traditional approaches to the creation of social movements. As keen scholar-activist Harry Cleaver states,[16]

> [T]he process of alliance building has created a new organizational form—a multiplicity of rhizomatically connected autonomous groups—that is connecting all kinds of struggles throughout North America that have previously been disconnected and separate (Cleaver, 1994a).

> The new organizational forms we see in action are not substitutes for old formulas—Leninist or social democratic. They provide something different: inspiring examples of workable solutions to the post-socialist problem of revolutionary organization and struggle (Cleaver, 1994b).

For these information-age activists, nonviolent but compelling action is crucial; to this end, they need rapid, far-reaching communications, as well as freedom of information and travel. Much of the netwar has

---

[15]It should be noted that there was a partial disjuncture between some demands of the *indigenas*, which were quite specific and immediate (e.g., electricity), and those of many intellectuals and NGO activists, which were general and sweeping (e.g., electoral reform). In a sense, the *indigenas* and the intellectuals spoke in different languages. The latter generally made for better press.

[16]Harry Cleaver (1994a) was among the first to identify and discuss the advent of new network designs and to show (1994b) how the NGOs' responses to Chiapas grew out of networking by groups opposed to NAFTA and by groups concerned with the rights of indigenous peoples. Cleaver (1995) expands on this. When journalist Joel Simon (1995) wrote an article proposing that netwar might be an interesting concept for understanding this model of conflict, a brief storm of discussion followed its circulation on the Internet. See the interesting article by Jason Wehling (1995) and other texts at www.teleport.com/~jwehling/OtherNetwars.html.

thus been waged through the media—both old media like newspapers, magazines, and television, and new media like faxes, email, and computer conferencing systems. (Old-fashioned face-to-face and telephone communications were important, too.)

Since word of the Zapatista insurrection first spread via the new media, activists have made heavy use of the Internet and such conferencing systems as Peacenet and Mexico's nascent La Neta (which came online in 1993) to disseminate information,[17] to mobilize their forces, and to coordinate joint actions. By the end of 1994, a remarkable number of web pages, email listserves, and gopher archives had been created on the Internet to convey the EZLN's and Marcos' statements for anyone to read and download,[18] to communicate the views and policy positions of various NGOs, and to show how to conduct what would later be termed "electronic civil disobedience."[19] The Zapatista movement gained an unprecedented transnational presence on the Net, and that presence endures and grows to this day.[20]

As the netwar got under way, two types of NGOs mobilized in regard to Chiapas, and both were important: (a) issue-oriented NGOs, and (b) infrastructure-building and network-facilitating NGOs. The former received most of the attention, but the latter were equally important. In a sense, the former correspond to the "content" and the latter to the "conduit"—or the "message" and "medium," respectively—of social activism.

Issue-oriented NGOs consist of those whose identities and missions revolve around a specific issue area, such as human rights, indigenous rights, peace, the environment, or trade and development. Numerous NGOs were active in each such issue area.

---

[17]And sometimes misinformation and disinformation, as discussed later.

[18]An early and famous site, regarded as the EZLN's unofficial home in cyberspace, was established by an American student, Justin Paulson, www.peak.org/ ~justin/ezln/, now located at www.ezln.org/.

[19]See Stefan Wray's web site (www.nyu.edu/projects/wray/ecd.html) for background and materials on electronic civil disobedience.

[20]The best general guide is Harry Cleaver's web site, *Zapatistas in Cyberspace: A Guide to Analysis and Resources,* www.eco.utexas.edu/faculty/Cleaver/zapsincyber.html.

Acting in tandem with these organizations were the second type: the network-facilitating and infrastructure-building NGOs. These are not defined by specific issues; rather, they assist other NGOs and activists, no matter what the issue. They specialize in facilitating communications; organizing demonstrations, caravans, and other events; and fostering education and exchange activities.

Of these, the most important from a technological and training standpoint is the Association for Progressive Communications (APC), a global network of computer networks that has many affiliates, such as the U.S.-based Peacenet and Conflictnet, and La Neta in Mexico. All are attached or have access to the Internet. The APC and its affiliates amount to a worldwide computer-conferencing and email system for activist NGOs. It enables them to consult and coordinate, disseminate news and other information, and put pressure on governments, including by mounting fax-writing and email campaigns. The APC also helps activist NGOs acquire the equipment and the training their members may need to get online.[21]

Of course, using the Internet to accomplish all this depends on there being good telecommunications systems for making Internet connections. In Mexico, such systems—including APC affiliates like La Neta, which came online with a weak presence in 1993, as well as direct connections available only at universities or through a few commercial providers, many of which are expensive for activists—were pretty reliable in Mexico City, other major cities, and at universities. Connecting to the Internet from a place like San Cristóbal de las Casas is another story; it can be done, but only at slow speeds and not very reliably. Faxes and telephone calls afford better communications.[22]

The APC itself did not have activists in Mexico specifically for Chiapas, but other important infrastructure-building NGOs did. These included an American NGO, Global Exchange; a Canadian networking NGO, Action Canada; and Mexico's CONPAZ. Again, cooperative connections existed among all such organizations. (At the same time, it

---

[21]For general background, see Frederick (1993a).

[22]The notion that Marcos uploads his statements to the Internet is apocryphal. He does reportedly have a laptop computer with him in the jungle, but uploading and downloading anything is accomplished by having diskettes taken to San Cristóbal.

should be noted that issue-oriented NGOs also serve as dissemina-tors of information to other NGOs. One of the most important and re-liable has been the "Miguel Agustín Pro" Center for Human Rights, which issues daily and weekly bulletins.)

Few transnational NGOs had or would install a permanent presence in Chiapas—a key exception was Global Exchange (not to mention some international governmental organizations, like the Internation-al Red Cross). Most had representatives who would come and go epi-sodically, with their timing often depending on meetings organized by the EZLN, activities organized by other NGOs, or on their own plans to visit and draw up a report. Nonetheless, the new communi-cations technologies enabled many NGOs to maintain a "virtual pres-ence" by being on mailing lists of supporters, signing petitions, par-ticipating in fax- and letter-writing campaigns, and circulating NGO-derived reports on the Internet and in other media. Such a "virtual presence" may be important to the conduct of a transnational social netwar.

Throughout, the fact that the Catholic Church, especially the diocese at San Cristóbal and church-related Mexican NGOs like the "Fray Bar-tolome de las Casas" Center for Human Rights, had a strong presence in Chiapas was crucial for the whole array of NGOs discussed above. The diocese and the NGOs related to it, soon to include CONPAZ, pro-vided a physical point of contact—a key node—for the transnational activists. (Such a node is missing in other states, like Guerrero and Oaxaca, where new conflicts are emerging.)

Thus the Zapatista networking conformed to what we would expect from a netwar. The activists' networking assumed informal, often ad hoc shapes. Participation shifted constantly, depending partly on the issues—although some NGOs did maintain a steady involvement and sought, or were accorded, leading roles. While the NGOs generally seemed interested in the collective growth of the networks, to create what would later be termed a "network of struggles," each still aimed to preserve its autonomy and independence and had its own particu-lar interests and strategies in mind. Clearly, the NGOs were—and are still—learning how to use this new approach to strategy, which re-quires that they develop and sustain a shared identity as a network and stress information operations.

There was impressive solidarity and harmony when a swarm took shape around a hot issue, such as demanding a halt to military operations or pressing for the release of an imprisoned Zapatista. At the same time, there was never complete solidarity and harmony among all members of the Zapatista networks at all times. According to our interviews, coordination was not always smooth. Problems and differences would get worked out most readily among activists present in the conflict zone, while the tone of debate might be quite different and more contentious in Mexico City. Indeed, some significant tensions existed, and surfaced, that had limiting effects.

For example, the EZLN's initial rhetoric in January 1994 was quite socialist in style and content, and it barely acknowledged the importance of *indigenista* issues like cultural rights and autonomy. In February, following Marcos's lead, a rebalancing occurred: The socialist rhetoric diminished, and demands for attention to indigenous rights came to the fore (see Van Cott, 1996, pp. 74–77; Nash, 1995). This reassured many indigenous-rights NGOs that were already supporting the EZLN. Yet some wanted to see even more Indian and less Marxist language used, and wanted the EZLN to join in building a pan-Indian movement—but the EZLN remained determined to keep its goals in a nationalist framework. From another perspective, some leftist activists were not comfortable with the EZLN's elevation of ethnicity as a factor; the Marxist left in particular regards economic class as the key factor, and ethnicity as a divisive rather than unifying factor, in social struggles.

Overall, however, many Mexican NGO activists gained confidence in their turn to networked approaches to communication, coordination, and mobilization, in regard not only to the conflict in Chiapas but also to other efforts to promote reform in Mexico. As Sergio Aguayo remarked (as a leader of Civic Alliance, a multi-NGO prodemocracy network that was created to monitor the August 1994 presidential election and later chosen in August 1995 by the EZLN to conduct a national poll, known as the National Consultation, about opinions of the EZLN):[23]

> We're seeing a profound effect on their [the NGOs'] self-esteem. They've proven to themselves that they can coordinate and do difficult tasks which have significant political implications.[24]

> [Furthermore,] if civic organizations have had so much impact, it is because they created networks and because they have received the support and solidarity of groups in the United States, Canada, and Europe.[25]

TRANSFORMATION OF THE CONFLICT

The physical—and electronic—swarming of activist NGOs into Mexico rapidly transformed the context and conduct of the Zapatista conflict. Within days, a traditional guerrilla insurgency changed into an information-age social netwar. The principal participants already had, or had shifted in the direction of, networked organizational structures—a point that is much truer for the EZLN and its NGO cohorts than for the Mexican government and army, but applies to the latter as well.

Within weeks, if not days, the conflict became less about "the EZLN" than about "the Zapatista movement" writ large and included a swarm of NGOs. This movement had no precise definition, no clear boundaries. To some extent, it had centers of activity for everything from the discussion of issues to the organization of protest demonstrations, notably in San Cristóbal de las Casas and Mexico City. It had organizational centers where issues got raised before being broadcast, such as the diocese in San Cristóbal and CONPAZ. And it drew on a core set of NGOs. Yet it had no formal organization, or headquarters, or leadership, or decisionmaking body. The movement's mem-

---

[23]Sergio Aguayo has been one of the keenest analysts of the rise of NGOs in Mexico. For example, see Sergio Aguayo Quezada, "Los modos del Marcos," *La Jornada*, January 10, 1996, as received via an email list.

[24]As quoted in Scott (1995).

[25]From Sergio Aguayo, "Citizens Chip Away at the Dinosaur," *Los Angeles Times*, August 15, 1996, p. B9.

bership (assuming it can be called that) was generally ad hoc and in flux; it could shift from issue to issue and from situation to situation, partly depending on which NGOs had representatives physically visiting the scene at the time, which NGOs were mobilizable from afar and how (including electronically), and what issues were involved. Evidently, some NGOs took a constant interest in the Zapatista movement; others showed solidarity only episodically, especially if it was not high on their agenda of concerns. In short, the Zapatista movement writ large was a sprawling, swirling, amorphous collectivity— and in a sense, its indefinition was part of its strength.[26]

As "information operations" came to the fore, the insurgents further decentralized organizationally and deemphasized combat operations in favor of gaining tighter links with the NGOs. Meanwhile, the latter utilized, and advocated that others utilize, nonviolent strategies for using varied new and old media to pressure the Mexican government to rein in its military response and accede to negotiations.

After 12 days of hard, sometimes brutal fighting in January, the government did indeed halt its initial counteroffensive. Since Mexican military forces were proving quite effective against the Zapatistas, the government's forbearance remains a puzzle. The cessation of combat operations cannot be explained by traditional state-centered theories wherein, for example, it might be thought that fear of recrimination from the U.S. government would constrain Mexican behavior. In this case, there was no overt U.S. support for the suppression of the EZLN, although there may have been some tacit or indirect support. Despite tacit external support from other governments, the Mexican government found itself unable to deflect the initiatives of the EZLN and the NGOs.

As the netwar developed, it actually impelled two Mexican presidents to halt combat operations and turn to political dialogue and negotiations: The first, as noted above, was President Salinas in January 1994, after which negotiations took place at the main cathedral in San Cristóbal de las Casas. Then a year later, in February 1995, his successor,

---

[26]The literature available on the Zapatista movement so far simply does not provide for a precise definition of "the Zapatista movement."

President Ernesto Zedillo (1994–2000), four days after ordering the army to expand its presence in the conflict zone and go arrest the EZLN leaders, called a halt and agreed to a new round of negotiations, now at San Andrés Larráinzar. Both turns of events surprised government officials, army officers, and the public at large. The halt in January 1994 also came as a surprise to the EZLN, whose leaders expected to wage war for months before seeing any possibility of negotiations. The government even agreed to treat the EZLN home base in the rain forest as a "free zone" essentially under the EZLN's own rule, for the time being.

What led President Salinas, and later Zedillo, to halt military operations and agree to dialogue and negotiations? Varied propositions have been raised for explaining their decisions: e.g., confidence that the army had gained the upper hand, or worries about a backlash among foreign creditors and investors, damage to Mexico's image in the media, infighting among Mexico's leaders, or a widespread aversion to violence among the Mexican public. Our analysis, however, is that in both instances, the transnational activist netwar—particularly the information operations stemming from it—was a key contributing factor. It lay behind many of the other explanations, including arousing media attention and alarming foreign investors. This activism was made possible by networking capabilities that had emerged only recently as a result of the information revolution. In this conflict, global civil society proved itself for the first time as a key new actor in relations between states and vis-à-vis other nonstate actors. The NGOs were able to accomplish this because of their information operations. Mexican officials admit that they were overwhelmed by the "information war" in the early days of the conflict.

BEYOND MEXICO

As noted earlier, the Zapatista case has been hailed from the beginning as the world's first "postmodern" insurgency or movement. As such, it has generated enormous comment outside as well as inside Mexico, and much of that has involved whether, and how, this case offers an information-age model of social struggle that can be further developed and replicated elsewhere.

That view is not without critics. For example, writing from a rather traditional leftist position, Daniel Nugent (1995) has decried the postmodern label by pointing out that the EZLN remains quite traditional and premodern in many respects:

> It is difficult to see how a rebel army of peasants, aware of itself as the product of five hundred years of struggle, that quotes from the Mexican constitution to legitimate its demand that the president of Mexico immediately leave office, that additionally demands work, land, housing, food, health, education, independence, liberty, democracy, justice, and peace for the people of Mexico, can be called a "postmodern political movement." How can the EZLN move beyond the politics of modernity when their vocabulary is so patently modernist and their practical organization so emphatically pre-modern? Their democratic command structure is a slow-moving form of organization—requiring as it does direct consultation and discussion with the base communities in five or six different languages—which is difficult to reconcile with postmodernist digital simultaneity. Do their demands include a modem and VCR in every jacale or adobe hut in Mexico? No. Is their chosen name "The Postmodern Army of Multinational Emancipation" or "Cyberwarriors of the South"? No.

But his points draw sharp dividing lines between what is deemed premodern, modern, or postmodern. The marvel, according to Chris Hables Gray (1997, pp. 5–6), in opening his book *Postmodern War*, is that the Zapatistas represent a hybrid of all three eras, and in a sense to be a hybrid is to be postmodern:

> Theirs is a hybrid movement, with the traditional virtues of peasant rebellions augmented by media-savvy spokespeople who use the internet and the tabloid press with the shamelessness of athletic shoe companies. . . . [Marcos] is clearly part of a sophisticated attempt by the Zapatistas to break their political isolation with a strange combination of small unit attacks, national mobilizations, and international appeals. . . . Victory, for Marcos, isn't achieving state power, it is reconfiguring power.

Irrespective of whether the postmodern label is applied, there is no denying that information plays a seminal, decisive role in this movement. As Manuel Castells (1997, p. 79) points out, in an important, wide-ranging discussion about how the information age may affect the nature of social conflict around the world,

> The success of the *Zapatistas* was largely due to their communication strategy, to the point that they can be called the *first informational guerrilla movement*. They created a media event in order to diffuse their message, while desperately trying not to be brought into a bloody war. . . . The *Zapatistas'* ability to communicate with the world, and with Mexican society, propelled a local, weak insurgent group to the forefront of world politics.

And his points are not unique to the Zapatistas. As a result of the information revolution, many new social movements—Castells also discusses environmental, religious fundamentalist, women's liberation, and American militia movements—are being redefined by the rise of a "networking, decentered form of organization and intervention" (Castells, 1997, p. 362). What is important about these networks is not just their ability to organize activities, but also to produce their own "cultural codes" and then disseminate them throughout societies:

> Because our historical vision has become so used to orderly battalions, colorful banners, and scripted proclamations of social change, we are at a loss when confronted with the subtle pervasiveness of incremental changes of symbols processed through multiform networks, away from the halls of power (Castells, 1997, p. 362).

The Mexican case is so seminal that Harry Cleaver (1998, pp. 622–623) speaks of a "Zapatista effect" that may spread contagiously to other societies:

> Beyond plunging the political system into crisis in Mexico, the Zapatista struggle has inspired and stimulated a wide variety of grassroots political efforts in many other countries. . . . [I]t is perhaps not exaggerated to speak of a "Zapatista Effect" reverberating through social movements around the world—homologous to, but ultimately much more threatening to, the New World Order of neoliberalism than the "Tequila Effect" that rippled through emerging financial markets in the wake of the Peso Crisis of 1994.

Anti-Maastricht marches in Europe and the roles played by Zapatista-inspired Italian radicals are among the examples he cites. But his analytical point is broader than any single example: A new "electronic fabric of struggle" is being constructed, helping to interconnect and inspire activist movements around the world (Cleaver, 1995 and 1998).[27]

We should note that there is some intellectual circularity in our presentation here. Most of the writings that we cite and quote from as evidence for the rise of netwar are by authors (e.g., Castells, Cleaver, Gray) who cite and quote from our original work proposing the netwar concept (especially Arquilla and Ronfeldt, 1993 and 1996). However, this circularity does not invalidate our using their writings as evidence for the spread of netwar. Instead, it confirms, as have discussions at the two Intercontinental Encounters organized by the Zapatistas, that the "network" meme[28] is taking hold in intellectual and activist circles and diffusing to new places around the world.

Thus, Chiapas provides the first of what may become a plethora of social netwars in the years ahead. Each may have its own characteris-

---

[27]Further evidence for this point appeared with news reports that a coalition of transnational civil-society NGOs, including the Council of Canadians and the Malaysia-based Third World Network, making use of the Internet and other media, had "routed" international negotiations that were supposed to lead to a Multilateral Agreement on Investment (MAI):

> The success of that networking was clear this week when ministers from the 29 countries in the Organization for Economic Cooperation and Development admitted that the global wave of protest had swamped the deal.

Some of the Canadians involved in this network had previously been active in anti-NAFTA networking. See Madelaine Drohan, "How the Net Killed the MAI: Grassroots Groups Used Their Own Globalization to Derail Deal," *The Globe and Mail*, April 29, 1998—as posted on the Internet.

[28]Dawkins (1989) originated the notion of memes as a postgenetic basis for continued human evolution, in order to convey his point that cultural as well as biological bodies are based on units of "self-replicating patterns of information" (p. 329). In his view (p. 192),

> Just as genes propagate themselves in the gene pool by leaping from body to body via sperm or eggs, so memes propagate themselves in the meme pool by leaping from brain to brain via a process which, in the broad sense, can be called imitation.

Lynch (1996) discusses how memes spread through "thought contagion."

tics, depending on the country and region in which it occurs. Chiapas, partly because it is an early case, may turn out to be a special case; so we should beware of generalizing from it. Yet it is portentous.

The case of Chiapas instructs that netwar depends on the emergence of "swarm networks"[29] and that swarming best occurs where dispersed NGOs are internetted and collaborate in ways that exhibit "collective diversity" and "coordinated anarchy." The paradoxical tenor of these phrases is intentional. The swarm engages NGOs that have diverse, specialized interests; thus, any issue can be rapidly singled out and attacked by at least some elements of the swarm. At the same time, many NGOs can act, and can see themselves acting, as part of a collectivity in which they share convergent ideological and political ideals and similar concepts about nonviolent strategy and tactics. While some NGOs may be more active and influential than others, the collectivity has no central leadership or command structure; it is multiheaded—impossible to decapitate.[30] A swarm's behavior may look uncontrolled, even anarchic at times, but it is shaped by extensive consultation and coordination that are made feasible by rapid communications among the parties to the swarm.[31]

The Zapatista case hints at the kind of doctrine and strategy that can make social netwar effective for transnational NGOs. The following are three key principles. (1) Make civil society the forefront—work to build a "global civil society," and link it to local NGOs. (2) Make "information" and "information operations" a key weapon—demand freedom of access and information,[32] capture media attention, and use all manner of information and communications technologies. Indeed, in a social netwar where a set of NGO activists challenge a gov-

---

[29]For elaboration, see Arquilla and Ronfeldt (1997), and Chapter Ten in this volume.

[30]However, particular leaders can make a difference. The development of many NGOs is at such an early stage that a leader's abilities and preferences can make a big difference in how a specific NGO behaves. Brysk (1992) makes this point well and provides examples.

[31]Of course, there may be significant divisions and factions within a network that affect its overall shape and behavior. Intranetwars may arise that alter or limit the network's capacity.

[32]On efforts to create an international charter on NGOs' rights to information and communications, see Frederick (1993c), among other sources.

ernment or another set of activists over a hot public issue, the battle tends to be largely about information—about who knows what, when, where, how, and why. (3) Make "swarming" a distinct objective and capability for trying to overwhelm a government or other target actor. Although, as noted above, swarming is a natural outcome of information-age, network-centric conflict, it should be a deliberately developed dimension of doctrine and strategy, not just a happenstance.

Where all this is feasible, netwarriors may be able to put strong pressure on state and market actors, without aspiring to seize power through violence and force of arms. In some instances, this may pose a potential threat to some U.S. interests. But in other cases, like Mexico's, a social netwar may amount to a challenge rather than a threat—it may even have some positive consequences, especially for spurring social and political reforms. Indeed, in its more positive aspects, the Zapatista netwar has not been bad for Mexico (or for U.S. interests), even though it has heightened uncertainty in Mexico and abroad regarding Mexico's stability and future prospects.

POSTSCRIPT (SUMMER 2001)

That was the case in 1998. 1999 and 2000 were mostly quiet years for the Zapatista movement and all related actors. Marcos rarely spoke out. The EZLN did not mount new operations. The Mexican army confined it to a small zone. Mexican officials kept a sharp eye on foreign activists. And many NGO activists turned their attention to other matters in Mexico and elsewhere. Whatever its potential negative consequences might be, the Zapatista movement writ large continued to have varied positive consequences in this period. In Chiapas, it stimulated the Mexican army to respond innovatively, refining the organization and performance of small units and networking them across the zone. For Mexico as a whole, the movement may have contributed, directly and indirectly, to the improved climate for democratic competition and electoral transparency that brought a new party—the National Action Party (PAN)—to power in December 2000.

Since assuming office, President Vicente Fox has energized a new peace initiative, released many imprisoned Zapatistas, and withdrawn army forces from some positions in the rebel zone. Subcomandante Marcos and the EZLN responded by expressing both hope and doubt, highlighted by a dramatic two-week march from Chiapas to Mexico City. Thus 2001 began with a traditional, theatrical political give-and-take between the government and the EZLN. Yet, the prospect of a renewed social netwar lingered in the background, fed by fresh disagreements between the Fox administration and the Zapatistas over indigenous-rights legislation.

BIBLIOGRAPHY

Arquilla, John, and David Ronfeldt, "Cyberwar Is Coming!" *Comparative Strategy,* Vol. 12, No. 2, Summer 1993, pp. 141–165. Available as RAND reprint RP-223.

Arquilla, John, and David Ronfeldt, *The Advent of Netwar,* Santa Monica, Calif.: RAND, MR-789-OSD, 1996.

Arquilla, John, and David Ronfeldt (eds.), *In Athena's Camp: Preparing for Conflict in the Information Age,* Santa Monica, Calif.: RAND, MR-880-OSD/RC, 1997.

Asprey, Robert, *War in the Shadows,* New York: Morrow, 1994.

Brysk, Alison, "Acting Globally: International Relations and Indian Rights in Latin America," paper presented at the XVII International Congress of the Latin American Studies Association, Los Angeles, September 24–27, 1992.

Castells, Manuel, *The Information Age: Economy, Society and Culture,* Vol. II, *The Power of Identity,* Malden, Mass.: Blackwell Publishers, 1997.

Castro Soto, Oscar, "Elementos Para un Analisis de Coyuntura y una Posible Estrategia desde las Clases Populares y las Organizaciones no Gubernamentales," in Mario B. Monroy (ed.), *Pensar Chiapas, Repensar México: Reflexiones de las ONGs Mexicanas,* Mexico: Convergecia de Organismos Civiles por la Democracia, August 1994.

Cleaver, Harry, "The Chiapas Uprising and the Future of Class Struggle in the New World Order," for *RIFF-RAFF,* Padova, Italy, February 1994a (online at gopher://lanic.utexas.edu:70/11/la/ Mexico/).

Cleaver, Harry, "Introduction," in Editorial Collective, ¡Zapatistas! Documents of the New Mexican Revolution, Brooklyn: Autonomedia, 1994b (online at gopher://lanic.utexas.edu:70/11/la/Mexico/Zapatistas/).

Cleaver, Harry, "The Zapatistas and the Electronic Fabric of Struggle," 1995, www.eco.utexas.edu/faculty/Cleaver/zaps.html, printed in John Holloway and Eloina Pelaez (eds.), Zapatista! Reinventing Revolution in Mexico, Sterling, Va.: Pluto Press, 1998, pp. 81–103.

Cleaver, Harry, "The Zapatista Effect: The Internet and the Rise of an Alternative Political Fabric," Journal of International Affairs, Vol. 51, No. 2, Spring 1998, pp. 621–640.

Collier, George, "Roots of the Rebellion in Chiapas," Cultural Survival Quarterly, Vol. 18, No. 1, Spring 1994a, pp. 14–18.

Collier, George, with Elizabeth Lowery Quaratiello, BASTA! Land and the Zapatista Rebellion in Chiapas, A Food First Book, Oakland, Calif.: Institute for Food and Development Policy, 1994b.

Dawkins, Richard, The Selfish Gene, New York: Oxford University Press, 1989.

Diaz del Castillo, Bernal, The Conquest of New Spain [1568], Baltimore: Penguin, 1963.

Fox, Jonathan, "The Difficult Transition from Clientelism to Citizenship: Lessons from Mexico," World Politics, Vol. 46, No. 2, January 1994, pp. 151–184.

Fox, Jonathan, and Luis Hernandez, "Mexico's Difficult Democracy: Grassroots Movements, NGOs and Local Government," Alternatives, Vol. 17, 1992, pp. 165–208.

Frederick, Howard, "Computer Networks and the Emergence of Global Civil Society," in Linda Harasim (ed.), Global Networks: Computers and International Communication, Cambridge, Mass.: MIT Press, 1993a, pp. 283–295.

Frederick, Howard, North America NGO Networking on Trade and Immigration: Computer Communications in Cross-Border Coalition-Building, Santa Monica, Calif.: RAND, DRU-234-FF, 1993b.

Frederick, Howard, *Global Communication and International Relations*, Belmont, Calif.: Wadsworth Publishing Co., 1993c.

Fuentes, Carlos, "Chiapas: Latin America's First Post-Communist Rebellion," *New Perspectives Quarterly*, Vol. 11, No. 2, Spring 1994, pp. 54–58.

Gossen, Gary H., "Comments on the Zapatista Movement," *Cultural Survival Quarterly*, Vol. 18, No. 1, Spring 1994, pp. 19–21.

Gray, Chris Hables, *Postmodern War: The New Politics of Conflict*, New York: The Guilford Press, 1997.

Griffith, Samuel, *Mao Tse-Tung on Guerrilla Warfare*, New York: Praeger, 1961.

Guevara, Che, *Guerrilla Warfare* [1960], Lincoln: University of Nebraska Press, 1985. Translated by J. P. Morray.

Harvey, Neil, "Rebellion in Chiapas: Rural Reforms, Campesino Radicalism, and the Limits to Salinismo," *Transformation of Rural Mexico*, Number 5, Ejido Research Project, La Jolla, Calif.: Center for U.S.-Mexican Studies, 1994, pp. 1–43.

Hernandez, Luis, "The Chiapas Uprising," *Transformation of Rural Mexico*, Number 5, Ejido Research Project, La Jolla, Calif.: Center for U.S.-Mexican Studies, 1994a, pp. 44–56.

Hernandez, Luis, "The New Mayan War," *NACLA: Report on the Americas*, Vol. 27, No. 5, March/April 1994b, pp. 6–10.

Kelly, Kevin, *Out of Control: The Rise of Neo-Biological Civilization*, New York: A William Patrick Book, Addison-Wesley Publishing Company, 1994.

Lynch, Aaron, *Thought Contagion: How Belief Spreads Through Society*, New York: Basic Books, 1996.

Mathews, Jessica, "Power Shift," *Foreign Affairs*, Vol. 76, No. 1, January/February 1997, pp. 50–66.

Nash, June, "The Reassertion of Indigenous Identity: Mayan Responses to State Intervention in Chiapas," *Latin American Research Review*, Vol. 30, No. 3, 1995, pp. 7–41.

Nugent, Daniel, "Northern Intellectuals and the EZLN," *Monthly Review,* Vol. 47, No. 3, July–August 1995 (as circulated on the Internet).

Prescott, W. H., *A History of the Conquest of Mexico* [1843], New York: Heritage, 1949.

Reygadas Robles Gil, Rafael, "Espacio Civil por la Paz," in Mario B. Monroy (ed.), *Pensar Chiapas, Repensar México: Reflexiones de las ONGs Mexicanas,* Mexico: Convergecia de Organismos Civiles por la Democracia, August 1994.

Ronfeldt, David, *Tribes, Institutions, Markets, Networks: A Framework About Societal Evolution,* Santa Monica, Calif.: RAND, P-7967, 1996.

Ronfeldt, David, Cathryn Thorup, Sergio Aguayo, and Howard Frederick, *Restructuring Civil Society Across North America in the Information Age: New Networks for Immigration Advocacy Organizations,* Santa Monica, Calif.: RAND, DRU-599-FF, 1993.

Ross, John, *Rebellion from the Roots: Indian Uprising in Chiapas,* Monroe, Me.: Common Courage Press, 1995.

Scott, David C., "NGOs Achieve Credibility in Mexico," *Crosslines Global Report,* October 31, 1995 (http://burn.ucsd.edu/archives/chiapas-l/1995.11/msg00013.html).

Simon, Joel, "Netwar Could Make Mexico Ungovernable," Pacific News Service, March 13, 1995.

Slaughter, Anne-Marie, "The New World Order," *Foreign Affairs,* Vol. 76, No. 5, September/October 1997, pp. 183–197.

Stephens, John Lloyd, *Incidents of Travel in Central America, Chiapas and Yucatan* [1841], New York/London: Harper/Century, 1988.

Taber, Robert, *The War of the Flea,* New York: Citadel, 1970.

Tello Díaz, Carlos, *La Rebelión de las Cañadas,* Mexico City: Cal y Arena, 1995.

Thomas, Hugh, *Conquest: Montezuma, Cortes, and the Fall of Old Mexico,* New York: Simon and Schuster, 1993.

Van Cott, Donna Lee, *Defiant Again: Indigenous Peoples and Latin American Security,* McNair Paper 53, Washington, D.C.: Institute for National Strategic Studies, October 1996.

Wehling, Jason, "'Netwars' and Activists Power on the Internet," March 25, 1995 (as circulated on the Internet in the abridged version, "'Netwars': Politics and the Internet," August 7, 1995. The full version is posted at www.teleport.com/~jwehling/OtherNetwars.html).

Chapter Seven

NETWAR IN THE EMERALD CITY: WTO PROTEST STRATEGY AND TACTICS

Paul de Armond

Editors' abstract. In a free society, netwar can run wild—sometimes literally. The Battle of Seattle is the best case of this to date. De Armond (Public Good Project) offers an eyewitness account, analyzing all players and their strategies and revealing how and why the Direct Action Network did so well. This struggle featured a rich mix of activists and anarchists, from around the world, who were intent upon disrupting a gathering of governmental and international institutional actors that were assembling to launch the World Trade Organization. The chapter is largely condensed from a longer paper titled "Black Flag Over Seattle," Albion Monitor, No. 72, March 2000, www.monitor.net/monitor/ seattlewto/index.html. Reprinted by permission.

Seattle, like many American cities, has self-appointed nicknames. One of Seattle's nicknames is "The Emerald City," a reference to its perpetually soggy evergreen vegetation and to the mythical Land of Oz. On November 30, 1999, Seattleites awoke to the reality of an emerging global protest movement. This movement was not created in Seattle. Other protests with similar motives, participants, and strategies had been happening in the United States and around the world for a considerable time. What made the "N30" protests remarkable was the shock that we, like Dorothy and Toto, were no longer in Kansas.

For the next year, roving protests continued the agitation that exploded in Seattle. In the United States, Boston (Biodevastation), Washington, D.C. (A16), numerous cities on May Day (M1), Milwaukee (animal rights), Detroit and Winsor, Ontario (OAS), Philadelphia (Republican Convention), and Los Angeles (Democratic Convention) were visited by what protesters called the "spirit of Seattle." Around the world, protests took place in Bangkok, London, Prague, Melbourne, and other cities.

On N30, all that lay in the future. Previous protests, particularly the J18/"Seize the Streets" protests in London and other cities around the world on June 18, 1999, foreshadowed the N30 demonstrations in Seattle. The J18 protest was ignored, dismissed, or misinterpreted. Seattle was where the protests broke through the infosphere and into the notice of the world. Oz did not fall, but the walls were breached.

Networked forms of social organization distinguish the new protest movement. Dubbed "netwar" by David Ronfeldt and John Arquilla, this style of conflict depends heavily on information and communications technology, nonhierarchical organization, and tactics that are distinctly different from previous forms of civil-society conflicts. Understanding what happened in the Emerald City on N30 requires identifying the numerous actors, outlining their strategies and tactics, and knowing the sequence of events as the protests unfolded.

PROTEST BACKGROUND

The central fact of the Seattle protests is the utter surprise and confusion during the initial confrontation on Tuesday morning. "It was a classic example of two armies coming into contact and immediately experiencing the total collapse of their battle plans," said Daniel Junas, a Seattle political researcher.[1]

---

[1]Most quotations are from news coverage in the *Seattle Times* and the *Seattle Post-Intelligencer* series on the WTO protests which ran during December 1999 and January 2000. The complete WTO coverage by these two newspapers is available on the web at http://seattletimes.nwsource.com/wto/ and http://seattlep-i.nwsource.com/wto/. Quotes from Daniel Junas and Jeff Boscole are from personal conversations with the author. The chronology of events was assembled from the WTO documentary *Four Days in Seattle* aired by KIRO TV on December 10, 1999.

For an anarchist view of the Black Bloc, see Tom Trouble, Black Bloc Participant Interview by Active Transformation, http://csf.colorado.edu/forums/pfvs/2000/msg03110.html. The police perspective is drawn from: Mike Carter and David Postman, "There Was Unrest Even at the Top During WTO Riots," *Seattle Times*, December 16, 1999, and Brett Smith and Dan Raley, "Police Officer Blames City's Poor Planning," *Seattle Post-Intelligencer,* December 4, 1999. One participant's experience in the AFL-CIO march is described by Greta Gaard, "'Shut Down the WTO!' Labor and Activists Create Change," *Every Other Weekly,* Bellingham, Wash., Dec. 16–Dec. 29, 1999.

The street action falls into three distinct phases. First, the Direct Action Network (DAN) protesters seized and held a handful of strategic intersections, immobilizing the police. Second, the police strategy fragmented over two contradictory goals: suppressing the DAN protests and allowing the labor parade. Third, the labor parade failed in its goal of controlling and diverting the DAN protesters away from the Convention Center. The influx of reinforcements who abandoned the labor parade and joined the DAN protests left the streets more firmly in control of the protesters, despite the use of tear gas by police from around 10 a.m. By approximately 3 p.m. Tuesday, the battle was decided and the Direct Action Network prevailed in its goal of shutting down the conference.

After that time, the outcome was certain. The battle continued for three days, spreading into other areas of the city. By Thursday, the World Trade Organization ground to an inconclusive halt, and the police ceased attacking civilians, thereby recognizing a conclusion that had been reached before darkness fell on Tuesday.

The Players: WTO Opponents

DAN represents an emerging species of political organization based on networks rather than institutions. The primary networked organizations in DAN were a coalition of such groups as the Rainforest Action Network, Art & Revolution, and the Ruckus Society. Through DAN, these groups coordinated nonviolent protest training, communications, and collective strategy and tactics through a decentralized process of consultation/consensus decisionmaking.

The strategy and tactics of these new—and primarily information-based—networks of nongovernmental organizations evolved from trends represented by the ad hoc mobilization committees of the Vietnam protest era, the "alternative summits" at recent world environmental and human rights conferences, and the loose coalitions that formed in opposition to U.S. policy during the Gulf War. Networks, as opposed to institutions, are shaped by decentralized command and control structures, are resistant to "decapitation" attacks targeting leaders, and are amorphous enough to weld together coalitions with significantly different agendas while concentrating forces on a single symbolic target. Conflicts involving networks blur distinctions between offense and defense.

The overall strategic goal of the Direct Action Network was to "shut down" the World Trade Organization meeting in Seattle. The main instrument for doing so was the fielding of a few dozen "affinity groups"—small units into which the activists organized on their own. These affinity groups were organized at DAN training sessions in the weeks prior to the protests. The central training was conducted by the Ruckus Society and was attended by approximately 250 people, who then became the hard core of protestors in the "first wave"—i.e., those who were willing to risk violent confrontation with the police and arrest once the demonstrations began. Through a variety of independent but strategically congruent actions, this first wave was to be followed by a "second wave" of other affinity groups and supporters who were still militant but less willing to risk arrest and injury—all summing up to a street blockade in the vicinity of the WTO conference. The numerically small affinity groups anchored the protests and provided a catalytic nucleus of blockades around which crowd actions were directed. The Direct Action Network's goals and consultative strategy were sufficiently broad to encompass all of the protesters' grievances.

The second major WTO opponent was American organized labor, the AFL-CIO. The AFL-CIO is a hierarchical institution emphasizing unitary, top-down command. There is little participation by rank and file in union decisionmaking, although ceremonial elections are sometimes held to legitimize leadership decisions. Essentially nationalist in outlook, the AFL-CIO has policy goals that are directed more at American politics and less at international issues. Simply stated, the AFL-CIO's strategic target was supporting and legitimizing President Clinton's actions at the conference through purely symbolic displays as a loyal opposition. The AFL-CIO helped attract thousands of people to Seattle. Its main adherents had little interest in joining with DAN's; but during the second and third days of the protests, a spill-over from the AFL-CIO crowds into DAN's street actions added to a "third wave" of protest that ultimately overwhelmed the police.

The Players: World Trade Organization and Allies

On the other side of the conflict, the World Trade Organization and its allies composed a much more divided picture. The purpose of the WTO conference was to produce a new framework for the next round of negotiations on international trade. To a lesser extent, the WTO deliberations would broaden the scope of existing trade agreements to include developing countries. Prior to the Seattle conference, three major trading blocks have dominated the WTO: the western hemisphere block organized around the NAFTA treaties, the European Economic Community (EEC), and the Asian industrialized nations. The Seattle talks were the first to include developing countries. Even in the absence of protests outside the meeting, the tensions inside made it very likely that the Seattle round of negotiations would be off to a very rocky start.

The American posture consisted of blocking agreements while giving the appearance of support. President Clinton's strategy was concentrated around his appearance at the conference, rather than the success of the conference itself. If the talks failed to produce a new framework, then the existing agreements (which heavily favored the shared interests of industrialized countries over developing countries) would continue to provide the basis for international negotiations. In relation to the protests, the federal strategy hinged on getting Clinton into the conference.

The City of Seattle, as host of the conference and lead jurisdiction, was the center of responsibility for containing the demonstrations. Aside from this hospitality, Mayor Schell's political concerns were complex. First of all, the primary reason for Seattle hosting the WTO conference was to promote regional trade interests: principally timber and forest products, wheat, and a variety of high-tech industries, of which Microsoft and Boeing are the best known examples. Second, Schell is a liberal with strong ties to the Democratic Party and its main source of financial support, the AFL-CIO. Third and last, Schell is deeply beholden to the progressive Democrats and environmentalists who are a key political constituency in Seattle, although mostly excluded from the Democratic Party by the labor interests. Schell's attempts to satisfy all of these interests were so riddled with contradictions that he became unable to control events and was ultimately left to twist slowly in the wind.

The direct point of contact between the Direct Action Network and the WTO was the Seattle Police Department (SPD). Under the leadership of Chief Norm Stamper, the SPD has become a national laboratory for a progressive philosophy of law enforcement known as "community policing." Recently, the relations between the police and Mayor Schell's administration have not been good. The road to community policing has been rough and rocky, particularly in light of the resistance from rank and file cops.

The total size of the Seattle Police Department is roughly 1,800 officers, of whom about 850 are available for street duty throughout the city. Of these, 400 were assigned to the WTO demonstrations. Seattle has about the same ratio of police to population as Chicago, but Seattle's smaller size limited the number of officers it could field against the protesters—unless, of course, the SPD entered into some sort of joint WTO operation with other police agencies in the region. By Wednesday, the second day of the protests, more than 500 state and regional police, plus some 200 National Guard were deployed.

The largest two outside police forces available to Seattle are the King County Sheriff's Department and the Washington State Patrol. King County Sheriff Dave Reichert is a conservative Republican and political foe of Mayor Schell. This reflects the long-standing division between Seattle and the King County government. The suburban fringe surrounding Seattle is the traditional political battleground in which statewide elections are fought. The outlying areas go to the Republicans and the heavily urbanized areas go to the Democrats. The suburbs swing back and forth between the two. The State Patrol and National Guard are responsible to Gov. Gary Locke, a nominal Democrat who rose to the governorship through the King County Council. None of these outside agencies are supporters of community policing policies, which meant that assistance entailed Chief Stamper presiding over a joint command divided by fundamental policy differences.

One consideration weighing against the employment of outside police on Tuesday was the strong possibility that they would attack the union parade and city residents. The delayed deployment of outside police reinforcements prevented contact with the union parade. Once the union supporters boarded their buses and left town, the augmented police hit the streets. Then the police attacks on city residents began and continued through Tuesday and Wednesday night. Unified police command was not established until Thursday, after the Wednesday night debacle on Capitol Hill—which included police attacks on media and elected officials.

The Players: Wild Cards

There are two more players who deserve examination, especially since one ended up dominating the national media coverage. Neither of these two groups was numerous nor strategically significant in terms of the overall outcome of the WTO protests. However, both ended up in effective control of the informational conflict in which the media was both the battleground and the prize.

The first of these groups was the so-called "Anarchists from Eugene," more correctly known as the "Black Blocs." The total number of Black Bloc participants numbered between one and two hundred people, slightly less than DAN's "lockdown" affinity groups. The appearance of Black Blocs at protests is a relatively recent phenomenon. The purpose of Black Blocs is to show a visible presence of the more radical anarchist factions. A Black Bloc consists of protesters who wear black, carry anarchist flags and banners, and take a more confrontational approach to protest.

In an interview in *Active Transformation*, an anarchist journal, one participant in the Seattle Black Blocs explained it this way:

> Anarchists were not isolated in the black block. There were anarchists involved in every possible way. There were anarchist labor activists, puppeteers, non-violent lockdown blockaders, marching musicians, medics, communication people, media people, whatever—as well as a group of about two hundred in black masks who had prepared, also in affinity groups, to do as much symbolic physical damage to multi-national capitalism as possible. I have seen black blocks used in protests in the U.S. a lot but never so successfully. It is important to note that the black block was not the result of

some conspiracy. It too happened quite spontaneously, with people who came from all over the country—with similar desires.

The media's tag line of "Anarchists from Eugene" is one of those lazy half-truths that sums up to a conscious lie. The half-truth is that people from Eugene participated in the Black Blocs. The other unreported half of the truth is that people from Seattle and the surrounding region—not affiliated with the Black Blocs—committed much of the vandalism and nearly all of the looting. These people were not part of the Black Blocs, nor were their actions directed or controlled by the Black Blocs. The lie was that the Black Bloc caused the police violence in the streets, when actually the police attacks on the crowds began several hours before the window-breaking spree.

The primary target of the Black Blocs was neither the WTO nor the businesses whose windows were broken. The Black Blocs were in Seattle to radicalize the protest and prevent the nascent movement from being absorbed by the AFL-CIO umbrella group.

The second wild card was a segment of the Seattle Police Department that actively sought to disrupt the chain of command and forcibly turn the initial confrontation with demonstrators into chaos. One clear sign of eroding police discipline was the circulation of mutinous talk regarding the "softness" of the official strategy for dealing with the demonstrators. During an October crowd-control training session, Assistant Chief Ed Joiner answered questions about protester violence by saying that there was nothing to worry about and the protests would be nonviolent. SPD Officer Brett Smith told the *Seattle Post-Intelligencer* that the FBI and Secret Service had briefed King County Sheriff's officers to "fully anticipate that five to six officers would be lost during the protests, either seriously injured or killed." By noon on Tuesday, the police chain of command was seriously eroding. From this moment on, more and more command responsibilities passed to officers in the streets. The breakdown in command continued through the next day, culminating in the events of Wednesday night. It was not until Thursday that a unified command was established and able to assert total control over police actions in the streets.

STRATEGIES

The geography of Seattle's downtown favors protesters. In the last decade, two major civil disturbances—accompanying first the Gulf War
protests, and later the "Rodney King" riots—have followed much the
same path over the same streets, as did the numerous protests during
the Vietnam War. Given sufficient numbers and even the most hare-
brained strategy, protesters have the ability to dominate the streets of
Seattle.

The outcome of the Seattle protests was mostly due to the failure, not
the success, of the respective strategies of the AFL-CIO, the Direct Action Network, and the Seattle Police. As is often the case in netwar
conflicts, victory goes to the side that decontrols most effectively. As
each of the strategies collapsed into confusion and disarray, the DAN
strategy proved to be the one that survived the chaos.

The AFL-CIO strategy was to hold a rally at the Seattle Center and
then march downtown (but not too far). Central to the AFL-CIO strategy was the notion that they could contain the majority of the demonstrators and keep them out of the downtown area. All the AFL-CIO
had to do was prevent any effective protests by groups not under their
control and allow the media to spin the tale of how labor caused a
sudden change in national policy. The AFL-CIO proved to be unequal
to the task of rounding up all the protesters and keeping them muz-
zled.

The Direct Action Network planned more effectively, and in the end
more realistically, with a "peoples convergence" consisting of three
waves (mentioned above) of blockaders enclosing the WTO conference site.

- The first wave consisted of 200–300 people in "lockdown" affinity
  groups—those who had opted for nonviolent civil disobedience
  and arrest. Their job was to penetrate the area close to the conference site, seize the dozen strategic intersections that controlled
  movement in the protest target, and hang on until reinforcements
  arrived. DAN estimated correctly the size of participation in this
  first wave.

- The second wave included several thousand protesters, also organized as affinity groups, who had opted for nonviolent demonstration and not being arrested. Their task was to protect the first wave from police violence and plug up the streets by sheer numbers and passive resistance. Many more people joined this second wave than DAN expected.

- The third wave was a march by several more thousand people in the People's Assembly, composed mostly of environmental and human rights groups who elected to participate in the street protests instead of the labor parade. This group entered downtown from the south at about 1 p.m. and marched to the Paramount Theatre inside the protest zone. The size of the third wave vastly exceeded DAN's expectations, as numerous marchers from the AFL-CIO parade merged into the protests downtown.

The first and second waves were organized around a dozen simultaneously converging affinity groups, swarming the protest target from all directions. Each affinity group blockaded a specific intersection. DAN expected the blockade would be maintained until police had arrested sufficient demonstrators to regain control of the streets. Much to DAN's surprise, the blockade was so effective that the expected arrests proved impossible.

Throughout the protests, the Direct Action Network protesters were able to swarm their opponents; seizing key intersections on Tuesday and penetrating the "no-protest" zone on Wednesday. DAN communications channels blanketed the Seattle area and had global reach via the Internet. Indeed, DAN's cohesion was partly owed to an improvised communications network of cell phones, radios, police scanners, and portable computers. Protesters in the street with wireless handheld computers were able to link into continuously updated web pages giving reports from the streets. Police scanners were used to monitor transmissions and provided some warning of changing police tactics. Cell phones were widely used. In addition to the organizers' all-points network, protest communications were leavened with individual protesters using cell phones, direct transmissions from roving independent media feeding directly onto the Internet, personal computers with wireless modems broadcasting live video, and a variety of other networked communications. Floating above the tear gas was a pulsing infosphere of enormous bandwidth, reaching around the planet via the Internet—although on the scene, at street level, the Internet played little role, because most communications among the affinity groups were face-to-face and via cell phone, unencrypted.

Institutions, such as corporate media, police, and the AFL-CIO, tend to depend on narrow communications—highly centralized and hierarchical. DAN's diffuse communications network allowed protesters to continuously adapt to changing conditions. The consultative form of decisionmaking enhanced the ability to coordinate large-scale actions. The police attempts to arrest "ringleaders" on Wednesday were fruitless, since leadership and communication were widely shared throughout the network of protest groups, and the communications network was continuously expanded and modified. On Tuesday, police cut off some of DAN's communications channels, but in a few hours a new and larger network based on new cell phones was functioning.

The competing strategies of the Direct Action Network and the AFL-CIO put the police in the classically disastrous position of dividing their limited forces and inviting defeat in detail. Working with the labor leaders, the police intended to use the AFL-CIO rally as a means of containing the crowds and keeping the majority of them away from the Convention Center. Much has been made of the decision to rely on a close perimeter defense of the Convention Center, but a larger perimeter and more police would have simply moved the line of battle and dispersed the police, as occurred on Wednesday.

The real question facing the police was whether they would be confronting a protest or a parade. The police put their money on the parade and lost. The labor parade as the dominant factor of the protests was the least likely of all outcomes, but the only one that the police had a chance of controlling.

Overestimating the importance of the parade and underestimating the numbers of the DAN mobilization resulted in the police plans that collapsed early Tuesday morning. The police relied on a "tripwire" outer perimeter to trigger the arrests around the Convention Center, backed up with an inner perimeter to prevent DAN protesters from entering the WTO conference. At the Paramount Theatre, the distance between these two tripwires was less than the width of a city street. When the crunch came, the outer tripwire (a flimsy barricade of rope) melted into the inner perimeter (a barricade of buses) in a matter of seconds. The police had prepared to defend a perimeter measured in feet and the protesters had arranged a blockade measured in city blocks.

Intelligence Failure

Underlying the failure of the police strategy for controlling the demonstrations was the fundamental failure of intelligence. The picture that law enforcement built of the developing protests was a catastrophe of wishful thinking, breathing their own exhaust, and the most classic of all blunders—mistaking tactics for strategy. The law enforcement agencies had the information necessary to appraise the situation. They lacked a comprehensive understanding of the strategy of the protests. Without that, the pieces of the intelligence puzzle were not going to fit into an accurate picture.

The wishful thinking centered on the alliance between the police and the AFL-CIO. The plan for the labor parade to engulf the protests and steer them into a marginal venue was never a real possibility. The Direct Action Network and their allies had no intention of turning the protest organizing over to the unions. If there was going to be an alliance between protesters and paraders, it was going to be on the protesters terms or not at all. City officials chose to believe the labor assurances of controlling the protesters. This led the police to drastically underestimate the number of protesters. Neither the police nor the unions foresaw the Direct Action Network being able to mount a successful protest, nor did they anticipate a blockade engulfing a dozen city blocks. Once false assumptions became the basis for planning, any evidence to the contrary was disregarded or misinterpreted.

The intelligence picture was further confused by the claims of federal law enforcement officials that the protests would be violent. The publicly released text of one FBI forecast was replete with hysterical predictions:

[E]lements within the protest community are planning to disrupt the conference . . . environmental or animal rights extremists or anarchist-induced violence . . . computer-based attacks on WTO-related web sites, as well as key corporate and financial sites. . . . Corporate sponsors . . . may be subject to surveillance efforts from these groups. . . . to identify the residences of key employees of sponsoring corporations. . . . These employees should remain alert for individuals who may be targeting them in furtherance of anti-WTO activities. . . . Recipients should remain sensitive to threats made by anti-WTO groups.

Nowhere in the FBI "Terrorist Threat Advisory" was the slightest inkling of what was going to be happening in the streets beyond the fact that the conference was going to be "disrupted." The competing strategies of the Direct Action Network and the AFL-CIO had been trumpeted loudly, widely, and in considerable detail in the press by the organizers, summing up to nonviolent civil disobedience, shutting down the conference, and an ineffectual parade designed to keep protesters away from the Convention Center. City officials at the top elected to pick and choose among information to support their plans. The frontline officers did the same, with opposite results. The rumors within the police department (fantasy or otherwise) about federal expectations of dead and wounded police added to the unreality.

Correlation of Forces

By Monday evening, November 30, the forces had aligned themselves. The Direct Action Network planned to shut down the WTO conference by swarming the streets. The AFL-CIO planned to hold a rally and parade in an effort to influence national trade policy—and the upcoming presidential elections. Police Chief Norm Stamper had decided the protests could be peacefully controlled by his own forces without outside assistance, knowing that the price of assistance could be the peace. The mayor allowed the AFL-CIO to control his actions on Tuesday, hoping against all evidence that the unions would swallow and control the protesters. The Seattle Police Department was tasked with preventing the protests while allowing the labor parade. The outside law enforcement agencies were champing at the bit to enter into the fray, but as long as the SPD maintained order, they had to sit on the sidelines. The FBI and Secret Service cried doom and gloom—while signing off on Mayor Schell and Chief Stamper's plan. The Black Blocs were milling around the edges, fondling their crowbars and dreaming of chaos.

What would happen next was anybody's guess, but the best guessers would win and the others would lose.

In the end, the advantage went to the Direct Action Network, since its strategy effectively enclosed the coordinated strategy of the AFL-CIO and the federal government. As will be seen, at the critical moment in the street actions, the balance shifted to the Direct Action Network as nonunion protesters and a few union members left the AFL-CIO parade and joined the street protests, effectively sealing the success of the Direct Action Network's day-long blockade.

CHRONOLOGY

At 5 a.m. Tuesday morning, Washington State Patrol Chief Annette Sandberg was having coffee at the Starbucks near the Convention Center. Nobody would be having coffee there later that evening because it would be smashed and looted. Sandberg saw demonstrators moving into strategic positions before any police had arrived. The converging columns of the Direct Action Network began to shut down Seattle.

The first DAN "arrest" affinity groups moved in on the strategic intersections in the vicinity of the Convention Center. Afterwards, these protesters said that they were surprised by the absence of any police presence on the streets. In many locations, the "arrest" groups arrived earlier than the "nonarrest" groups, which were supposed to protect the arrest groups from removal by the police. The news photographs of these initial "lock-down" groups have a surrealistic air to them. In the empty streets after dawn, groups of protesters lock themselves together with bicycle locks or tubes, covering their linked arms to prevent police from removing them individually. By 8 a.m., most of the key intersections had been seized by the protesters, now reinforced by their second wave.

King County Sheriff Dave Reichert says he got a telephone call at 8 a.m. from a county detective. "He said, 'Sheriff, we're trapped. . . . We have no backup,'" Reichert claimed. "I had officers barricaded in the hotel with a mob literally pounding on the glass, and there was nobody to help them. Nobody." Reichert wasn't on the scene, but already he was seeing "mobs." KIRO TV crews were at the same location and showed lines of grinning demonstrators holding hands and blocking the street—no "mob literally pounding on the glass."

As the number of protesters increased, the 400 police remained in their lines around the Convention Center or at their positions at the Memorial Stadium. The slow infiltration of demonstrators made it difficult for the police to gauge the intentions of the crowd. The Direct Action Network had already swarmed the police and shifted to a defensive strategy of holding on to the streets that they now controlled. The flimsy rope and netting barriers, the "tripwire" at the Paramount Theatre, went down as protesters walked toward the line of city buses next to the theater. The buses were a second line of defense, separating the police from the crowd. The police strategy relied on these "tripwires" to trigger the shift from a passive "wait and see" mode to more aggressive tactics. Unfortunately, the "tripwire" perimeters were now engulfed and isolated by the DAN affinity groups and the crowds that surrounded them.

Meanwhile, at the Memorial Stadium at the Seattle Center, the gates were opening for the AFL-CIO rally, which was scheduled to begin at 10 a.m. Chartered buses from around the region were on the road for some time, carrying a mixture of union members and protesters to Seattle. The AFL-CIO had done mass mailings throughout Washington State, sending postcards to nonunion supporters of a variety of liberal and progressive organizations. "Join the March of the Century," the cards read. Before the stadium opened, DAN had rewritten the AFL-CIO script. The labor parade, a "head fake to the left," was now a sideshow rather than the main event.

By 9:10 a.m., "crowd-control efforts were encountering difficulty," according to Washington State Patrol Chief Sandberg. She placed troopers throughout Western Washington on alert. The day was barely started and the police plan was already beginning to break down. The Secret Service, responsible for the security of federal and visiting government officials, discovered that the streets between the Convention Center, the adjacent hotels, and the Paramount Theater—a distance of up to five blocks along some routes—were closed by protesters. "It hadn't taken long for things not to be working very well," said Ronald Legan, the special agent in charge of the Seattle office of the Secret Service.

Police Go on the Offensive

Shortly after 10 a.m., the Seattle Police Department began using tear gas at the southern end of the triangular area blockaded by the Direct Action Network. The use of gas may have been a botched effort to open a pathway into the protest area from outside, since the gas was fired on Sixth Avenue, between University and Union Streets, immediately outside the Olympic, one of the delegate hotels. Police officials later explained that the gas was an attempt to expand and reconnect their now isolated perimeters inside the crowds.

With the release of the gas, mood in the streets rapidly changed. The police were successful in advancing against the crowd only over short distances. There were no instances where police charges were repulsed, or where the crowds counterattacked and cut off police. One major effort to reopen the street connecting the Paramount Theatre to the hotels moved the crowds back until running out of steam. In short, the police tactics were ineffective because of the enormous ratio of protesters to police.

The net effect of the use of gas and the police charges was to cause the crowds to surge from one point to another without allowing police to gain control of the streets. In the midst of the melee, the "lock-down" affinity groups remained in place, blocking intersections and anchoring the protest to the area around the Convention Center. Police gassed and pepper-sprayed the immobile groups, but could not arrest them and remove them from the area because of the continued blockade. These tactics were both ineffective in getting the blockaders to move and successful in infuriating the crowds who saw their main mission as the protection of these groups. The crowds were now frightened and angry, but determined to maintain control of the streets.

The overall strategic situation remained unchanged, despite the tactical chaos. The protesters numbers were sufficient to keep the blockade intact, although it was now a blockade of continuous movement. The police remained isolated inside the protest area without an open avenue to the outside through which arrestees could be removed. The area involved in the disorder—and that's what it clearly was after an hour of tear gas and chaos—spread down Pike and Pine Streets. The protests remained centered on the Convention Center. Although the crowds expanded into the surrounding blocks under the police attacks, they kept surging back to protect the "lock-down" affinity groups holding the key intersections.

Labor's Head-Fake Becomes a U-Turn

By 11 a.m., the rally at Memorial Stadium had been under way for an hour. Roughly 20,000 people half-filled the stadium. The union numbers were swelled by the anti-WTO organizations that had accepted the labor invitation to protest the WTO. These groups were a mixture of environmental, social justice, and human rights groups. Over the next two hours, the joint planning by the labor leaders and police to break the DAN blockade would irretrievably split the brief alliance between labor and the progressive left.

As the labor rally was getting under way, Assistant Chief Ed Joiner was turning down demands from his field commanders to declare a state of civil emergency that would cancel the parade. Joiner said he overruled a recommendation by Assistant Chief John Pirak to declare a state of emergency on Tuesday at about 11 a.m. Despite the fact that "we were getting hit with much larger numbers of protesters than we had anticipated," Joiner refused.

The veto, Joiner said, was made in consideration of plans for the AFL-CIO march toward downtown. "I felt declaring a state of emergency at that time, before the march ever got under way, was going to send a very strong public message that we already had major difficulties as a city," Joiner said.

Joiner believed the march would actually work in favor of his stretched police lines. The strategy, he said, was for the peaceful march to sweep the other demonstrators into its ranks and lead them out of the downtown area.

The final decision was to allow the AFL-CIO parade to proceed from the Seattle Center to downtown. This sealed the fate of the street actions as a victory for the Direct Action Network. If the march had been canceled and the additional protesters had been prevented from joining in the chaos downtown, the city stood a better chance of restoring order. Instead, the strategy of using the AFL-CIO to contain and neutralize the Direct Action Network protests was woefully misdirected.

The march was supposed to wheel away from downtown several blocks from the Convention Center, draw people away from the street protests, and move north to a "dispersal point" near Republican Avenue near Memorial Stadium. The police intended to move in behind the demonstrators and expand the perimeter around the hotels and Convention Center. Joiner said,

> I still believe we could have controlled what we were dealing with at that time had the march turned. It was not going to be clean. It would have been messy. But I think we would have been able to open a corridor to get delegates in and out.

As the parade approached downtown, AFL-CIO marshals began blocking progress toward the Convention Center, saying "The route has been changed. Circle around here." Police were massing several blocks to the south but were not visible to the people arriving from the Seattle Center. Several thousand people broke away from the march, just in time to run into the renewed police push to move people away from the Convention Center. The momentum of the thousands moving toward the Convention Center carried several blocks south, past the parade's planned pivot at 5th and Pike. Behind them, the leaders of the labor parade moved north from downtown and returned to the Seattle Center, unmolested by police.

Pause to Regroup

Assistant Chief Joiner's "messy" plan to force the Direct Action Network protesters out of the downtown area and into the AFL-CIO parade set in motion several different actions that had a dramatic effect on perceptions of the Battle in Seattle. To understand how these actions converged, it is necessary to step back in time to around noon, when Assistant Chief Joiner was turning down requests to declare a civil emergency and cancel the AFL-CIO parade.

From about noon on, the Multi-Agency Command Center in the Public Safety Building began filling with top-ranking officials from government and law enforcement. Federal officials were speaking loudly about the consequences of not regaining control of the streets. State Patrol Chief Annette Sandberg described the federal officials as in a "kind of panicky mode."

The police attacks on the protesters reached a peak shortly before the parade departed from the Seattle Center. According to police sources, nearly all of the available tear gas was expended before the parade approached downtown. In the preparations for the protest, Mayor Schell and Chief Stamper had laid in stocks of about $20,000 worth of gas. This was one-fifth the amount recommended by federal officials. According to the *Seattle Post-Intelligencer*, police officers "took matters into their own hands" to obtain new supplies of gas and pepper spray. Other information suggests that the new supplies were part of Joiner's "messy" postparade attack plans.

Officers sped to Auburn, Renton, and Tukwila police departments, as well as the King County Jail and Department of Corrections, emptying munitions stores and ferrying the supplies back to downtown. Other officers bought additional chemical agents from a local law enforcement supply business. Meanwhile, a police captain flew to Casper, Wyoming, to pick up a large quantity of gas, "stinger shells," and other paraphernalia from Defense Technology Corp., a subsidiary of Armor Holdings. The locally obtained gas and pepper spray were driven as close to the street action as possible. The munitions were transferred into gym bags and knapsacks that were then run through the streets by plainclothes detectives.

Other improvised preparations did not go as well as the deliveries of tear gas and pepper spray. The declaration of civil emergency was delayed until 3:24 p.m., preventing police reinforcements from other law enforcement agencies and the National Guard from being legally deployed until long after the AFL-CIO paraders had withdrawn to their buses. Assistant Chief Ed Joiner's "messy" plan was also impeded by the flat refusal of the Seattle Fire Department to turn fire hoses on demonstrators. The fire department decision resulted in the trucks being delivered to the SPD by out-of-uniform firemen who refused to operate the equipment on the grounds that people would be injured by the spray.

Black Bloc Runs Amok

While the police were regrouping and preparing to force the Direct Action Network protesters to join the AFL-CIO parade, several groups took advantage of the lull in the battle. They have all been lumped together into a nameless anarchist horde, but the fact remains that there were two distinct groups acting out different agendas, not one "organized" anarchist conspiracy as the myth would have it.

At approximately 1 p.m., the police temporarily stopped trying to push corridors through the protest area. Earlier, the Black Bloc anarchists had entered into an understanding with the Direct Action Network that they would refrain from vandalism at least as long as the streets remained peaceful. But meanwhile, the Black Bloc arrived downtown armed with hammers, crowbars, spray paint, M-80 firecrackers, and paint bombs. Their goal was a "propaganda of the deed," centering around vandalizing chosen stores—Nike, Starbucks, the Gap, Old Navy, and others—that they saw as fitting targets.

The Black Bloc anarchists were simply biding their time and waiting for an opportunity to vandalize these stores and then get away. They had been closely monitored by the police and FBI since the preceding day. Early Tuesday morning, the FBI had briefed Seattle police on the Black Bloc's whereabouts and activities. The close observation of the Black Bloc included undercover FBI agents dressed to blend in with the anarchists, right down to wearing masks to hide their faces.

According to KIRO TV, the Black Bloc rampage started on 6th Avenue between Pine Street and Olive Way. Vandals smashed the windows of a Starbucks coffee shop in the middle of the block, then moved north toward Olive Way. Turning west on Olive Way, they attacked the SeaFirst bank, then turned south on 5th Avenue. Two or three stores along this block were vandalized. Emerging onto Pine Street, the

Black Bloc turned again, moving west and attacking three or four more stores in the next two blocks. Reaching Third Avenue, the Black Bloc turned south and dispersed.

The *Seattle Times* reported that the vandalism centered mainly along Pike Street, between Third and Sixth Avenue. A map showing the location of vandalized and looted stores published in the *Times* overlaps the route of the Black Bloc only at the beginning and end. The majority of the vandalism occurred around 4th and Pike, a corner that the Black Bloc *avoided* while being videotaped by KIRO TV.

Large numbers of teenagers who were not part of the Black Bloc took advantage of the situation and likewise engaged in vandalism. It was this second group, estimated to number at least one hundred or more, who engaged in looting some of the broken store windows, as well as occupying the awning over the Nike store. In addition to the damage to commercial property, police cars and limousines were vandalized with spray paint and by having their tires slashed.

Jeff Boscole, an eyewitness who was on Sixth Avenue, described how the two groups could be distinguished by their dress and the different slogans that they spray painted on buildings and windows. According to Boscole, the Black Bloc graffiti consisted of legible political slogans, while the "wilding teenagers" were "tagging" with illegible individualized symbols that were not slogans.

The Black Bloc engaged in vandalism numbered no more than thirty to forty people. They all dressed similarly. Many were dressed in black and all were hooded or masked to prevent their identification. They moved at a brisk pace, occasionally stopping in small groups to break windows or spray paint anarchist and anticorporate slogans. Early in the raid, they twice attacked KIRO TV news crews, spraying the camera lenses with paint to stop the crews from taking pictures. After these attacks, news crews followed from half a block to avoid further attacks. The Black Bloc maintained cohesion and moved along its route in a determined manner, several times scuffling with the nonviolent protesters from the Direct Action Network. A handful of plainclothes police and FBI shadowed the group, reporting their movements. Police made no effort to halt the vandalism, but in several instances DAN protesters stopped or interfered with members of the

Black Bloc, while others chanted "no violence" to little avail. The vandalism and looting occurred in the area evacuated by police to create a buffer zone between the DAN protesters and the AFL-CIO parade. The center of the vandalized area coincides with the turning point of the parade.

Declaration of Emergency

At 12:45 p.m., Gov. Gary Locke authorized his chief of staff to begin preparing to call up the National Guard. An hour earlier, State Patrol Chief Annette Sandberg had ordered State Patrol troopers in Eastern Washington on higher alert and dispatched a 22-member Civil Disturbance Team from Spokane to drive the 400 miles to Seattle. Traveling at top speed, they would not arrive before dark.

Shortly after Locke set the National Guard in motion, his office in Olympia received a telephone call from a furious Secretary of State Madeline Albright. Albright demanded the governor immediately take action to release her from her hotel where she was trapped by the demonstrators.

Gov. Locke was able to claim that he was taking action—but all of these things would take time. Locke arrived at the Multi-Agency Command Center in the Public Safety Building at 2:50 p.m., about ten minutes ahead of the mayor. "Almost immediately upon arriving at the command center, there was no doubt in my mind that we needed to call up the National Guard," Locke said.

Mayor Schell spent most of the day at the WTO conference site, waiting for the opening ceremonies to begin. He did not arrive at the Multi-Agency Command Center until about 3 p.m., two hours after the ceremonies had been canceled.

Upon Schell's arrival, officials from the SPD, Secret Service, FBI, State Patrol, Department of Justice, State Department, King County, the governor's office, and the White House moved into a back room and engaged in a heated discussion. While the argument continued, U.S. Attorney General Janet Reno called the governor and insisted that the National Guard be called up.

After speaking with Reno, Locke met with the mayor. Schell then spoke with Assistant Police Chiefs Joiner and Pirak. "By that time, we had a chance to look at what was happening. The mayor immediately agreed and authorized [the emergency declaration]," said Joiner. "There was never any hesitation." The period between the mayor's arrival at the Multi-Agency Command Center and issuing the proclamation of civil emergency was less than half an hour. At 3:24 p.m., the mayor issued the emergency declaration.

The mayor's declaration of civil emergency set in motion the reinforcements from the King County Sheriff's Department, the Washington State Patrol, and local police departments from surrounding cities and towns. The arrival of the reinforcements in the streets occurred relatively slowly over the next three hours, impeded by the discord that dominated the relations between the Seattle police and King County Sheriff Dave Reichert.

Battle Resumes

By 3 p.m., the belated attempt by police to push the protesters away from the triangle of key intersections surrounding the Convention Center was in full motion. The Direct Action Network blockade was still intact, immobilizing the police and preventing movement through the strategic triangle surrounding the Convention Center. As a result, most of the police action took place south and west of the Convention Center. Starting from the south along Union and University Streets, the police moved north along Third to Seventh Avenue to sweep the demonstrators north into the route by which the labor leaders had already retreated.

The police sweep northwards compressed the crowds into the east-west corridor running along Pike and Pine Streets. Here, the police again stalled against the large size of the crowds. The compression halted the police movement for several hours, as dumpsters that had been pushed into the streets to block the center of intersections began to burn. These bonfires slowly spread in an irregular way as the crowds withdrew east, not north as the police wished, and moved up into Capitol Hill in the early evening.

By 5:30 p.m., the police lines—now increased by the arrival of Sheriff's deputies—had reached the corner of Fourth and Pike. The protesters began withdrawing east along Pike and Pine Streets, toward Capitol Hill, followed by police firing tear gas, stun grenades, and rubber and wooden projectiles, and accompanied in some instances by vehicles. The police did not maintain close contact with the crowds and followed—not drove—them into Capitol Hill. The turning movement of the police—from a northern push to an eastern one—was contrary to the plan outlined by Assistant Chief Ed Joiner. According to political researcher Daniel Junas, the Direct Action Network overheard police radio messages in which units in the East Precinct on Capitol Hill frantically demanded that the police downtown cease pushing demonstrators up the hill. The central command replied that they were pushing the crowds north (i.e., along the route of the AFL-CIO retreat from downtown) not east.

The protesters' withdrawal from downtown coincided with the arrival of additional police reinforcements, the declaration of a 7 p.m. curfew, and the fall of darkness. The WTO had announced the cancellation of activities around 1 p.m., although word of the cancellation did not become widespread until late afternoon. Based on videos and photographs of the move east up Pine Street, the protesters appear to have decided to leave downtown and were followed, not "swept," by police. As the police crossed the freeway, the demonstrators melted away. Residents of Capitol Hill began to be attacked by the newly arrived police units from King County and adjoining communities who followed the pursuit teams up Pine.

The police decision to follow up the hill, firing tear gas and rubber bullets, is inexplicable in terms of clearing downtown and appears to be contrary to the "messy" plan outlined by Assistant Chief Joiner. Like the initial deployment of tear gas, it is evidence of loss of control by the commanders. The hot pursuit of the protesters was the second instance where tactics at the street level ran contrary to the strategic direction of the commanders. The police decision not to disengage continued the disturbance late into the night. Failure by commanders to halt the attacks on residents of Capitol Hill would have serious repercussions a day later.

The loose contact between police and demonstrators permitted the last act of serious vandalism of the day. Police were not controlling the intersection at Sixth Avenue and Stewart Street, near the Westin Hotel. Protesters had built a bonfire in the center of the intersection. At approximately 7:15 p.m., a group of vandals smashed the window of the Starbucks coffee shop. This was the same coffee shop from which Washington State Patrol Chief Annette Sandberg saw the Direct Action Network affinity groups at 5 a.m., as they moved into position and seized the strategic intersections surrounding the WTO conference site. Events had come full circle.

Day Two

By the end of the first day, with the departure of the AFL-CIO parade participants, the Direct Action Network assumed total control of the protests in Seattle. After their one brief appearance, the Black Bloc presence in the streets subsided, effectively now under the control of the DAN nonviolence strategy. The media directed considerable attention to the Eugene contingent, and the Black Bloc created unprecedented attention for the philosophy of "autonomist" anarchism and John Zerzan, a Eugene anarchist philosopher who promotes "primitivism" and a withdrawal from technological society. Yet, the Direct Action Network strategy of nonviolent civil disobedience had succeeded against the Black Bloc's efforts to escalate the police violence, the AFL-CIO's strategy of controlling and marginalizing protests in favor of a symbolic parade, the attempts of the Seattle police to clear the streets with tear gas, and media efforts to frame the issue in terms of "violent protesters."

Then, at 7:30 a.m. on Wednesday morning, the police began mass arrests. Direct Action Network protesters began assembling at a few locations and others made their way into the downtown core. Some of the arrests occurred at Denny Park, well to the north of the downtown. Police handcuffed some demonstrators and put them on city buses to the temporary jail at the former Sand Point Naval Air Station. Other demonstrators had their signs taken away from them but were not arrested. These proceeded downtown.

Protesters converged on the Westlake Center, and arrests there began at approximately 8 a.m. As the morning wore on, it became apparent that Westlake Center, rather than the WTO conference location, was the focus of Wednesday's blockade. The Direct Action Network had correctly identified the shopping and business district as being the vulnerable point in the new police strategy. By 9 a.m. Westlake Center was clogged by a peaceful sit-down protest as protesters patiently waited for police to arrest them. The crowds, consisting of demonstrators waiting to join the sit-in and spectators from the business district, continued to swell. As one protester was arrested, more would leave the crowd and sit down. Once again, the netwar tactic of "swarming" the target by stealthy approach succeeded.

By 10 a.m. it was becoming evident that the police tactics were not going to halt the sit-in and that the police were creating a situation that they could not control. At 10:30 a.m., the police commander stepped between his men and the protesters. He walked to the seated protesters, leaned down and said, "We're outta here." He then motioned to his men to leave the area and the police withdrew in an orderly manner. The protesters, both seated and among the crowd, were jubilant. They had prevailed. The disengagement of the police at Westlake Center marked the end of mass arrests as a police tactic.

The preceding day, as police and federal security officials had milled around in an atmosphere of panic at the Multi-Agency Command Center in the Public Safety Building, Ronald Legan, the special agent in charge of the Seattle office of the Secret Service, laid down an ultimatum to Seattle officials about the presidential visit. Legan said,

> I remember saying that unless we get control of the streets, we would recommend that he not come. Now the problem there is that, with this president, he sets his own agenda and goes where he wants. And we did not want to have to battle a 30-car motorcade in and out of Seattle.

Seattle Assistant Chief Ed Joiner said he would not characterize the Tuesday discussion as "threatening . . . but it was clear that if the situation was going to be the following day what it was then, there was no way you could bring the president of the United States into Seattle."

Postpresidential Disorder

On the streets, Wednesday afternoon was a repeat of Tuesday. The police pulled back for the four hours that President Clinton was in public view, just as they had pulled back as the AFL-CIO parade approached downtown. In the words of one TV reporter, "The streets were strangely quiet." At 1 p.m., Washington Gov. Gary Locke gave a live interview on local television. Locke stated that order was restored to Seattle and told local shoppers to "come downtown"—inside the perimeter of the "no protest" zone. Unfortunately, the governor hadn't heard about police plans for a 4 p.m. crackdown to drive protesters out of the downtown core, a time that coincides with the downtown rush hour.

As Clinton's motorcade departed, the streets were once again blanketed in tear gas and police fired pepper spray at anyone who got in their way. At the Pike Place Market, tear gas was severe enough that produce merchants put out signs the next day announcing they were closed because their fruits and vegetables were contaminated by tear gas.

As on Tuesday, the police failed to move the crowds of protesters and the main axis of protest movement once again became Pike and Pine Streets. After two hours, police were able to move only two blocks up Pike from the market to Second Avenue. A protester blockade at Third and Pine stayed in place until protesters voluntarily dispersed at 6:45 p.m.

To celebrate their "control" of the now empty streets, a column of a dozen police cars raced through the empty downtown core with emergency lights flashing and sirens blaring. Police officials explained to reporters that this "wild weasel" operation was a "show of force to clear the streets." The news videos of the stream of cars is one of the more surreal images from the entire week. Things would get even stranger that night.

At about the same time as the "wild weasels" were racing through the streets, police assaulted Seattle City Councilman Richard McIver. McIver said city police officers yanked him from his car, pulled his arms behind his back and started to cuff him as he drove to a World

Trade Organization reception event at the nearby Westin Hotel. Councilman McIver said,

> I don't want to aid the hooligans who are raising hell and I don't want to take on specific officers. . . . But there are huge flaws with the officers when it comes to people of color. I'm 58 years old. I had on a $400 suit, but last night, I was just another nigger.

Street Battles for the Hell of It

The final incident of Wednesday night demonstrated that civilian control of law enforcement ceased to exist for a time. The "Battle of Capitol Hill" degenerated into a police riot, perhaps the only time during the WTO protests that police command totally lost control of their forces on the street.

As demonstrators withdrew from the downtown curfew area at around 7 p.m., a group of several hundred protesters moved north on Fourth Avenue, followed at a distance by police. The group withdrew in an orderly manner, stopping several times along the way to vote on where they were going. They moved east on Denny Way into Capitol Hill and reached Broadway and East Harrison Street at about 7:45 p.m., where they joined another group that was already at the intersection. By now the group numbered approximately 500. As they passed through the neighborhood, cheering residents and honks of support from motorists greeted them. The crowd marched back and forth along Broadway for about an hour, carrying banners and accompanied by a band playing music. The atmosphere was one of celebration rather than protest.

The crowd was predominantly residents of Capitol Hill, many of whom had been angered by police the previous night when bystanders and people on their way home from work had been indiscriminately attacked by the police who had pursued demonstrators up the hill. At about 9 p.m., police and National Guard forces began arriving in the area. By 9:30 p.m., police closed several blocks of Broadway between East Republican and East John. Tensions were high, as a result of the residents' resentment of the police presence and police fears of violence. KIRO TV reported that the people opposing police that night consisted entirely of Capitol Hill residents. Afterwards, police claimed there were reports of agitators carrying gasoline bombs and throwing rocks and bottles. They said some protesters charged officers. No gasoline bombs were thrown that evening and news videos show only police charging, not civilians attacking police.

Police began using pepper spray, tear gas, and concussion grenades shortly after 9:30 p.m., first at John Street and Broadway to the south, moving north, and near Harrison, moving south. More police blocked side streets, preventing the crowds from dispersing.

"The protesters looked completely calm to me. . . . They were not instigating this," said Erin Katz, a Capitol Hill resident who watched from behind police lines near Pine Street. "I heard absolutely no warning and they started to gas them."

For the next two and a half hours, police rampaged along Broadway. It was during this period that some of the week's worst instances of police misconduct occurred. National television repeatedly aired footage of a Tukwila officer kicking a young man in the groin and then immediately firing a shotgun within inches of the young man's torso. At a parking lot near Broadway, two journalism students were videotaping the action. A King County deputy went up to their car and motioned for the young women to roll down a window. When they did, the deputy pepper sprayed them both, shouting "Tape this, bitch!" This footage has also been repeatedly aired on national television. These and other incidents have resulted in civil suits filed against the Seattle police as the agency in charge, as well as the officers involved.

Around midnight, the disorder had drawn several local government officials, who tried to get the police and demonstrators to disengage. They included King County Councilman Brian Derdowski, City Councilmen Richard Conlin and Nick Licata, and Councilwoman-elect Judy Nicastro.

"Those council members tried to work through the chain of command of the Police Department and they were unable to get anybody," Derdowski said. For two hours, the civilian officials tried to get the police to cease attacking the crowd. Finally, around 2 a.m., the crowd began to leave. The police responded with volleys of gas and rubber projectiles. Derdowski said,

I asked the police to be professional and just take one step back. That would be the sign that these folks needed, and they would disperse. The police said they couldn't do that, so we went back and told the people that they needed to leave the area. And a lot of them did, but a few persisted. And they started singing Christmas carols. They sang "Jingle Bells," and when they started singing "Silent Night," the tear gas started. Something hit me in the back, and there was pandemonium there, and so we left the area.

Jail Blockade and Release

By Thursday, the success of the Direct Action Network protests was undeniable. The WTO conference was prevented from holding its opening ceremonies on Tuesday. On Wednesday, the conference began to come unraveled when President Clinton made repeated statements supporting the demonstrators—although it appears he was referring only to the AFL-CIO—and announced a U.S. policy initiative that guaranteed that major consensus at the WTO conference would be impossible. On Wednesday night, police attacked local residents in the sort of breakdown of command and discipline shown by defeated troops. Graffiti began appearing around Seattle reading: "Remember, We Are Winning!" On Thursday afternoon, police finally came to an accommodation with Direct Action Network protesters and provided a police escort for a march.

The focus of the Direct Action Network strategy now shifted from the WTO to support for those still in jail as a result of the mass arrests. For two days, vigils were held at the Public Safety Building, at times completely surrounding the building.

On Friday evening, after meeting with city officials, Direct Action Network legal staff announced an agreement with the city. Jailed protesters would now begin cooperating with the courts and properly identify themselves. Many had refused to provide their names and addresses, giving their names only as "Jane WTO," "John WTO," or in one case "Emiliano Zapata." Once processed for arraignment, they were released on personal recognizance. Nearly all of those jailed were released by Sunday. After the jailed protesters were released, Seattle City Attorney Mark Sidran issued a statement to the press denying that any agreement had been reached and promising to prosecute all cases. In January, all of the mass arrest cases were dismissed because police had not filled out arrest forms.

Police Officials Resign

The final act of the WTO protests was the announced departures of Seattle Police Chief Norm Stamper, strategic commander Assistant Chief Ed Joiner, Nancy McPherson, civilian director of the Community and Information Services, and Assistant Chief of Investigations Harve Fergusson. Those who made public statements regarding their resignations or retirements said that the decisions had been made before the WTO protests. Chief Stamper stated that one purpose of announcing his resignation was to "depoliticize" the investigations into police actions during the protests and "in making this announcement, I've taken my tenure off the table." The departures of the other police officials were virtually ignored in the media, although they represent the departure of three out of seven of the chief's highest-ranking assistants.

The Seattle police organizations launched a massive public-relations blitz. In one of the more bizarre actions, police officers began selling T-shirts to local merchants—as if the police had won some sort of a major victory. The shirts showed the Space Needle engulfed in a tornado, saying "Battle in Seattle WTO 99." Mike Edwards, president of the Seattle Police Officers Guild, said that money from the T-shirt sales would be used to buy merchandise from downtown merchants and that the items purchased would be given to charities. The guild also organized a rally to show support for the police. State Rep. Luke Esser, R-Bellevue, a conservative "law-and-order" advocate, issued a statement saying that he would be attending the police rally "commending those brave men and women for working around the clock in treacherous conditions to maintain law and order during the WTO riots." The *Seattle Times* ran a variety of pro-police articles, including one front-page headline announcing the retirement of a police dog.

CONCLUSION

The WTO protests in Seattle were the largest left-wing demonstrations in America since the Gulf War. They were also the most successful American political demonstrations of the decade, if success is measured by the degree of congruence between the protesters' goals and the effect on public policy issues.

The WTO protests succeeded in the streets through a combination of strategic surprise and tactical openness. The three key phases of the street actions leading to this success consisted of the Tuesday morning "swarm," which blockaded strategic intersections; the collapse of the police strategy to suppress the Direct Action Network protests while allowing the AFL-CIO parade; and the failure of the AFL-CIO parade to engulf the Direct Action Network protests.

Three things distinguished the N30 protests from the others that followed in other cities and countries. None of the later protests had an AFL-CIO contingent, a rampage of vandalism by anarchists, or a divided police command. The much-touted "Teamsters and turtles together" alliance evaporated immediately. The AFL-CIO shifted its target to China's admission to the WTO and severed what few ties had been made to environmental and human rights groups. The protest movement thoroughly rejected the property destruction tactics of the militant anarchist factions, having never embraced them in the first place. The Black Blocs were never an influential factor in future protests. Every police department expecting protests noted the fate of Seattle Police Chief Norm Stamper and made extensive preparations to contain, disrupt, and control the protestors. The protesters, for their part, did not evolve new tactics or repeat the strategic surprise of N30. The parallel to the strategic surprise of the January 1994 Zapatista attacks in Chiapas and the subsequent stalemate in Mexico is worth considering.

The most profound outcome of the WTO protests is the appearance of the netwar construct in American politics. The "Battle in Seattle" was fought not only in the streets, but also in the infosphere. Once the idea of an international left-wing opposition to globalization was demonstrated to be a political force, the informational offensive had succeeded. Strategic surprise occurs in the mind of the opponent.

The WTO protests were the first to take full advantage of the extremely dense and wide-reaching alternative media network via the Internet. The use of "media special forces" is one of the hallmarks of netwar and informational conflicts. The flexible and improvised communications infrastructure used by the Direct Action Network is a notable feature of the protests. One of the dictums of netwar is that netwar actors have a much greater interest in keeping communications working, rather than shutting them down. The dense and diversified communications used by the Direct Action Network could not have been significantly harmed by any action less than a total media and communications blackout in Seattle. Not only is such an action impossible because of the economic and social costs that would result, but a blackout of the required magnitude would be the netwar equivalent of unconditional surrender by the establishment. Both protesters and their opponents will have to come to terms with the implications of netwar and the struggle for information, understanding, and "topsight." Because the ultimate prize in a netwar conflict is understanding—not opinion—the quality of information (not quantity) determines the final outcome.

Since the N30 protests, a new hybrid of institution and network has multiplied with every protest. Beginning with the Independent Media Center in Seattle, each new protest has spawned a new "indymedia" organization. Producing newspapers, web sites, videos, radio programs, and a steady barrage of information, the indymedia network is an attempt to gain some sort of informational parity with the corporate-controlled mass media. As of this writing, www.indymedia.org lists over forty nodes: ten international web sites in various countries, plus two in Australia, six in Canada, twenty in the United States, and five specialized subsites for indymedia support operations. Significantly, the indymedia network has an "all points" connectivity. All of the sites are linked to the others and share information, links, technical support, and web design.

Netwar is nothing new as a form of conflict. What is new is the richer informational environment, which makes the organization of civil (and uncivil) society into networks easier, less costly, and more efficient. The essential conditions for victory in a social netwar conflict are also the conditions that make waging netwar possible: the shared understanding of a situation demanding direct action. In many ways, the victory of the Direct Action Network was implicit in the fact that so many people understood the conflict and were willing to act on that understanding. The streets of Seattle showed what democracy looks like.

EDITORS' POSTSCRIPT (SUMMER 2001)

Seattle was a seminal win. It sparked new netwars in the streets of Washington (A16), Los Angeles, and in a string of other cities where activists have persisted in their opposition to the World Trade Organization, the International Monetary Fund, and the general process of corporate globalization. One activist has reportedly boasted that protests could be mounted in any city around the world, at any time.

In the United States, netwar in the streets has fared badly since Seattle. Seattle was, in many ways, unique. First, the voluminous swarm of protesters who formed the third wave, drawn from the AFL-CIO participants, surprised both DAN and the law enforcement authorities. In addition, governmental authorities may have learned more from the Battle of Seattle than the activists did. In both the Washington and Los Angeles demonstrations, police were able to preempt or prevent almost all the tactical maneuvers of the activists. In these post-Seattle cases, protest organizers reverted to centralized control of operations—including by locating some command, media, and other functions in the same building—which made them vulnerable to counter-leadership targeting. The Battle of Seattle was won without a field general, and without a general staff. Post-Seattle actions have violated the key netwar principle of "leaderlessness."

Law enforcement, government authorities, and even the American Civil Liberties Union have conducted instructive after-action analyses of the Battle of Seattle. Exactly what lessons the AFL-CIO has drawn are not known, but the practical result has been its withdrawal from post-Seattle demonstrations—leaving NGO activists with less of a pool to draw on. By way of contrast, none of the protest organizations has rendered an after-action analysis of the strategies and tactics used in Seattle, even though the Internet teems with eyewitness accounts.

In all forms of protracted conflict, early confrontations are seedbeds of doctrinal innovation—on all sides. If governmental authorities learned much from their defeat in Seattle, perhaps we should also expect that social netwarriors will learn lessons from their defeats in Los

Angeles, Washington, and elsewhere. Indeed, the events of the summer of 2001 in Genoa indicate that the netwarriors are learning their own lessons—and are steadily willing to apply them in practice.

Part III

ONCE AND FUTURE NETWARS

Chapter Eight

ACTIVISM, HACKTIVISM, AND CYBERTERRORISM: THE INTERNET AS A TOOL FOR INFLUENCING FOREIGN POLICY

Dorothy E. Denning

Editors' abstract. Netwar is not mainly about technology—but good information technology sure makes a difference. In this chapter, Denning (Georgetown University) examines how activists, hacktivists, and cyberterrorists use the Internet, and what influence they have been able to exert on policymakers. Social activists seem the most effective of these netwar actors. Hacktivists and cyberterrorists have not posed much of a real threat to date—but this could change if they acquire better tools, techniques, and methods of organization, and if cyberdefenses do not keep pace. In this swiftly evolving area, today's tools and techniques are often soon outdated; yet Denning's analytic approach should prove conceptually sound for years to come. The original version of this paper was sponsored by the Nautilus Institute and presented at a conference on "The Internet and International Systems: Information Technology and American Foreign Policy Decision Making," The World Affairs Council, San Francisco, December 10, 1999 (www.nautilus.org/info-policy/workshop/papers/denning.html). Reprinted by permission.

The conflict over Kosovo has been characterized as the first war on the Internet. Government and nongovernment actors alike used the Net to disseminate information, spread propaganda, demonize opponents, and solicit support for their positions. Hackers used it to voice their objections to both Yugoslav and NATO aggression by disrupting service on government computers and taking over their web sites. Individuals used it to tell their stories of fear and horror inside the conflict zone, while activists exploited it to amplify their voices and reach a wide, international audience. And people everywhere used it to discuss the issues and share text, images, and video clips that were not available through other media. In April 1999, the *Los Angeles Times* wrote that the Kosovo conflict was "turning cyberspace into an ethereal war zone where the battle for the hearts and minds is being waged through the use of electronic images, online discussion group postings, and hacking attacks."[1] Anthony Pratkanis, professor of psychology at the University of California, Santa Cruz, and author of *Age of Propaganda: The Everyday Use and Abuse of Persuasion,* observed,

> What you're seeing now is just the first round of what will become an important, highly sophisticated tool in the age-old tradition of wartime propaganda . . . . The war strategists should be worried about it, if they aren't yet.[2]

Just how much impact did the Internet have on foreign policy decisions relating to the war? It clearly had a part in the political discourse taking place, and it was exploited by activists seeking to alter foreign policy decisions. It also affected military decisions. While NATO targeted Serb media outlets carrying Milosevic's propaganda, it intentionally did not bomb Internet service providers or shut down the satellite links bringing the Internet to Yugoslavia. Policy instead was to keep the Internet open. James P. Rubin, spokesman for the U.S. State Department, said "Full and open access to the Internet can only help the Serbian people know the ugly truth about the atrocities and crimes against humanity being perpetrated in Kosovo by the Milosevic regime."[3] Indirectly, the Internet may have also affected public support for the war, which in turn might have affected policy decisions made during the course of the conflict.

The purpose of this chapter is to explore how the Internet is altering the landscape of political discourse and advocacy, with particular em-

---

[1]Ashley Dunn, "Crisis in Yugoslavia—Battle Spilling over onto the Internet," *Los Angeles Times*, April 3, 1999.

[2]Quoted in Rick Montgomery, "Enemy in Site—It's Time to Join the Cyberwar," *Daily Telegraph* (Australia), April 19, 1999, p. 19.

[3]David Briscoe, "Kosovo-Propaganda War," *Associated Press*, May 17, 1999.

phasis on how it is used by those wishing to influence foreign policy. Emphasis is on actions taken by nonstate actors, including both individuals and organizations, but state actions are discussed where they reflect foreign policy decisions triggered by the Internet. The primary sources used in the analysis are news reports of incidents and events. These are augmented with interviews and survey data where available. A more scientific study would be useful.

The chapter is organized around three broad classes of activity: activism, hacktivism, and cyberterrorism. The first category, activism, refers to normal, nondisruptive use of the Internet in support of an agenda or cause. Operations in this area includes browsing the web for information, constructing web sites and posting materials on them, transmitting electronic publications and letters through email, and using the Net to discuss issues, form coalitions, and plan and coordinate activities. The second category, hacktivism, refers to the marriage of hacking and activism. It covers operations that use hacking techniques against a target's Internet site with the intent of disrupting normal operations but not causing serious damage. Examples are web sit-ins and virtual blockades, automated email bombs, web hacks, computer break-ins, and computer viruses and worms. The final category, cyberterrorism, refers to the convergence of cyberspace and terrorism. It covers politically motivated hacking operations intended to cause grave harm such as loss of life or severe economic damage. An example would be penetrating an air traffic control system and causing two planes to collide. There is a general progression toward greater damage and disruption from the first to the third category, although that does not imply an increase of political effectiveness. An electronic petition with a million signatures may influence policy more than an attack that disrupts emergency 911 services.

Although the three categories of activity are treated separately, the boundaries between them are somewhat fuzzy. For example, an email bomb may be considered hacktivism by some and cyberterrorism by others. Also, any given actor may conduct operations across the spectrum. For example, a terrorist might launch viruses as part of a larger campaign of cyberterrorism, all the while using the Internet to collect information about targets, coordinate action with fellow conspirators, and publish propaganda on web sites. Thus, while this chapter distinguishes activists, hacktivists, and terrorists, an individual can play all three roles.

The following sections discuss and give examples of activity in each of these three areas. The examples are drawn from the Kosovo conflict, cryptography policy, human rights in China, support for the Mexican Zapatistas, and other areas of conflict. The examples are by no means exhaustive of all activity in any of these areas, but are intended only to be illustrative. Nevertheless, they represent a wide range of players, targets, and geographical regions.

The main conclusion here is that the Internet can be an effective tool for activism, especially when it is combined with other communications media, including broadcast and print media and face-to-face meetings with policymakers. It can benefit individuals and small groups with few resources as well as organizations and coalitions that are large or well-funded. It facilitates such activities as educating the public and media, raising money, forming coalitions across geographical boundaries, distributing petitions and action alerts, and planning and coordinating events on a regional or international level. It allows activists in politically repressive states to evade government censors and monitors.

With respect to hacktivism and cyberterrorism, those who engage in such activity are less likely to accomplish their foreign policy objectives than those who do not employ disruptive and destructive techniques. They may feel a sense of empowerment, because they can control government computers and get media attention, but that does not mean they will succeed in changing policy. The main effect is likely to be a strengthening of cyberdefense policies, both nationally and internationally, rather than accommodation to the demands of the actors.

ACTIVISM

The Internet offers a powerful tool for communicating and coordinating action. It is inexpensive to use and increasingly pervasive, with an estimated 300 million people online as of May 2000.[4] Groups of any size, from two to millions, can reach each other and use the Net to promote an agenda. Their members and followers can come from any geographical region on the Net, and they can attempt to influence foreign policy anywhere in the world. This section describes five modes of using the Internet: collection, publication, dialogue, coordination of action, and direct lobbying of decisionmakers. While treated separately, the modes are frequently used together, and many of the examples described here illustrate multiple modes.

Collection

One way of viewing the Internet is as a vast digital library. The web alone offers several billion pages of information, and much of the information is free. Activists may be able to locate legislative documents, official policy statements, analyses and discussions about issues, and other items related to their mission. They may be able to find names and contact information for key decisionmakers inside the government or governments they ultimately hope to influence. They may be able to identify other groups and individuals with similar interests and gather contact information for potential supporters and collaborators. There are numerous tools that help with collection, including search engines, email distribution lists, and chat and discussion groups. Many web sites offer their own search tools for extracting information from databases on their sites.

One advantage of the Internet over other media is that it tends to break down barriers erected by government censors. For example, after Jordanian officials removed an article from 40 print copies of *The Economist* on sale in Jordan, a subscriber found a copy online, made photocopies, and faxed it to 1,000 Jordanians. According to Daoud Kuttab, head of the Arabic Media Internal Network (AMIA), the government would have been better off leaving the print version intact. "We found this very exciting," he said. "For the first time the traditional censorship that exists within national borders was bypassed." Kuttab said AMIA opened Jordanian journalists to the non-Arab world

---

[4]Nua Internet Surveys, www.nua.ie. The site is updated regularly with the latest estimate.

and use of the web as a research tool. "In the Jordanian media, we have been able to detect a much more open outlook to the world as well as to Arab issues," he said.[5]

The Internet itself is not free of government censorship. According to Reporters Sans Frontiers, 45 countries restrict their citizens' access to the Internet, typically by forcing them to subscribe to a state-run Internet service provider, which may filter out objectionable sites.[6] Authoritarian regimes recognize the benefits of the Internet to economic growth, but at the same time feel threatened by the unprecedented degree of freedom of speech.

Chinese authorities block access to web sites that are considered subversive to government objectives. This has been only partially effective, however, and Chinese activists have found ways of slipping information past the controls. For example, the editors of *VIP Reference*, a Washington-based electronic magazine with articles and essays about democratic and economic evolution inside China, email their electronic newsletter directly to addresses inside mainland China. The email is sent from a different address every day to get past email blocks. It is also delivered to random addresses, compiled from commercial and public lists, so that recipients can deny having deliberately subscribed. As of January 1999, about 250,000 people received the pro-democracy publication, including people inside the government who did not want it. Chinese officials were not, however, complacent. When 30-year-old Shanghai software entrepreneur Lin Hai sold 30,000 email addresses to *VIP Reference*, he was arrested and later sentenced to two years in prison. In addition, authorities fined him 10,000 yuan (a little over $1,000) and confiscated his computer equipment and telephone. Lin was said to be the first person convicted in China for subversive use of the Internet. He claimed he was only trying to drum up business and was not politically active.[7]

---

[5]Alan Docherty, "Net Journalists Outwit Censors," *Wired News*, March 13, 1999.

[6]*The Twenty Enemies of the Internet*, press release, Reporters Sans Frontiers, August 9, 1999.

[7]Maggie Farley, "Dissidents Hack Holes in China's New Wall," *Los Angeles Times*, January 4, 1999. Adrian Oosthuizen, "Dissidents to Continue E-Mail Activity Despite Court Verdict," *South China Morning Post*, February 2, 1999.

During the Kosovo conflict, people in Yugoslavia had full access to the Internet, including Western news sites. In April 1999, the *Washington Post* reported that according to U.S. and British officials, the NATO governments controlled all four Internet access providers in Yugoslavia and kept them open for the purpose of spreading disinformation and propaganda. The *Post* also said that Belgrade, with a population of 1.5 million, had about 100,000 Internet connections.[8] Individuals without their own connections could get access at Internet cafes.

Even though Serbs had access to Western news reports, both through the Internet and through satellite and cable television, many did not believe what they saw and heard from Western media. They considered coverage on Western television stations such as CNN and Sky News to be as biased as that on the Yugoslav state-run station, citing instances when Western reports of Serbian atrocities turned out to be wrong. Alex Todorovic, a Serbian-American who spent time in Belgrade during the conflict observed, "By and large, Serbs mistrust the rest of the world's media. CNN, for example, is considered the official voice of Washington."[9] Some Yugoslav surfers did not even bother looking at Western news sites on the Internet. When asked if she visited web sites of Western news stations, one 22-year-old student replied, "No, I don't believe in their information, so why should I upset myself?"[10] Thus, it is not clear whether open Internet access in Yugoslavia undermined Milosevic's objectives. Further, given that people living in Yugoslavia personally witnessed and felt the effects of the NATO bombing and either disbelieved reports or heard little about Serb atrocities against the ethnic Albanians in Kosovo, it is not surprising that an anti-NATO discourse ran throughout Belgrade. As one pharmacist observed, "I have two children. The people who are bombing my kids are my only enemy right now."[11]

In addition to information relating to a particular policy issue, the web offers cyberactivists various information that can help them use the Net effectively. For example, NetAction offers a training guide for

---

[8]Michael Dobbs, "The War on the Airwaves," *Washington Post*, April 19, 1999.

[9]Alex Todorovic, "I'm Watching Two Different Wars," *Washington Post*, April 18, 1999.

[10]Dobbs, 1999.

[11]Todorovic, 1999.

the virtual activist. The guide provides information on the use of email for outreach, organizing, and advocacy; web-based outreach and advocacy tools; membership and fundraising; netiquette and policy issues; and various resources.[12]

Publication

The Internet offers several channels whereby advocacy groups and individuals can publish information (and disinformation) to further policy objectives. They can send it through email and post it to newsgroups. They can create their own electronic publications or contribute articles and essays to those of others. They can put up web sites, which can serve as a gathering place and source of information for supporters, potential supporters, and onlookers.

One reason the Internet is popular among activists is its cost advantage over traditional mass media. It is easier and cheaper to post a message to a public forum or put up a web site than it is to operate a radio or television station or print a newspaper. Practically anyone can afford to be a web publisher. In addition, the reach of the Internet is global. A message can potentially reach millions of people at no additional cost to the originator. Further, activists can control their presentation to the world. They decide what is said and how. They do not have to rely on the mass media to take notice and tell their story "right."

Kosovo. During the Kosovo conflict, organizations and individuals throughout the world used their web sites to publish information related to the conflict and, in some cases, to solicit support. Nongovernment organizations with Kosovo-related web pages included the press, human rights groups, humanitarian relief organizations, churches, and women's groups.

Government web sites on Kosovo tended to feature propaganda and materials that supported their official policies. An exception was the U.S. Information Agency (USIA) web site, which presented a survey of news stories from around the world, some of which were critical of

---

[12]*NetAction's Virtual Activist Training Guide,* www.netaction.org/training.

NATO actions.[13] Jonathan Spalter, USIA chief information officer, commented that "The measure of our success is the extent to which we are perceived not as propaganda but anti-propaganda."[14] The British government's Foreign Office used their web site, in part, to counter Serb propaganda. Concerned that the Yugoslav public was getting a highly distorted view of the war, Foreign Secretary Robin Cook posted a message on the office's web site intended for the Serbs. The message said that Britain has nothing against the Serbs but was forced to act by the scale of Yugoslav President Slobodan Milosevic's brutality.[15] British Defence Secretary George Robertson said the Ministry of Defence (MoD) had translated its web site into Serbian to counter censorship of the news by Belgrade.[16]

The Yugoslav media was controlled by the Serbian government and served to promote Milosevic's policies. Yugoslavia had an independent, prodemocracy radio station, B92, but it was raided by the police in the early days of the Kosovo conflict and turned over to a government-appointed station manager.[17] B92 had had a run-in with the government earlier, in late 1996, when government jammers tried to keep it from airing news broadcasts. At that time, however, B92 prevailed, in part by encoding their news bulletins in RealAudio format and posting them on a web site in Amsterdam. Radio Free Europe acquired tapes of the news programs and rebroadcasted them back to the Serbs, circumventing the jammers, who then gave up.[18] But when the government took over B92's facility in 1999, B92's then-managers ceded to the government and also discontinued posting materials on their web site, which had offered viewers a reliable source of information about the conflict. This was considered a great loss to Yugoslavia's

---

[13]See www.usia.gov.

[14]David Briscoe, "Kosovo-Propaganda War," *Associated Press*, May 17, 1999.

[15]"Conflict in the Balkans—Cook Enlists Internet to Send Serbs Message," *Daily Telegraph*, London, April 2, 1999, p. 9.

[16]Rebecca Allison, "Belgrade Hackers Bombard MoD Website in 'First' Internet War," *PA News*, March 31, 1999.

[17]Leander Kahney, "Yugoslavia's B92 Goes Dark," *Wired News*, April 2, 1999.

[18]Bob Schmitt, "An Internet Answer to Repression," *Washington Post*, March 31, 1997, p. A21.

prodemocracy movement and general public, which had rallied be-hind Belgrade's top-rated news station.

A few individuals inside Yugoslavia posted to the Internet firsthand accounts of events as they were being witnessed or shortly thereafter. Their stories told of fear and devastation, the latter caused not only by the Serb military, but also by NATO bombs. By all accounts, the situation inside Yugoslavia was horrible for citizens everywhere, whether Serbian or ethnic Albanian. The stories may have inspired activists and influenced public opinion, but it is not clear what if any effect they had on government decisionmaking.

New-media artists used the web to voice their opinions on the Balkans conflict. In late March, artist and high school teacher Reiner Strasser put up a site called Weak Blood, which featured works of visual poetry, kinetic imagery, and interactive art, all making an antiviolence statement. Strasser vowed to add one or two pieces a day "as long as bombs are falling and humans are massacred" in the region.[19]

Some Serbs with Internet access sent emails to American news organizations calling for an end to the NATO bombing. Many of the messages contained heated rhetoric that was anti-NATO and anti-U.S. One letter directed to the Associated Press ended, "To be a Serb now is to be helpless . . . to listen to the euphemistic and hypocritical phrases as 'peace-making mission,' 'moral imperative.'" Other messages contained human stories about how their lives were affected. Tom Reid, London correspondent to the Washington Post, said he received 30–50 messages a day from professors at universities and activists all over Yugoslavia. The general tenor of the messages was the same, "'Please remember there are human beings under your bombs,'" he said.[20]

The Serbs used email distribution lists to reach tens of thousands of users, mostly in the United States, with messages attacking the NATO bombing campaign. One message read

---

[19]Matthew Mirapaul, "Kosovo Conflict Inspires Digital Art Projects," The New York Times (Cybertimes), April 15, 1999.

[20]Larry McShane, "Yugoslavs Condemn Bombs over E-mail to U.S. Media," Nando Times, April 17, 1999, www.nandotimes.com.

In the last nine days, NATO barbarians have bombed our schools, hospitals, bridges, killed our people but that was not enough for them now they have started to destroy our culture monuments which represents [sic] the core of existence of our nation.

Most recipients were annoyed by this unwanted "spam," which the *Wall Street Journal* dubbed "Yugospam."[21]

Dennis Longley, a professor in the Information Security Research Centre at Australia's Queensland University of Technology, said they received a suspicious email from Serbia. The message had two paragraphs. The first was the usual friendly greetings, while the second was a rant about NATO that read like pure propaganda, characterizing NATO as a "terrorist organization" that "brought nothing but a gigantic humanitarian disaster to Kosovo," while attributing the cause of the problem to "Albanian terrorist and separatist actions, not the repression by the government security forces." The second paragraph exhibited a style unlike the first and a standard of English well below that of the sender, leading them to speculate that Serb authorities had modified the email.[22] If that is so, one is left wondering how much other anti-NATO talk hitting the Net was the work of the Yugoslav government.

Of course, not all of the messages coming out of the Balkans were anti-NATO. Shortly after the Kosovo conflict began, I found myself on a list called "kcc-news," operated by the Kosova[23] Crisis Center from the Internet domain "alb-net.com." The messages included Human Rights Flashes from Human Rights Watch, Action Alerts from the Kosova Task Force, and other appeals for support in the war against the Serbs. One message contained a flier calling for "sustained air strikes until total Serb withdrawal" and "ground troops to STOP GENOCIDE now." The flier included links to web pages that documented Serb atrocities and aggression.

---

[21]Ellen Joan Pollock and Andrea Petersen, "Unsolicited E-Mail Hits Targets in America in First Cyberwar," *Wall Street Journal*, April 8, 1999.

[22]Dennis Longley, personal communication, July 15, 1999.

[23]The task force uses the spelling "Kosova" in its name and in all references to Kosovo.

Even though the Yugoslav government did not prohibit Internet activity, fear of government reprisals led some to post their messages through anonymous remailers so they could not be identified. This allowed for a freer discourse on Internet discussion groups and contributed to the spread of information about the situation inside Belgrade and Kosovo. Microsoft Corp. initiated a section called "Secret Dispatches from Belgrade" on the web site of its online magazine *Slate*. An anonymous correspondent gave daily reports of both alleged Serb atrocities and civilian suffering inflicted by NATO bombs.[24]

After human rights organizations expressed concern that the Yugoslav government might be monitoring Internet activity and cracking down on anyone expressing dissenting views, Anonymizer Inc., a provider of anonymous web browsing and email services, launched the Kosovo Privacy Project web site. The site, which went online in April 1999, offered surfers anonymous email and instant, anonymous access to Voice of America, Radio Free Europe, and about 20 other web sites. According to *Federal Computer Week*, Anonymizer planned to add NATO and other Western government information sites to the Kosovo list, and to launch similar projects for human rights situations in other parts of the world, for example, China.[25] However, the effectiveness of the Kosovo project was never established. In August 1999, *USA Today* reported that activists said the project was little noticed inside Kosovo, where traditional media seemed unaware while the fighting knocked out Internet trunk lines in short order.[26]

Internet Policy Issues. The Internet has raised numerous policy issues in such areas as privacy, encryption, censorship, electronic commerce, international trade, intellectual property protection, taxation, Internet governance, cybercrime, and information warfare, all of which have a foreign policy dimension. As the issues surfaced and took on some urgency, existing industry and public-interest groups began to address them. In addition, both national and international

---

[24]Rick Montgomery, "Enemy in Site—It's Time to Join the Cyberwar," *Daily Telegraph*, Australia, April 19, 1999.

[25]Daniel Verton, "Net Service Shields Web Users in Kosovo," *Federal Computer Week*, April 19, 1999.

[26]Will Rodger, "Online Human-Rights Crusaders," *USA Today*, August 25, 1999.

advocacy groups sprung up specifically devoted to Internet issues. They all operate web sites, where they publish policy papers and information about issues, events, and membership. Many also send out email newsletters and alerts.

In the area of encryption policy, for example, the major players include Americans for Computer Privacy (ACP), the Center for Democracy and Technology (CDT), Cyber-Rights & Cyber Liberties, the Electronic Frontier Foundation (EFF), the Electronic Privacy Information Center (EPIC), the Global Internet Liberty Campaign (GILC), and the Internet Privacy Coalition. The ACP has perhaps the largest group of constituents, being composed of 40 trade associations, over 100 companies, and more than 3,000 individual members.[27] GILC is one of the most global, with member organizations from Europe, North America, Australia, and Asia.

In July 1999, nine leading U.S.-based Internet companies joined forces to become the voice of the Internet on such issues as privacy, consumer protection, and international trade. The industry group, called NetCoalition.com, includes America Online, Amazon.com, eBay, Lycos, Yahoo!, DoubleClick, Excite@Home, Inktomi, and Theglobe.com. The companies represent seven of the top ten Internet sites, and more than 90 percent of the world's Internet users visit one of the sites at least once a month. The group plans to focus on 150 Internet-related bills that were introduced in Congress.[28]

Hackers and Terrorists. The Internet is used extensively as a publication medium by hackers (including hacktivists) and terrorists. Hackers publish electronic magazines and put up web sites with software tools and information about hacking, including details about vulnerabilities in popular systems (e.g., Microsoft Windows) and how they can be exploited, programs for cracking passwords, software packages for writing computer viruses, and scripts for disabling or breaking into computer networks and web sites. In March 1997, an article in

---

[27]See www.computerprivacy.org.

[28]"Internet Heavies Back New Net-Policy Group," IDG, July 14, 1999.

The New York Times reported that there were an estimated 1,900 web sites purveying hacking tips and tools, and 30 hacker publications.[29]

Terrorist groups use the Internet to spread propaganda. Back in February 1998, Hizbollah was operating three web sites: one for the central press office (www.hizbollah.org), another to describe its attacks on Israeli targets (www.moqawama.org), and the third for news and information (www.almanar.com.lb).[30] That month, Clark Staten, executive director of the Emergency Response & Research Institute (ERRI) in Chicago, testified before a U.S. Senate subcommittee that "even small terrorist groups are now using the Internet to broadcast their message and misdirect/misinform the general population in multiple nations simultaneously." He gave the subcommittee copies of both domestic and international messages containing anti-American and anti-Israeli propaganda and threats, including a widely distributed extremist call for *jihad* (holy war) against America and Great Britain.[31] In June 1998, *U.S. News & World Report* noted that 12 of the 30 groups on the U.S. State Department's list of terrorist organizations are on the web. As of August 1999, it appears that virtually every terrorist group is on the web, along with a mishmash of freedom fighters, crusaders, propagandists, and mercenaries.[32] Forcing them off the web is impossible, because they can set up their sites in countries with free-speech laws. The government of Sri Lanka, for example, banned the separatist Liberation Tigers of Tamil Eelam, but they have not even attempted to take down their London-based web site.[33]

---

[29]Steve Lohr, "Go Ahead, Be Paranoid: Hackers Are out to Get You," *The New York Times*, March 17, 1997.

[30]John Arquilla, David Ronfeldt, and Michele Zanini, "Networks, Netwar, and Information-Age Terrorism," in Ian O. Lesser et al., *Countering the New Terrorism*, RAND, 1999, p. 66. The authors cite "Hizbullah TV Summary 18 February 1998," *Al-Manar Television World Wide Webcast*, FBIS-NES-98-050, February 19, 1998, and "Developments in Mideast Media: January–May 1998," Foreign Broadcast Information Service (FBIS), May 11, 1998.

[31]Clark L. Staten, testimony before the Subcommittee on Technology, Terrorism and Government Information, U.S. Senate Judiciary Committee, February 24, 1998.

[32]Bob Cromwell's site at Purdue has an excellent collection of links: http://RVL4.ecn.purdue.edu/~cromwell/lt/terror.html.

[33]Kevin Whitelaw, "Terrorists on the Web: Electronic 'Safe Haven,'" *U.S. News & World Report*, June 22, 1998, p. 46. The State Department's list of terrorist organizations is at www.state.gov/www/global/terrorism/index.html.

Dialogue

The Internet offers several venues for dialogue and debate on policy issues. These include email, newsgroups, web forums, and chat. Discussions can be confined to closed groups, for example through email, as well as open to the public. Some media sites offer web surfers the opportunity to comment on the latest stories and current issues and events. Government officials and domain experts may be brought in to serve as catalysts for discussion, debate issues, or answer questions. Discussion can even take place on web sites that themselves lack such facilities. Using Gooey software from the Israeli company Hypernix, for example, visitors to a web site can chat with other Gooey users currently at the site.[34]

Internet discussion forums are frequently used to debate, blast, and maybe even attempt to influence government policies. Encryption policy, for example, is discussed on the email lists "cypherpunks" and "ukcrypto" and on several newsgroups, including alt.privacy and sci.crypt.

The ukcrypto list was created in early 1996 by two academics, Ross Anderson (Cambridge) and Paul Leyland (Oxford), and one person then in government, Brian Gladman (NATO SHAPE), who was acting outside his official capacity. Motivated by a concern that a lack of public discussion and debate in the United Kingdom on cryptography issues was allowing the government to set policies that they believed were not in the interests of the United Kingdom and its citizens, they formed the list with the objective of affecting cryptography policy. They were concerned both with domestic policy, particularly proposals to restrict the use of cryptography by U.K. citizens, and foreign policy, particularly export controls. As of May 1999, the list had 300 subscribers, including government officials responsible for U.K. policy and persons in other countries, including the United States. Many of the key contributors held influential positions in other policymaking fora. The focus is on U.K. policy issues, but items of international interest are also discussed, including export controls adopted under the Wassenaar Arrangement (31 countries participate);

---

[34]Chris Oaks, "Every Web Site a Chat Room," *Wired News*, June 14, 1999.

policy changes adopted by France, the United States, and other countries; policy statements from the European Union and other organizations; and some technical issues.[35]

Gladman believes the list has made four contributions: (1) educating
many about the policy issues and encouraging journalists and writers
to write about them; (2) bringing individual and industry views closer
together and allowing U.K. industry to see more clearly that agreeing
with their government may not be a good thing if private citizens do
not support government policy; (3) encouraging the more progressive
voices in government to speak out and argue from within government
that their views represent those of the public; and (4) bringing groups
together that were previously campaigning separately. "The most significant contribution of ukcrypto is not direct," Gladman said. "It is
the contribution that it has made in promoting an educated community of commentators and a forum for the review of what government
is doing that is fully open."

On the downside, some postings on ukcrypto may alienate the very
government officials the authors hope to influence. According to
Gladman,

> discussions on the list can become slinging matches that quickly put
> those in government on the defensive and hence inclined to dis-
> count what is being said. It would be more effective if we had a way
> of focusing on the issues and not the personalities.[36]

But Andrew Brown gave ukcrypto high marks, crediting it with most
of the thought and coordination behind the successful campaign to
keep strong cryptography legal and widely available. He wrote in *New
Statesman,*

> There, for the past two years, the civil servants responsible for policy
> have actually been available, more or less, to the people who dis-
> agree with them. . . . They have had to justify their actions, not to the

---

[35]Brian Gladman, personal correspondence, May 4, 1999, augmented by my own observations from subscribing to the list since the beginning.
[36]Ibid.

public, but to a small group of geographically dispersed experts. . . . It's a kind of updated version of Lions *v* Christians.[37]

Nigel Hickson, one of the principal players in the policy debates from the U.K. Department of Trade and Industry (DTI), agrees the Internet and ukcrypto in particular have played a role in shaping U.K. cryptography policy.[38] But he was also critical of the list:

> Whilst ukcrypto has undoubtedly had an influence on the development of U.K. encryption policy, it has tended to polarise the debate into extremes. This may be because there tends to be a large "silent majority" on the list who do not directly contribute because of commercial or policy reasons.[39]

Besides participating in ukcrypto, the DTI has published draft consultation documents on the web for comment. Many of the comments they receive arrive through electronic mail. DTI has also met with industry groups and participated in non-Internet forums, such as conferences and seminars. These have also helped shape policy decisions.

There are Usenet newsgroups and other interactive forums that focus on practically every conceivable topic relating to foreign (and domestic) policy. Whether these are effective or not in terms of influencing policy is another matter. After studying the effects of the Net on the American political system, Richard Davis, a political science professor at Brigham Young University and author of *The Web of Politics*, observed that

> In Usenet political discussions, people talk past one another, when they are not verbally attacking each other. The emphasis is not problem solving, but discussion dominance.[40]

---

[37]Andrew Brown, "Editors Wanted," *New Statesman*, April 26, 1999.

[38]Nigel Hickson, private conversation, April 29, 1999.

[39]Nigel Hickson, private communication, July 28, 1999.

[40]Richard Davis, *The Web of Politics*, Oxford University Press, 1999, p. 177.

Davis also found interactivity on the Internet to be primarily an illusion:

> Interest groups, party organizations, and legislators seek to use the web for information dissemination, but they are rarely interested in allowing their sites to become forums for the opinions of others.[41]

Coordination of Action

Advocacy groups can use the Internet to coordinate action among members and with other organizations and individuals. Action plans can be distributed by email or posted on web sites. Services are cheaper than phone and fax (although these services can also be delivered through the Internet), and faster than physical delivery (assuming Internet services are operating properly, which is not always the case). The Internet lets people all over the world coordinate action without regard to constraints of geography or time. They can form partnerships and coalitions or operate independently.

One web site was created to help activists worldwide coordinate and locate information about protests and meetings. According to statements on Protest.Net, the web site serves "to help progressive activists by providing a central place where the times and locations of protests and meetings can be posted." The site's creator said he hoped it would "help resolve logistical problems that activists face in organizing events with limited resources and access to mass media."[42] The site features news as well as action alerts and information about events.

The power of the Internet to mobilize activists is illustrated by the arrest of Kurdish rebel leader Abdullah Ocalan. According to Michael Dartnell, a political science professor at Concordia University, when Turkish forces arrested Ocalan, Kurds around the world responded with demonstrations within a matter of hours. He attributed the swift action in part to the Internet and web. "They responded more quickly than governments did to his arrest," he said. Dartnell contends the Internet and advanced communication tools are changing the way peo-

---

[41]Davis, 1999, p. 178.

[42]See www.protest.net.

ple around the world play politics. Antigovernment groups are establishing alliances and coalitions that might not have existed before the technology was introduced.[43]

The force of the Internet is further illustrated by the day of protest against business that took place on June 18, 1999. The protests, which were set up to coincide with a meeting of the G8 in Cologne, Germany, was coordinated by a group called J18 from a web site inviting people to plan individual actions focusing on disrupting "financial centres, banking districts and multinational corporate power bases." Suggested activity included marches, rallies, and hacking. In London, up to 2,000 anticapitalists coursed through the city shouting slogans and spray-painting buildings.[44] According to the *Sunday Times*, teams of hackers from Indonesia, Israel, Germany, and Canada attacked the computers of at least 20 companies, including the Stock Exchange and Barclays. More than 10,000 attacks were launched over a five-hour period.[45]

During the Kosovo conflict, the Kosova Task Force used the Internet to distribute action plans to Muslims and supporters of Kosovo. A March 1999 Action Alert, for example, asked people to organize rallies in solidarity with Kosovo at local federal buildings and city halls on April 3 at 11 a.m.; organize public funeral prayers; make and encourage others to make daily calls or send email to the White House asking for Kosovo independence, sustained air strikes until there was total Serb withdrawal from Kosovo, and arming of ethnic Albanians in Kosovo; and make and encourage others to make calls to their representatives and senators. An April 18 alert asked every community in the United States to establish a Kosova Room for action and information. Each room was to be equipped with a bank of phones for making 1,000 calls to the White House and Congress in support of resolution #HCR 9, calling for independence of Kosovo.

---

[43]Martin Stone, "Prof to Build Archive of Insurgency Groups," *Newsbytes*, March 3, 1999.

[44]Edward Harris, "Web Becomes a Cybertool for Political Activists," *Wall Street Journal*, August 5, 1999, B11; Barbara Adam, "J18 Hackers Could Target Australian Companies on Friday," *Australian Associated Press*, June 16, 1999.

[45]Jon Ungoed-Thomas and Maeve Sheehan, "Riot Organisers Prepare to Launch Cyber War on City," *Sunday Times*, August 15, 1999.

The International Campaign to Ban Landmines (ICBL), a loose coalition of over 1,300 groups from more than 75 countries, has made extensive use of the Internet in efforts to stop the use, production, stockpiling, and transfer of antipersonnel landmines, and to increase international resources for humanitarian mine clearance and victim assistance. According to ICBL's Liz Bernstein, the Net has been the dominant form of communication since 1996.[46] It has been used to coordinate events and committee functions, distribute petitions and action alerts, raise money, and educate the public and media. Although most direct lobbying is done through face-to-face meetings and letters, email has facilitated communications with government policymakers. Bernstein said the Net "has helped the nature of the campaign as a loose coalition, each campaign setting their own agenda yet with common information and communication."[47] Ken Rutherford, cofounder of Land Mine Survivors Network, noted that the Internet also helped establish bridges from North America and Europe to Asia and Africa, and helped enable quick adoption of the 1997 landmine treaty.[48] It became international law on March 1, 1999, and, as of September 16, 1999, has been signed by 135 countries and ratified by 86. In 1997, the Nobel Peace Prize was awarded to the ICBL and its then coordinator, Jody Williams.[49]

Human rights workers increasingly use the Internet to coordinate their actions against repressive governments. One tool that has become important in their battles is encryption, because it allows activists to protect communications and stored information from government interception. Human rights activists in Guatemala, for example, credited their use of Pretty Good Privacy (PGP) with saving the lives of witnesses to military abuses.[50] Encryption is not the ultimate solution, however, since governments can outlaw its use and arrest those who do not comply.

---

[46]Liz Bernstein, private communication, October 4, 1999.

[47]Ibid.

[48]Ken Rutherford, private communication, October 6, 1999.

[49]See also the ICBL web site at www.icbl.org and the web site of the Land Mine Survivors Network at www.landminesurvivors.org.

[50]Alan Boyle, "Crypto Can Save Lives," *ZDNet*, January 26, 1999. PGP provides both file and electronic-mail encryption.

Terrorists also use the Internet to communicate and coordinate their activities. Back in 1996, the headquarters of terrorist financier Osama bin Laden in Afghanistan was equipped with computers and communications equipment. Egyptian "Afghan" computer experts were said to have helped devise a communication network that used the web, email, and electronic bulletin boards.[51] Hamas activists have been said to use chat rooms and email to plan operations and coordinate activities, making it difficult for Israeli security officials to trace their messages and decode their contents.[52]

The U.S. government's program to establish an Advanced Encryption Standard (AES) illustrates how government can use the Internet to invite and coordinate participation in a decisionmaking process of international significance. The Department of Commerce National Institute of Standards and Technology (NIST) set up a web site with information about the AES program and AES conferences, a schedule of events, candidate encryption algorithms (more than half from outside the United States), documentation and test values, and links to public analysis efforts all over the world. The site contains an electronic discussion forum and Federal Register call for comments. Public comments are posted on the site and NIST representatives contribute to the online discussions and answer questions.[53] Because the AES will offer a foundation for secure electronic commerce and privacy internationally, involving the international community from the beginning will help ensure its success and widespread adoption. Cryptographers from all over in the world have been participating.

NIST's use of the Internet to aid a decision process seems to be unusual. While most government sites provide an email address for making contact, they do not support discussion forums or even ac-

---

[51]John Arquilla, David Ronfeldt, and Michele Zanini, "Networks, Netwar, and Information-Age Terrorism," in Ian O. Lesser et al., *Countering the New Terrorism*, RAND, 1999, p. 65. The authors cite "Afghanistan, Saudi Arabia: Editor's Journey to Meet Bin-Laden Described," *London al-Quds al-'Arabi*, FBIS-TOT-97-003-L, November 27, 1996, p. 4, and "Arab Afghans Said to Launch Worldwide Terrorist War," 1995.

[52]Ibid. The authors cite "Israel: U.S. Hamas Activists Use Internet to Send Attack Threats," *Tel Aviv IDF Radio*, FBIS-TOT-97-001-L, October 13, 1996, and "Israel: Hamas Using Internet to Relay Operational Messages," *Tel Aviv Ha'aretz*, FBIS-TOT-98-034, February 3, 1998, p. 1.

[53]The NIST AES web site is at csrc.nist.gov/encryption/aes/aes_home.htm.

tively solicit comments on specific pending policy decisions. However, to the extent that government agencies invite or welcome email messages and input through electronic discussion groups, the Internet can serve the democratic process. Because it is easier to post or send a message on the Internet than to send a written letter, professionals and others with busy schedules may be more inclined to participate in a public consultation process or attempt to influence policy when policymakers are readily accessible through the Internet.

Lobbying Decisionmakers

Whether or not government agencies solicit their input, activists can use the Internet to lobby decisionmakers. One of the methods suggested by the Kosova Task Force for contacting the White House, for example, was email. Similarly, a Canadian web site with the headline "Stop the NATO Bombing of Yugoslavia Now!" urged Canadians and others interested in stopping the war to send emails and/or faxes to the Canadian Prime Minister, Jean Chretien, and all members of the Canadian Parliament. A sample letter was included. The letter concluded with an appeal to "stop aggression against Yugoslavia and seek a peaceful means to resolve the Kosovo problem."[54]

Email has been credited with halting a U.S. banking plan aimed to combat money laundering. Under the "Know Your Customer" policy, banks would have been required to monitor customer's banking patterns and report inconsistencies to federal regulators. Recognizing the value of the Internet to its deliberations, the Federal Deposit Insurance Corporation (FDIC) put up a web site, published an email address for comments, and printed out and tabulated each message. By the time the proposal was withdrawn, they had received 257,000 comments, 205,000 (80 percent) of which arrived through email. All but 50 of the letters opposed the plan. FDIC's chair, Donna Tanoue, said it was the huge volume of email that drove the decision to withdraw the proposal. "It was the nature and the volume [of the comments]," she

---

[54]See www.aeronautix.com/nato/yugoslavia.html.

said. "When consumers can get excited about an esoteric bank regulation, we have to pay attention."[55]

Most of the email was driven by an online advocacy campaign sponsored by the Libertarian Party. About 171,000 (83 percent) of the email messages were sent through the party's web site. The party advertised its advocacy campaign in talk radio interviews and by sending a notice to its email membership list.[56] One could argue that the results were due more to the efforts of a large nongovernment organization than to a grassroots response from the citizens.

Indeed, many email campaigns have been driven by nongovernment organizations. The organizations send email alerts on issues to electronic mailing lists, offer sample letters to send members of Congress and other decisionmaking bodies, and, in some cases, set up email boxes or web sites to gather signatures for petitions. The petition process can be automated, making it possible to gather huge volumes of signatures across a wide geographic area with little effort and cost. One web site, e-The People, offers hundreds of petitions to choose from and 170,000 email addresses of government officials.[57]

Computer Professionals for Social Responsibility (CPSR) organized an Internet petition campaign in early 1994 to protest the U.S. government's proposal to adopt the Clipper encryption chip as a standard.[58] The chip offered strong encryption but would have given law enforcement agencies the capability to decrypt a subject's messages when conducting a court-ordered wiretap against the subject. Despite numerous safeguards to ensure that government agencies could not violate the privacy of users of the chip,[59] Clipper was strongly opposed for privacy (and other) reasons, and the general sentiment expressed

---

[55]Rebecca Fairley Raney, "Flood of E-Mail Credited with Halting U.S. Bank Plan," *The New York Times (Cybertimes)*, March 24, 1999.

[56]Ibid.

[57]Edward Harris, "Web Becomes a Cybertool for Political Activists," *Wall Street Journal*, August 5, 1999, B11. The web site is at www.e-thepeople.com.

[58]The persons organizing the campaign went on to form the Electronic Privacy Information Center shortly thereafter.

[59]For example, each chip was uniquely keyed and decryption was not possible without getting the keys to the subject's chip from two separate government agencies.

on Internet newsgroups and email discussion lists was strongly anti-Clipper. CPSR announced its petition through email and set up an email address whereby people could sign on. Tens of thousands of signatures were collected, but it is not clear the petition had much impact. The government moved forward with the standard anyway, although Clipper eventually met its death.[60]

Because of the low cost of operation, individuals can run their own advocacy campaigns. For example, during the heart of the impeachment process against President Clinton, Joan Blades and Wes Boyd, a husband and wife team in Berkeley, founded MoveOn.org and put up a web site inviting citizens to sign a one-sentence petition: "The Congress must immediately censure President Clinton and move on to pressing issues facing the country." In just four months, the petition gathered a half-million signatures. Another petition that read "In the Year 2000 election, I will work to elect candidates who courageously address key national issues and who reject the politics of division and personal destruction" was sent to every member of the House and Senate. MoveOn.org received pledges of $13 million and more than 650,000 volunteer hours for congressional candidates in the 2000 election who supported its position.[61] It is difficult to assess the effects of the site on the impeachment process, but it may have amplified public opinion polls, which showed the American public supported Clinton and wanted Congress to turn to other issues.

While activists can attempt to influence policymakers through email, it is not clear that most policymakers listen (the FDIC, which asked for comments, was an exception). Richard Davis found that

> the Internet has not lived up to its promise as a forum for public expression to elected officials. In fact, while publicly encouraging e-mail, members are becoming increasingly disenchanted with it. If the most idealistic members originally envisioned e-mail as the impetus for intelligent communication with constituents, they have

---

[60]For an interesting discussion of the Internet campaign against Clipper, see Laura J. Gurak, *Persuasion and Privacy in Cyberspace,* Yale University Press, 1997.

[61]Chris Carr, "Internet Anti-Impeachment Drive Yields Big Pledges of Money, Time," *Washington Post,* February 7, 1999. The site is at www.moveon.org.

seen e-mail deteriorate into a mass mailing tool for political activists. . . . [M]embers may even discount e-mail communication."[62]

According to the *Wall Street Journal*, Senator Charles Schumer's office
gives first priority to old-fashioned letters. Persons sending an email
to his account get back an automatic response telling them to submit
a letter if they want a personal reply.[63]

The most successful advocacy groups are likely to be those that use
the Internet to augment traditional lobbying methods, including personal visits to decisionmakers and use of broadcast media to reach
the public. These operations can be time consuming and expensive,
favoring groups that are well-funded. They also require a network of
long-term and trusted relationships with policymakers, sponsors,
and voters. This supports Davis's conclusion that the promise of the
Internet as a forum for participatory democracy is unlikely to be realized. Davis found that existing dominant players in American politics—the media, interest groups, candidates, and policymakers—are
adapting to the Internet to retain preeminence; and that the Internet
is not an adequate tool for public political movement.[64]

HACKTIVISM

Hacktivism is the convergence of hacking with activism, where "hacking" is used here to refer to operations that exploit computers in ways
that are unusual and often illegal, typically with the help of special
software ("hacking tools"). Hacktivism includes electronic civil disobedience, which brings methods of civil disobedience to cyberspace. This section explores four types of operations: virtual sit-ins
and blockades, automated email bombs, web hacks and computer
break-ins, and computer viruses and worms. Because hacking
incidents are often reported in the media, operations in this category
can generate considerable publicity for both the activists and their
causes.

---

[62]Davis, 1999, p. 135.

[63]Edward Harris, "Web Becomes a Cybertool for Political Activists," *Wall Street Journal*,
August 5, 1999, B11.

[64]Davis, 1999, p. 168.

Virtual Sit-Ins and Blockades

A virtual sit-in or blockade is the cyberspace version of a physical sit-in or blockade. The goal in both cases is to call attention to the protestors and their cause by disrupting normal operations and blocking access to facilities.

With a sit-in, thousands of activists simultaneously visit a web site and attempt to generate so much traffic against the site that other users cannot reach it. A group calling itself Strano Network conducted one of the first such demonstrations as a protest against French government policies on nuclear and social issues. On December 21, 1995, they launched a one-hour Net'Strike attack against the web sites operated by various government agencies. At the appointed hour, participants from all over the world were instructed to point their browsers to the government web sites. According to reports, at least some of the sites were effectively knocked out for the period.[65]

In 1998, the Electronic Disturbance Theater (EDT) took the concept of electronic civil disobedience a step further. They organized a series of web sit-ins, first against Mexican President Zedillo's web site and later against President Clinton's White House web site, and the web sites of the Pentagon, the School of the Americas, the Frankfurt Stock Exchange, and the Mexican Stock Exchange. The purpose was to demonstrate solidarity with the Mexican Zapatistas.[66] According to EDT's Brett Stalbaum, the Pentagon was chosen because "we believe that the U.S. military trained the soldiers carrying out the human rights abuses." For a similar reason, the School of the Americas was selected.[67] The Frankfurt Stock Exchange was targeted, Stalbaum said,

> because it represented capitalism's role in globalization utilizing the techniques of genocide and ethnic cleansing, which is at the root of the Chiapas' problems. The people of Chiapas should play a key role

---

[65]Information provided to the author from Bruce Sterling; Winn Schwartau, *Information Warfare*, 2nd ed., Emeryville, Calif.: Thunder's Mouth Press, 1996, p. 407.

[66]For an in-depth analysis of the Zapatista's "netwar," see David Ronfeldt, John Arquilla, Graham E. Fuller, and Melissa Fuller, *The Zapatista "Social Netwar" in Mexico*, Santa Monica, Calif.: RAND, MR-994-A, 1998.

[67]Niall McKay, "Pentagon Deflects Web Assault," *Wired News*, September 10, 1998.

in determining their own fate, instead of having it pushed on them through their forced relocation (at gunpoint), which is currently financed by western capital.[68]

To facilitate the strikes, the organizers set up special web sites with automated software. All participants had to do was visit one of the FloodNet sites. When they did, their browser would download the software (a Java Applet), which would access the target site every few seconds. In addition, the software let protesters leave a personal statement on the targeted server's error log. For example, if they pointed their browsers to a nonexistent file such as "human_rights" on the target server, the server would return and log the message "human_rights not found on this server." Stalbaum, who wrote the software, characterized FloodNet as "conceptual net art that empowers people through active/artistic expression."[69]

EDT estimated that 10,000 people from all over the world participated in the sit-in on September 9, 1998, against the sites of President Zedillo, the Pentagon, and the Frankfurt Stock Exchange, delivering 600,000 hits per minute to each. According to *Wired News'* Niall McKay, the Pentagon "apparently struck back." Stalbaum recalled the incident:

> They [Pentagon programmers] were redirecting any requests coming by way of the EDT Tactical FloodNet to a page containing an Applet called "HostileApplet." This Applet . . . instantly put all the FloodNet protestors' browsers into an infinite loop by opening a small window which tried to reload a document as fast as [possible]. . . . I had to restart my [computer] to recover control.

President Zedillo's site did not strike back on this occasion, but at a June sit-in, that site used software that caused the protestors' browsers to open window after window until their computers crashed. The Frankfurt Stock Exchange reported that it was aware of the protest but believed that the protest had not affected its services. The exchange normally gets about 6 million hits a day. Overall, EDT considered the

---

[68]Brett Stalbaum, private correspondence, July 23, 1999.

[69]Brett Stalbaum, "The Zapatista Tactical FloodNet," www.thing.net/~rdom/ecd/ZapTact.html.

attack a success. "Our interest is to help the people of Chiapas to keep receiving the international recognition that they need to keep them alive," said Stalbaum.[70]

When asked about the effects of its web strikes, EDT's Ricardo Dominguez responded,

> Digital Zapatismo is and has been one of the most politically effective uses of the Internet that we know of since January 1, 1994. It has created a distribution network of information with about 100 or more autonomous nodes of support. This has enabled the EZLN [Zapatista National Liberation Army] to speak to the world without having to pass through any dominant media filter. The Zapatistas were chosen by *Wired* as one of the 25 most important people online in 1998. . . . The Zapatista network has, also, held back a massive force of men and the latest drug war technologies from annihilating the EZLN in a few days.

Regarding FloodNet, specifically, he said the main purpose of the Electronic Disturbance Theatre's Zapatista FloodNet performance

> is to bring the situation in Chiapas to [the] foreground as often as possible. The gesture has created enough ripples with the Pentagon and the Mexican government that they have had to respond using both online and offline tactics. Thus, these virtual sit-ins have captured a large amount of traditional media attention. You would not be interviewing us if this gesture had not been effective in getting attention to the issues on a global scale.[71]

EDT also conducted web sit-ins against the White House web site to express opposition to U.S. military strikes and economic sanctions against Iraq. In the "Call for FloodNet Action for Peace in the Middle East," EDT articulated the group's philosophy.

---

[70]Niall McKay, "Pentagon Deflects Web Assault," *Wired News*, September 10, 1998; Brett Stalbaum, personal communication, January 30, 1999.

[71]Ricardo Dominguez, personal communication, February 2, 1999.

We do not believe that only nation-states have the legitimate authority to engage in war and aggression. And we see cyberspace as a
means for non-state political actors to enter present and future are-
nas of conflict, and to do so across international borders.[72]

Animal right's activists have also used EDT's FoodNet software to protest the treatment of animals. Over 800 protestors from more than 12
countries joined a January 1999 sit-in against web sites in Sweden.[73]
And on June 18, 1999, FloodNet was one of the tools used in the anticapitalist attack coordinated by J18.[74]

Whether web sit-ins are legal is not clear. Mark Rasch, former head of
the Department of Justice's computer crime unit, said that such attacks run the risk of violating federal laws, which make it a crime to
distribute a program, software code, or command with the intent to
cause damage to another's site. "It may be an electronic sit-in, but
people get arrested at sit-ins," he said.[75] A related question is the le-
gality of using a denial-of-service (DoS) counteroffensive. In the case
of the Pentagon, the response most likely would be considered lawful,
because it is permissible for a nation to take "proportional" actions to
defend against an attack that threatens its security.

The tools used to conduct web sit-ins have evolved so that protestors
need not all congregate on a central site at once. Instead, they can
download the software anytime or get it via email and then run it at
the appointed hour. The software remains, however, fundamentally
different from that used in standard DoS and distributed DoS attacks,
such as those launched against Yahoo! and other e-commerce web

---

[72]See www.aec.at/infowar/NETSYMPOSIUM/ARCH-EN/msg00633.html.

[73]*Day of Net Attacking Against Vivisection,* communiqué from the Animal Liberation
Front, December 31, 1998. *The First Ever Animal Liberation Electronic Civil Disobedi-
ence Virtual Sit-In on the SMI Lab Web Site in Sweden,* notice from Tactical Internet Re-
sponse Network, http://freehosting.at.webjump.com/fl/floodnet-webjump/smi.html.
ECD Report—SMI Shuts Down Their Computer Network!!!, www.aec.at/infowar/
NETSYMPOSIUM/ARCH-EN/msg00678.html, January 15, 1999.

[74]Jon Ungoed-Thomas and Maeve Sheehan, "Riot Organisers Prepare to Launch Cyber
War on City," *Sunday Times,* August 15, 1999.

[75]Carl Kaplan, "For Their Civil Disobedience, the 'Sit-In' Is Virtual," *Cyberlaw Journal,
The New York Times on the Web,* May 1, 1998. The law is Title 18 U.S.C. section 1030
(a)(5)(A).

sites in February 2000. It does not compromise any systems or spoof source addresses, and it usually does not shut down the target. Further, thousands or tens of thousands of people must hit the target at once to have any effect, a phenomena sometimes referred to as "swarming." Swarming ensures that the action being protested is reprehensible to more than just a single person or small group.

The Electrohippies Collective, another group of hacktivists, says that its web sit-ins must substitute the deficit of speech by one group with a broad debate on policy issues, and that the event used to justify the sit-in must provide a focus for the debate. The group also espouses a philosophy of openness and accountability.[76] The group conducted a sit-in in conjunction with the WTO protests in Seattle in late 1999. In April 2000, it planned a sit-in against the genetics industry but backed out when visitors to their web site voted a lack of support. The sit-in was to be part of a broader "E-Resistance is Fertile" campaign, which also included a low-level email lobbying campaign.[77]

Individuals acting alone or in small groups have used DoS tools to disable Internet servers. During the Kosovo conflict, Belgrade hackers were credited with conducting such attacks against NATO servers. They bombarded NATO's web server with "ping" commands, which test whether a server is running and connected to the Internet. The effect of the attacks was to cause line saturation of the targeted servers.[78]

Email Bombs

It is one thing to send one or two messages to government policymakers, even on a daily basis. But it is quite another to bombard them with thousands of messages at once, distributed with the aid of automated tools. The effect can be to completely jam a recipient's incoming email box, making it impossible for legitimate email to get through. Thus, an email bomb is also a form of virtual blockade. Al-

---

[76]See www.gn.apc.org/pmhp/ehippies.

[77]Ibid.

[78]Rebecca Allison, "Belgrade Hackers Bombard MoD Website in 'First' Internet War," *PA News*, March 31, 1999.

though email bombs are often used as a means of revenge or harassment, they have also been used to protest government policies.

In what some U.S. intelligence authorities characterized as the first known attack by terrorists against a country's computer systems, ethnic Tamil guerrillas were said to have swamped Sri Lankan embassies with thousands of electronic mail messages. The messages read "We are the Internet Black Tigers and we're doing this to disrupt your communications."[79] An offshoot of the Liberation Tigers of Tamil Eelam, which had been fighting for an independent homeland for minority Tamils, was credited with the 1998 incident.[80]

The email bombing consisted of about 800 emails a day for about two weeks. William Church, editor for the Centre for Infrastructural Warfare Studies (CIWARS), observed that

> the Liberation Tigers of Tamil are desperate for publicity and they got exactly what they wanted . . . considering the routinely deadly attacks committed by the Tigers, if this type of activity distracts them from bombing and killing then CIWARS would like to encourage them, in the name of peace, to do more of this type of "terrorist" activity.[81]

The attack, however, was said to have had the desired effect of generating fear in the embassies.

During the Kosovo conflict, protestors on both sides email bombed government sites. According to *PA News*, NATO spokesman Jamie Shea said the NATO server had been saturated at the end of March by one individual who was sending 2,000 messages a day.[82] Fox News reported that when California resident Richard Clark heard of attacks against NATO's web site by Belgrade hackers, he retaliated by sending an email bomb to the Yugoslav government's site. Clark said that a few

---

[79]"E-Mail Attack on Sri Lanka Computers," *Computer Security Alert*, No. 183, Computer Security Institute, June 1998, p. 8.

[80]Jim Wolf, "First 'Terrorist' Cyber-Attack Reported by U.S.," Reuters, May 5, 1998.

[81]*CIWARS Intelligence Report*, May 10, 1998.

[82]Rebecca Allison, "Belgrade Hackers Bombard MoD Website in 'First' Internet War," *PA News*, March 31, 1999.

days and 500,000 emails into the siege, the site went down. He did not claim full responsibility but said he "played a part." That part did not go unrecognized. His Internet service provider, Pacific Bell, cut off his service, saying his actions violated their spamming policy.[83]

An email bombing was conducted against the San Francisco–based Internet service provider Institute for Global Communications (IGC) in 1997 for hosting the web pages of the *Euskal Herria Journal*, a controversial publication edited by a New York group supporting independence of the mountainous Basque provinces of northern Spain and southwestern France. Protestors claimed IGC "supports terrorism" because a section on the web pages contained materials on the terrorist group Fatherland and Liberty, or ETA, which was responsible for killing over 800 people during its nearly 30-year struggle for an independent Basque state. The attack against IGC began after members of the ETA assassinated a popular town councilor in northern Spain.[84]

The protestors' objective was censorship. They wanted the site pulled. To get their way, they bombarded IGC with thousands of bogus messages routed through hundreds of different mail relays. As a result, mail was tied up and undeliverable to IGC's email users, and support lines were tied up with people who couldn't get their mail. The attackers also spammed IGC staff and member accounts, clogged their web page with bogus credit card orders, and threatened to employ the same tactics against organizations using IGC services. The only way IGC could stop the attack was by blocking access from all of the relay servers.[85]

IGC pulled the site on July 18, but not before archiving a copy so that others could put up mirrors. Within days of the shutdown, mirror sites appeared on half a dozen servers on three continents. Chris Ellison, a spokesman for the Internet Freedom Campaign, an English

---

[83]Patrick Riley, "E-Strikes and Cyber-Sabotage: Civilian Hackers Go Online to Fight," Fox News, April 15, 1999.

[84]Rebecca Vesely, "Controversial Basque Web Site Resurfaces," *Wired News*, August 28, 1997; "Two More Basque Politicians Get ETA Death Threats," Reuters, San Sebastian, Spain, December 16, 1997.

[85]"IGC Censored by Mailbombers," letter from Maureen Mason and Scott Weikart, IGC, www.infowar.com.

group that was hosting one of the mirrors, said they believe "the Net should prove an opportunity to read about and discuss controversial ideas." A New York–based journal maintained its objective was to publish "information often ignored by the international media, and to build communication bridges for a better understanding of the conflict."[86] An article by Yves Eudes in the French newspaper *Le Monde* said the email bomb attack against the IGC site represented an "unprecedented conflict" that "has opened up a new era of censorship, imposed by direct action from anonymous hackers."[87]

About a month after IGC threw the controversial Basque *Euskal Herria Journal* off its servers, Scotland Yard's Anti-Terrorist Squad shut down Internet Freedom's U.K. web site for hosting the journal. According to a press release from Internet Freedom, the squad claimed to be acting against terrorism. Internet Freedom said it would move its news operations to its U.S. site.[88]

The case involving *Euskal Herria Journal* illustrates the power of hacktivists on the Internet. Despite IGC's desire to host the controversial site, they simply could not sustain the attack and remain in business. They could have ignored a few email messages demanding that the site be pulled, but they could not ignore an email bombing. The case also illustrates the power of the Internet as a tool for free speech. Because Internet venues for publication are rich and dispersed throughout the world, it is extremely difficult for governments and hacktivists alike to keep content completely off the Internet. It would require extensive international cooperation, and, even then, a site could operate out of a safe haven that did not sign on to international agreements.

---

[86]Rebecca Vesely, "Controversial Basque Web Site Resurfaces," *Wired News*, August 28, 1997.

[87]Yves Eudes, "The Zorros of the Net," *Le Monde*, November 16, 1997.

[88]*Anti-Terrorist Squad Orders Political Censorship of the Internet*, press release from Internet Freedom, September 1997.

Web Hacks and Computer Break-Ins

The media is filled with stories of hackers gaining access to web sites and replacing some of the content with their own. Frequently, the messages are political, as when a group of Portuguese hackers modified the sites of 40 Indonesian servers in September 1998 to display the slogan "Free East Timor" in large black letters. According to *The New York Times*, the hackers also added links to web sites describing Indonesian human rights abuses in the former Portuguese colony.[89] Then in August 1999, Jose Ramos Horta, the Sydney-based Nobel laureate who represents the East Timor independence movement outside Indonesia, warned that a global network of hackers planned to bring Indonesia to a standstill if Jakarta sabotaged the ballot on the future of East Timor. He told the *Sydney Morning Herald* that more than 100 hackers, mostly teenagers in Europe and the United States, had been preparing the plan.[90]

In June 1998, a group of international hackers calling themselves Milw0rm hacked the web site of India's Bhabha Atomic Research Center (BARC) and put up a spoofed web page showing a mushroom cloud and the text "If a nuclear war does start, you will be the first to scream . . . ." The hackers were protesting India's recent nuclear weapons tests, although they admitted they did it mostly for thrills. They said that they also downloaded several thousand pages of email and research documents, including messages between India's nuclear scientists and Israeli government officials, and had erased data on two of BARC's servers. The six hackers, whose ages range from 15 to 18, hailed from the United States, England, the Netherlands, and New Zealand.[91]

Another way in which hacktivists alter what viewers see when they go to a web site is by tampering with the Domain Name Service so that

---

[89]Amy Harmon, "'Hacktivists' of All Persuasions Take Their Struggle to the Web," *The New York Times*, October 31, 1999.

[90]Lindsay Murdoch, "Computer Chaos Threat to Jakarta," *Sydney Morning Herald*, August 18, 1999, p. 9.

[91]James Glave, "Crackers: We Stole Nuke Data," *Wired News*, June 3, 1998; Janelle Carter, "Hackers Hit U.S. Military Computers," *Associated Press*, Washington, D.C., June 6, 1998; "Hackers Now Setting Their Sights on Pakistan," *Newsbytes*, June 5, 1998.

the site's domain name resolves to the Internet protocol address of some other site. When users point their browsers to the target site, they are redirected to the alternative site.

In what might have been one of the largest mass homepage takeovers, the antinuclear Milw0rm hackers were joined by Ashtray Lumberjacks hackers in an attack that affected more than 300 web sites in July 1998. According to reports, the hackers broke into the British Internet service provider (ISP) EasySpace, which hosted the sites. They altered the ISP's database so that users attempting to access the sites were redirected to a Milw0rm site, where they were greeted with a message protesting the nuclear arms race. The message concluded with ". . . use your power to keep the world in a state of PEACE and put a stop to this nuclear bullshit."[92]

Several web sites were hacked during the Kosovo conflict. According to Fox News, the *Boston Globe* reported that an American hacking group called Team Spl0it broke into government web sites and posted statements such as "Tell your governments to stop the war." Fox also said that the Kosovo Hackers Group, a coalition of European and Albanian hackers, had replaced at least five sites with black and red "Free Kosovo" banners.[93] The Bosnian Serb news agency SRNA reported that the Serb Black Hand hackers group had deleted data on a U.S. Navy computer, according to the Belgrade newspaper *Blic.* Members of the Black Hand group and Serbian Angel planned daily actions that would block and disrupt military computers operated by NATO countries, *Blic* wrote.[94] Black Hand had earlier claimed responsibility for crashing a Kosovo Albanian web site. "We shall continue to remove [ethnic] Albanian lies from the Internet," a member of the group told *Blic.*[95]

---

[92]Jim Hu, "Political Hackers Hit 300 Sites," *CNET,* July 6, 1998. The Milw0rm page is shown at www.antionline.com.

[93]Patrick Riley, "E-Strikes and Cyber-Sabotage: Civilian Hackers Go Online to Fight," Fox News, April 15, 1999.

[94]"Serb Hackers Reportedly Disrupt U.S. Military Computers," SRNA (Bosnian Serb news agency), March 28, 1999.

[95]"Serb Hackers Declare Computer War," *Associated Press*, October 22, 1998.

In the wake of NATO's accidental bombing of China's Belgrade embassy in May 1999, angry Chinese allegedly hacked several U.S. government sites. *Newsbytes* reported that the slogan "down with barbarians" was placed in Chinese on the home page of the U.S. Embassy in Beijing, while the Department of Interior web site showed images of the three journalists killed during the bombing, crowds protesting the attack in Beijing, and a fluttering Chinese flag.[96] According to the *Washington Post*, Interior spokesman Tim Ahearn said their computer experts had traced their hacker back to China. The newspaper also reported that the Department of Energy's home page read:

> Protest U.S.A.'s Nazi action! Protest NATO's brutal action! We are Chinese hackers who take no cares about politics. But we can not stand by seeing our Chinese reporters been killed which you might have know. Whatever the purpose is, NATO led by U.S.A. must take absolute responsibility. You have owed Chinese people a bloody debt which you must pay for. We won't stop attacking until the war stops![97]

NATO did not, of course, declare an end to the war because of the hacking. The effect on foreign policy decisions, if any at all, likely paled in comparison to the bombing itself. Following the accident, China suspended high-level military contacts with the United States.[98]

Acting in the name of democracy and human rights, hackers have targeted Chinese government computers. One group, called the Hong Kong Blondes, allegedly infiltrated police and security networks in an effort to monitor China's intelligence activities and warn political targets of imminent arrests.[99] According to OXblood Ruffin, "foreign minister" of the Cult of the Dead Cow, the Blondes are an under-

---

[96]Martyn Williams, "Federal Web Sites Under Attack After Embassy Bombing," *Newsbytes*, May 10, 1999.

[97]Stephen Barr, "Anti-NATO Hackers Sabotage 3 Web Sites," *Washington Post*, May 12, 1999.

[98]"China Suspends Contacts with U.S.," *Associated Press*, Beijing, May 9, 1999.

[99]Niall McKay, "China: The Great Firewall," *Wired News*, December 1, 1998. See also Sarah Elton, "Hacking in the Name of Democracy in China," *The Toronto Star*, July 4, 1999.

ground group of Chinese dissidents who aim to destabilize the Chinese government. They have threatened to attack both Chinese state-owned organizations and Western companies investing in the country.[100]

The *Los Angeles Times* reported that a California computer science student who calls himself Bronc Buster and his partner Zyklon cracked the Chinese network, defacing a government-run web site on human rights and interfering with censorship. The hackers said they came across about 20 firewall servers blocking everything from Playboy.com to Parents.com, and that they disabled the blocking on five of the servers. He said they did not destroy any data, but only moved files.[101]

Bronc Buster belonged to a group of 24 hackers known as the Legion of the Underground (LoU). In a press conference on Internet Relay Chat (IRC) in late December 1998, an LoU member declared cyberwar on the information infrastructures of China and Iraq. He cited civil rights abuses and said LoU called for the complete destruction of all computer systems in China and Iraq.[102]

The declaration of cyberwar prompted a coalition of other hacking groups to lash out against the campaign in early 1999. A letter co-signed by *2600*, the Chaos Computer Club, the Cult of the Dead Cow (CDC), !Hispahak, L0pht Heavy Industries, Phrack, Pulhas, and several members of the Dutch hacking community denounced the cyberwar, saying "Declaring 'war' against a country is the most irresponsible thing a hacker group could do. This has nothing to do with hacktivism or hacker ethics and is nothing a hacker could be proud of." Reid Fleming of the CDC said "One cannot legitimately hope to improve a nation's free access to information by working to disable its data networks."[103]

---

[100]Neil Taylor, "CDC Says Hackers Are Threat," *IT Daily*, August 26, 1999.

[101]Maggie Farley, "Dissidents Hack Holes in China's New Wall," *Los Angeles Times*, January 4, 1999.

[102]See www.hacknews.com/archive.html?122998.html.

[103]Letter of January 7, 1999.

By the time the letter went out on January 7, LoU had already issued a statement that day saying that the declaration of war on IRC did not represent the position of the group.

> The LoU does not support the damaging of other nations' computers, networks or systems in any way, nor will the LoU use their skills, abilities or connections to take any actions against the systems, networks or computers in China or Iraq which may damage or hinder in any way their operations.[104]

Bronc Buster said the IRC declaration was issued by a member before he left and never came back.[105]

In August 1999, a cyberwar erupted between hackers in China and Taiwan. Chinese hackers defaced several Taiwanese and government web sites with pro-China messages saying Taiwan was and would always be an inseparable part of China. "Only one China exists and only one China is needed," read a message posted on the web site of Taiwan's highest watchdog agency.[106] Taiwanese hackers retaliated and planted a red and blue Taiwanese national flag and an anti-Communist slogan—"Reconquer, Reconquer, Reconquer the Mainland"—on a Chinese high-tech Internet site. The cyberwar followed an angry exchange by Chinese and Taiwanese in response to Taiwan's President Lee Teng-hui's statement that China must deal with Taiwan on a "state-to-state" basis.[107]

One of the consequences of hacking is that victims might falsely attribute an assault to a foreign government rather than the small group of activists that actually conducted it. This could strain foreign relations or lead to a more serious conflict.

The Chinese government has been accused of attacking a U.S. web site devoted to the Falun Gong meditation sect, which has been outlawed by Chinese authorities. Bob McWee, a sect practitioner in Mid-

---

[104]Statement of January 7, 1999.

[105]James Glave, "Confusion Over 'Cyberwar,'" *Wired News*, January 12, 1999.

[106]"Pro-China Hacker Attacks Taiwan Government Web Sites," *Reuters*, August 9, 1999.

[107]Annie Huang, "Hackers' War Erupts Between Taiwan, China," *Associated Press*, Taipei, Taiwan, August 9, 1999.

dleton, Maryland, said his site had been under a persistent electronic assault in July 1999. In addition to a continuous denial-of-service attack, someone had tried breaking into his server. He said he was able to trace the penetration attempt to the Internet Monitoring Bureau of China's Public Security Ministry.[108] According to the *South China Morning Post*, additional attacks took place in April 2000 against at least five Falun Gong sites: three in the United States and two in Canada. The group's main web site, www.Falundafa.org, had received an anonymous tip-off warning that the police software security bureau had offered to pay a computer company money to hack into the sites.[109] If these attack did indeed originate with the Chinese police, this would have major foreign policy implications. It would suggest that the Chinese government views web sites operating on foreign soil as legitimate targets of aggression when those sites support activities prohibited on home soil.

Web hacks and computer break-ins are extremely common, and targets include commercial and educational computers as well as government ones. The results of *Information Security* magazine's "1999 Industry Survey" showed that the number of companies experiencing penetrations jumped from 12 percent in 1997 to 23 percent in 1998 (almost double).[110] About 26 percent of respondents to the "ERRI/EmergencyNet News Local/County/State Computer 'Hacking' Survey" said they thought they had been the victims of an unauthorized intrusion or attack on their computer systems.[111] And 30 percent of respondents to the "1999 CSI/FBI Computer Crime and Security Survey" reported intrusions from outsiders.[112] Most of the attacks, however, were probably not motivated by politics (hacktivism), but rather by thrills, curiosity, ego, revenge, or financial gain. In the area of web

---

[108]"Beijing Tries to Hack U.S. Web Sites," *Associated Press*, July 30, 1999. McWee's web site is at www.falunusa.net.

[109]"Web Sites of Falun Gong Hit," *South China Morning Post*, April 14, 2000.

[110]See www.infosecuritymag.com/articles/1999/julycover.shtml/.

[111]Clark Staten, private email, July 19, 1999.

[112]Richard Power, "1999 CSI/FBI Computer Crime and Security Survey," *Computer Security Issues & Trends*, Vol. V, No. 1, Winter 1999.

hacks alone, Attrition.org recorded more than 1,400 cases of vandalism by July 1999 for the year.[113]

Computer Viruses and Worms

Hacktivists have used computer viruses and worms to spread protest message and damage target computer systems. Both are forms of malicious code that infect computers and propagate over computer networks. The difference is that a worm is an autonomous piece of software that spreads on its own, whereas a virus attaches itself to other files and code segments and spreads through those elements, usually in response to actions taken by users (e.g., opening an email attachment). The boundary between viruses and worms, however, is blurry and not important to the discussion here.

The first protest to use a worm occurred about a decade ago, when antinuclear hackers released a worm into the U.S. National Aeronautics and Space Administration's SPAN network. On October 16, 1989, scientists logging into computers at NASA's Goddard Space Flight Center in Greenbelt, Maryland, were greeted with a banner from the WANK worm (see Figure 8.1).

At the time of the attack, antinuclear protestors were trying to stop the launch of the shuttle that carried the Galileo probe on its initial leg to Jupiter. Galileo's 32,500-pound booster system was fueled with radioactive plutonium. John McMahon, protocol manager with NASA's SPAN office, estimated that the worm cost them up to half a million dollars of wasted time and resources. It did not have its intended effect of stopping the launch. The source of the attack was never identified, but some evidence suggested that it might have come from hackers in Australia.[114]

Computer viruses have been used to propagate political messages and, in some cases, cause serious damage. In February 1999, the *London Sunday Telegraph* reported that an Israeli teen had become a national hero after he claimed to have wiped out an Iraqi government

---

[113]Ted Bridis, "Hackers Become an Increasing Threat," *Associated Press*, July 7, 1999.
[114]Ibid.

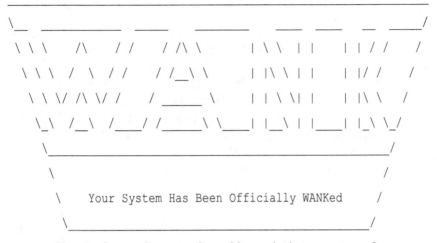

Figure 8.1—WANK Worm

web site. "It contained lies about the United States, Britain and Israel, and many horrible statements against Jews," 14-year-old Nir Zigdon said.[115] "I figured that if Israel is afraid of assassinating Saddam Hussein, at least I can try to destroy his site. With the help of some special software I tracked down the site's server to one of the Gulf states."[116] The Tel Aviv hacktivist then sent a computer virus in an email attachment to the site. "In the e-mail message, I claimed I was a Palestinian admirer of Saddam who had produced a virus capable of wiping out Israeli websites," Zigdon said. "That persuaded them to open the message and click on the designated file. Within hours the site had

---

[115]Tom Gross, "Israeli Claims to Have Hacked Saddam Off the Net," *London Sunday Telegraph*, February 7, 1999.
[116]Ibid.

been destroyed. Shortly afterwards I received an e-mail from the site manager, Fayiz, that told me to 'go to hell.'"[117]

During the Kosovo conflict, businesses, public organizations, and academic institutes received virus-laden emails from a range of Eastern European countries, according to mi2g, a London-based Internet software company. "The contents of the messages are normally highly politicised attacks on NATO's unfair aggression and defending Serbian rights using poor English language and propaganda cartoons," the press release said. It went on to say "The damage to the addressee is usually incorporated in several viruses contained within an attachment, which may be plain language or anti-NATO cartoon."[118] In an earlier press release, mi2g warned that "The real threat of cyber warfare from Serbian hackers is to the economic infrastructure of NATO countries and not to their better prepared military command and control network."[119]

It is extremely difficult, perhaps impossible, for an organization to prevent all viruses, because users unwittingly open email attachments with viruses and spread documents with viruses to colleagues. Although antiviral tools can detect and eradicate viruses, the tools must be kept up-to-date across the enterprise, which may have tens of thousands of computers, and they must be installed and used properly. While viruses bearing political messages may not seem to pose a serious problem, an organization hit by one may have to shut down services to eradicate it from its network.

Viruses, especially those carrying destructive payloads, are a potentially potent tool in the hands of cyberterrorists. Other tools of hacktivism, including computer network attacks, could likewise be put to highly destructive ends. This is the topic discussed next.

---

[117]Ibid.

[118]*mi2g Cyber Warfare Advisory Number 2*, April 17, 1999, M2 Communications, April 19, 1999.

[119]M2 Communications, April 8, 1999.

CYBERTERRORISM

In the 1980s, Barry Collin, a senior research fellow at the Institute for Security and Intelligence in California, coined the term "cyberterrorism" to refer to the convergence of cyberspace and terrorism.[120] Mark Pollitt, special agent for the FBI, offers a working definition:

> Cyberterrorism is the premeditated, politically motivated attack against information, computer systems, computer programs, and data which result in violence against noncombatant targets by subnational groups or clandestine agents.[121]

Politically motivated attacks that cause serious harm, such as severe economic hardship or sustained loss of power or water, might also be characterized as cyberterrorism.

The Threat

As previously discussed, terrorist groups are using the Internet extensively to spread their message and to communicate and coordinate action. However, there have been few, if any, computer network attacks that meet the criteria for cyberterrorism. The 1998 email bombing by the Internet Black Tigers against the Sri Lankan embassies was perhaps the closest thing to cyberterrorism that has occurred so far, but the damage cause by the flood of email, for example, pales in comparison to the deaths of 240 people from the physical bombings of the U.S. embassies in Nairobi and Dar es Salaam in August of that year.

Is cyberterrorism the way of the future? For a terrorist, it would have some advantages over physical methods. It could be conducted remotely and anonymously, it would be cheap, and it would not require the handling of explosives or a suicide mission. It would likely garner extensive media coverage, since journalists and the public alike are fascinated by practically any kind of computer attack. One highly ac-

---

[120]Barry Collin, "The Future of Cyberterrorism," *Crime and Justice International*, March 1997, pp. 15–18.

[121]Mark M. Pollitt, "Cyberterrorism: Fact or Fancy?" *Proceedings of the 20th National Information Systems Security Conference*, October 1997, pp. 285–289.

claimed study of the risks of computer systems began with a paragraph that concludes "Tomorrow's terrorist may be able to do more with a keyboard than with a bomb."[122] However, there are also drawbacks to terrorists using cyberweapons over physical ones. Because systems are complex, it may be harder to control an attack and achieve a desired level of damage. Unless people are injured, there is also less drama and emotional appeal. Further, terrorists may be disinclined to try new methods unless they see their old ones as inadequate.[123]

In a 1997 paper, Collin describes several possible scenarios for cyberterrorism. In one, a cyberterrorist hacks into the processing control system of a cereal manufacturer and changes the levels of iron supplement. A nation of children get sick and die. In another, a cyberterrorist attacks the next generation of air traffic control systems. Two large civilian aircraft collide. In a third, a cyberterrorist disrupts banks, international financial transactions, and stock exchanges. Economic systems grind to a halt, the public loses confidence, and destabilization is achieved.[124]

Analyzing the plausibility of Collin's hypothetical attacks, Pollitt concludes that there is sufficient human involvement in the control processes used today that cyberterrorism does not *at present* pose a significant risk in the classical sense. In the cereal contamination scenario, for example, he argues that the quantity of iron (or any other nutritious substance) that would be required to become toxic is so large that assembly line workers would notice. They would run out of iron on the assembly line and the product would taste different and not good. In the air traffic control scenario, humans in the loop would

---

[122]National Research Council, *Computers at Risk*, Washington, D.C.: National Academy Press, 1991.

[123]Kevin Soo Hoo, Seymour Goodman, and Lawrence Greenberg, "Information Technology and the Terrorist Threat," *Survival*, Vol. 39, No. 3, Autumn 1997, pp. 135–155. See also unpublished RAND research by Martin Libicki, James Mulvenon, and Zalmay Khalilzad, on information warfare, which examines these issues in some detail.

[124]Barry Collin, "The Future of Cyberterrorism," *Crime and Justice International*, March 1997, pp. 15–18. In terms of scenario building, RAND's longstanding "The Day After . . . in Cyberspace" research effort has done much pioneering work. See, for example, Roger C. Molander, Andrew S. Riddile, and Peter A. Wilson, *Strategic Information Warfare: A New Face of War*, Santa Monica, Calif.: RAND, 1996.

notice the problems and take corrective action. Pilots, he says, are trained to be aware of the situation, to catch errors made by air traffic controllers, and to operate in the absence of any air traffic control at all.[125] Pollitt does not imply by his analysis that computers are safe and free from vulnerability. To the contrary, his argument is that despite these vulnerabilities, because humans are in the loop, a cyberattack is unlikely to have such devastating consequences. He concludes:

> As we build more and more technology into our civilization, we must ensure that there is sufficient human oversight and intervention to safeguard those whom technology serves.

There is little concrete evidence of terrorists preparing to use the Internet as a venue for inflicting grave harm. However, in February 1998, Clark Staten, executive director of the Emergency Response & Research Institute in Chicago, testified that it was believed that

> members of some Islamic extremist organizations have been attempting to develop a "hacker network" to support their computer activities and even engage in offensive information warfare attacks in the future.[126]

And in November, the *Detroit News* reported that Khalid Ibrahim, who claimed to be a member of the militant Indian separatist group Harkat-ul-Ansar, had tried to buy military software from hackers who had stolen it from U.S. Department of Defense computers they had penetrated. Harkat-ul-Ansar, one of the 30 terrorist organizations on the State Department list, declared war on the United States following the August cruise-missile attack on a suspected terrorist training camp in Afghanistan run by Osama bin Laden, which allegedly killed nine of their members. The attempted purchase was discovered when an 18-year-old hacker calling himself Chameleon attempted to cash a $1,000 check from Ibrahim. Chameleon said he did not have the soft-

---

[125]Mark M. Pollitt, "Cyberterrorism: Fact or Fancy?" *Proceedings of the 20th National Information Systems Security Conference*, October 1997, pp. 285–289.

[126]Clark L. Staten, testimony before the Subcommittee on Technology, Terrorism and Government Information, U.S. Senate Judiciary Committee, February 24, 1998.

ware and did not give it to Ibrahim, but Ibrahim may have obtained it or other sensitive information from one of the many other hackers he approached.[127]

Given that there are no instances of cyberterrorism, it is not possible to assess the effects of such acts. It is equally difficult to assess potential damage, in part because it is hard to predict how a major computer network attack, inflicted for the purpose of affecting national or international policy, would unfold. So far, damage from attacks committed for reasons other than terrorism—for example, to seek revenge against a former employer—have generally been confined to immediate targets. No lives have been lost.

Cyberdefense

The main effect of cyberthreats on foreign and domestic policy relates to defending against such acts, particularly attacks against critical infrastructures. At the international level, several countries, including the United States, have been addressing such issues as mutual legal assistance treaties, extradition, the sharing of intelligence, and the need for uniform computer crime laws so that cybercriminals can be successfully investigated and prosecuted even when their crimes cross international borders, as they so often do. This effort is not focused on either cyberterrorism or hacktivism, but rather addresses an array of actions that includes all forms of hacking and computer network attacks, computer and telecommunications fraud, child pornography on the Net, and electronic piracy (software, music, etc.). It also covers state-sponsored cyberwarfare operations that use hacking and computer network attacks as a military weapon.

At the initiative of the Russian Federation, the U.N. General Assembly adopted a resolution related to cybercrime, cyberterrorism, and cyberwarfare in December 1998. Resolution 53/70, Developments in the Field of Information and Telecommunications in the Context of International Security, invites member states to inform the secretary-general of their views and assessments on (a) the issues of information security, (b) definition of basic notions related to information

---

[127]"'Dangerous' Militant Stalks Internet," *Detroit News*, November 9, 1998.

security, and (c) advisability of developing international principles that would enhance global information and telecommunications systems and help combat information terrorism and criminality.[128]

The United States has taken several steps to better protect its critical infrastructures. In July 1996, President Clinton announced the formation of the President's Commission on Critical Infrastructure Protection (PCCIP) to study the critical infrastructures that constitute the life support systems of the nation, determine their vulnerabilities to a wide range of threats, and propose a strategy for protecting them in the future. Eight infrastructures were identified: telecommunications, banking and finance, electrical power, oil and gas distribution and storage, water supply, transportation, emergency services, and government services. In its final report, issued in October 1997, the commission reported that the threats to critical infrastructures were real and that, through mutual dependence and interconnectedness, they could be vulnerable in new ways. "Intentional exploitation of these new vulnerabilities could have severe consequences for our economy, security, and way of life."[129]

The PCCIP noted that cyberthreats have changed the landscape:

> In the past we have been protected from hostile attacks on the infrastructures by broad oceans and friendly neighbors. Today, the evolution of cyberthreats has changed the situation dramatically. In cyberspace, national borders are no longer relevant. Electrons don't stop to show passports. Potentially serious cyberattacks can be conceived and planned without detectable logistic preparation. They can be invisibly reconnoitered, clandestinely rehearsed, and then mounted in a matter of minutes or even seconds without revealing the identity and location of the attacker.[130]

---

[128]G.A. Res. 53/70, U.N. GAOR, 53rd Sess., U.N. Doc. A/RES/53/70.

[129]*Critical Foundations: Protecting America's Infrastructures,* The Report of the President's Commission on Critical Infrastructure Protection, October 1997, report summary, www.pccip.gov.

[130]Ibid.

In assessing the threat from both physical and cyberattacks, the PCCIP concluded that

> Physical means to exploit physical vulnerabilities probably remain the most worrisome threat to our infrastructures *today*. But almost every group we met voiced concerns about the new cyber vulnerabilities and threats. They emphasized the importance of developing approaches to protecting our infrastructures against cyberthreats *before* they materialize and produce major system damage.[131]

The recommendations of the PCCIP led to Presidential Decision Directive 63, which established the National Infrastructure Protection Center (NIPC), the Critical Infrastructure Assurance Office (CIAO), the National Infrastructure Assurance Council (NIAC), and private-sector Information Sharing and Assessment Centers (ISACs).[132] The Department of Defense also established a joint task force—Computer Network Defense. The National Security Council's "National Plan for Information System Protection," issued in January 2001, provides an update on the most recent developments in this area.

That critical systems are potentially vulnerable to cyberattacks was underscored by a June 1997 exercise, code-named Eligible Receiver, conducted by the National Security Agency (NSA). The objective was to determine the vulnerability of U.S. military computers and some civilian infrastructures to a cyberattack. According to reports, two-man teams targeted specific pieces of the military infrastructure, including the U.S. Pacific Command in Hawaii, which oversees 100,000 troops in Asia. One person played the role of the attacker, while another observed the activity to ensure that it was conducted as scripted. Using only readily available hacking tools that could easily be obtained from the Internet, the NSA hackers successfully gained privileged access on numerous systems. They concluded that the military infrastructure could be disrupted and possible troop deployments hindered. The exercise also included written scenarios against

---

[131]Ibid.

[132]*Protecting America's Critical Infrastructures: PDD 63*, The White House, May 22, 1998. See also the White Paper *The Clinton Administration's Policy on Critical Infrastructure Protection: Presidential Decision Directive 63*, May 22, 1998, and "National Infrastructure Assurance Council," Executive Order, The White House, July 14, 1999.

the power grid and emergency 911 system, with resulting service disruptions. For the latter, they postulated that by sending sufficient emails to Internet users telling them the 911 system had a problem, enough curious people would phone 911 at once to overload the system. No actual attacks were made against any civilian infrastructures.[133]

The vulnerability of commercial systems to cyberattacks is repeatedly demonstrated by survey results such as those mentioned earlier. There is no evidence that nongovernment systems are any more or less vulnerable than government ones, or that the security posture of either group, as a whole, is generally improving—despite the availability and use of a growing supply of information security tools.

CONCLUSIONS

The Internet is clearly changing the landscape of political discourse and advocacy. It offers new and inexpensive methods for collecting and publishing information, for communicating and coordinating action on a global scale, and for reaching out to policymakers. It supports both open and private communication. Advocacy groups and individuals worldwide are taking advantage of these features in their attempts to influence foreign policy.

Several case studies show that when the Internet is used in normal, nondisruptive ways, it can be an effective tool for activism, especially when it is combined with other media, including broadcast and print media, and face-to-face meetings with policymakers. As a technology for empowerment, the Net benefits individuals and small groups with few resources as well as organizations that are large or well-funded. It facilitates such activities as educating the public and media, raising money, forming coalitions across geographical boundaries, distributing petitions and action alerts, and planning and coordinating events on a regional or international level. It allows activists in politically repressive states to evade government censors and monitors.

---

[133]*CIWARS Intelligence Report*, Centre for Infrastructural Warfare Studies, June 21, 1998; "Pentagon Computer Systems Hacked," *Info Security News*, June 1998; Douglas Pasternak and Bruce B. Auster, "Terrorism at the Touch of a Keyboard," *U.S. News & World Report*, July 13, 1998, p. 37.

In the area of hacktivism, which involves the use of hacking tools and techniques of a disruptive nature, the Internet will serve mainly to draw attention to a cause, since such incidents are regularly reported by news media. Whether that attention has the desired effect of changing policy decisions related to the issue at hand is much less certain. Hacktivists may feel a sense of empowerment, because they can control government computers and get media attention, but that does not mean they will succeed in changing policy. So far, anecdotal evidence suggests that for the majority of cases, they will not.

With regard to cyberterrorism, that is, the use of hacking tools and techniques to inflict grave harm such as loss of life, few conclusions can be drawn about its potential effect on foreign policy, because there have been no reported incidents that meet the criteria. What can be said is that the threat of cyberterrorism, combined with hacking threats in general, is influencing policy decisions related to cyberdefense at both a national and international level. If we look at terrorism in general for insights into the potential effects of cyberterrorism, we find that the effect of terrorism on the foreign policy issues at hand is similarly difficult to assess, but here again, the threat of terrorism, particularly chemical, biological, and nuclear terrorism, is having a significant effect on national defense policy.

Chapter Nine

THE STRUCTURE OF SOCIAL MOVEMENTS: ENVIRONMENTAL ACTIVISM AND ITS OPPONENTS

Luther P. Gerlach

Editors' abstract. Get ready for the "SPIN cycle." Gerlach (University of Minnesota) provides an excellent summary on the organizational and strategic dynamics that characterize all manner of "segmented, polycentric, integrated networks" found in American social movements. This is one of the few studies that discusses social movements from a thoroughgoing network perspective. We believe that many of his observations also apply across the range of "uncivil-society" actors. This chapter stems from his contribution[1] to Jo Freeman's and Victoria Johnson's edited volume, Waves of Protest *(1999), Lanham, Mass.: Rowman and Littlefield, a study of social movements since the 1960s. Reprinted by permission.*

In the late 1960s Virginia H. Hine and I examined the structure of several social movements. We found that the most common type of organization was neither centralized and bureaucratic nor amorphous, but one that was a segmentary, polycentric, and integrated network (acronym SPIN) (Gerlach and Hine, 1970, 1973; Gerlach, 1971/1983).

- Segmentary: Composed of many diverse groups, which grow and die, divide and fuse, proliferate and contract.

- Polycentric: Having multiple, often temporary, and sometimes competing leaders or centers of influence.

- Networked: Forming a loose, reticulate, integrated network with multiple linkages through travelers, overlapping membership,

---

[1]Luther Gerlach thanks Jo Freeman for her skillful assistance in editing this paper.

joint activities, common reading matter, and shared ideals and opponents.

We proposed that this segmentary, polycentric, and networked organization was more adapted to the task of challenging and changing society and culture than was centralized organization. At the time, even social movement participants did not fully appreciate the strengths of SPIN organization, believing that anything other than a centralized bureaucracy was either disorganized or an embryonic organization. Since then a consensus has emerged that SPINs have many benefits, and not just for social movements. This chapter revisits and supplements our analysis. Although examples abound from many movements since the 1960s, I will feature examples from the environmental movement (once called the ecology movement), and the Wise Use (property rights) movement, which opposes environmental activism. First we will examine each characteristic of a SPIN.

SEGMENTARY

Social movements have many organizationally distinct components that change through fission, fusion, and new creation. A typical SPIN is composed of semiautonomous segments. New segments are created by splitting old ones, by appending new segments, or by splitting and adding new functions. Segments overlap and intertwine complexly, so that many people are members of several segments at the same time. A person may be a leader in one segment and a follower in another.[2] When we examined what we then called the "participatory ecology" movement in 1969–1970, we found that movement groups included the following:

*   Regional and local branches of bureaucratically structured national and international institutions that had been founded many years previously. These included the Sierra Club (1892), the Audubon Society (1905), the Wilderness Society (1935), the National Wildlife Federation (1936), and the Isaac Walton League (1922).

---

[2]In being so differentiated in structure and role, these movement segments are unlike the segments of classical segmentary lineage systems in "tribal" Africa, which are like each other and unspecialized.

- Recently formed alternatives to these established institutions, notably the Environmental Defense Fund (1967), Friends of the Earth (1968), and Zero Population Growth (1968).

- A plethora of groups even more radical in ideology and/or tactics, with names such as the People's Architects, the Food Conspiracy, Ecology Action, Ecology Freaks, and Ecology Commandoes.

- A mushrooming array of small and local groups that people were forming in communities across the country to challenge the construction in their neighborhoods of power plants, jetports, dams, incinerators and other industrial facilities, and real estate development projects.

The ecology movement continued to move, grow, change, and promote change. Sometime in the late 1970s people began to refer to it as environmentalism. In the late 1980s and early 1990s the European term "the Greens" became popular. Today, what remains is usually called the environmental movement or the Green movement.

Why Groups Divide

Some groups divide when their participants differ over ideology and tactics. Some deliberately spin off new cells. Others are created by new people inspired by movement ideology or provoked by similar conditions. We have identified four factors that contribute to this process of segmentation.

1. *Personal power* is often a component of movement belief systems. In charismatic religious movements, participants believe that they can have direct access to God and this will empower them. The environmental movement exhorts participants to "think globally, act locally," and not to be dissuaded by those who claim that there is not enough expert agreement to support action. In all of these movements, individuals and small or local groups each feel the need to take the initiative in achieving those movement goals the person or group considers important. They don't wait to be asked. This helps produce divisions among persons and groups over ideology and tactics. It also motivates these participants to recruit others to support their competing ventures. For example, the Earth Liberation Front (ELF), a radical group that advocates and takes credit for the sabotage

of development projects that threaten the natural environment, advertises on its web site that anyone can begin taking such action and do so in its name. This, of course, could have the effect of producing many new saboteurs.

2. *Preexisting cleavages* derived from socioeconomic differences, factionalism, and personal conflicts are often brought into a group and increase its fissiparous tendencies.

3. *Competition* among movement members, especially leaders, for economic, political, social, and psychological rewards. Benefits include followers, media attention, influence, funds from foundations and government, and the satisfaction of knowing that they are advancing movement goals. Competition causes factions, realigns followers, and intensifies efforts to recruit new participants and broaden the base of support.

4. *Ideological differences* are a major source of new groups. A strongly committed movement participant experiences an intensity of concern for ideological purity that people ordinarily feel only for threats to their personal or family well-being. Thus, for instance, environmental groups have regularly divided over disagreements about how much conventional culture and society must be changed to protect the environment, and how militant the tactics must be to achieve such changes. Some groups also split over how much to couple environmental protection with other issues, such as social reform or multinational corporations.

Most division occurs during the growth phase of a movement and contributes to its expansion, but it may occur at any time. Although they vary, new groups are often small and decentralized. Many make and implement decisions through consensus, while others are driven by strong individuals, if only temporarily. Sometimes existing organizations that are large and bureaucratic become movement groups, often undergoing profound changes in the process. Groups that are more likely to split pursue radicalism, reject authority, and/or reject organization. Despite the frequency of fission, most groups are dismayed when they split. But some embrace it. Redwood Summer (RS) was launched by Earth First! in 1989–1990 to protest the logging of California redwoods. Its coalition included peace and justice as well as environmental and local watershed management groups; soon it became a separate entity. Participants were expected to take what they learned back to their own states and form distinct groups beholden to no one but themselves (Pickett, 1990, p. 8).

Having a variety of groups permits a social movement to do different things and reach out to different populations. Some of the participants in RS switched from protesting in the field to working for a referendum on protection of California forests. They then organized, temporarily, as the Environmental Protection Information Center, which functioned essentially as a coordinator of a new array of groups working to get voter support for their proposal. Moreover, although in the mid-1990s Earth First! renounced the use of violence to disrupt logging and resource development, other activists, notably Earth Liberation Front, emerged to advocate sabotage of projects that to them represent environmental risk.

Sometimes movements beget countermovements, which are themselves segmentary. Environmental activism prompted the "Wise Use" or property rights movement. Loggers, mill workers, ranchers, farmers, miners, natural resource developers, snowmobile and dirt-bike riders, property owners, libertarians, populists, political conservatives, and some religious fundamentalists organized into many different and often localized groups. They include the Center for the Defense of Free Enterprise, which takes a comprehensive approach to property use; the National Inholders/Multiple-Use Land Alliance, best known for demanding access to national parks and federal lands; the Blue Ribbon Coalition of recreation and off-road vehicle users in Pocatello, Idaho; the Women's Mining Coalition in Montana; the Pulp and Paper Workers Union; and the Pacific Legal Foundation in Sacramento, California. While coming from different places, they share the view that environmentalists and federal or state regulation of property in the name of ecosystem management threaten their interests. In the late 1980s and 1990s, they organized against such environmental legislation as wetland protection and the Endangered Species Act. They demanded multiple use of federal and state lands or the return of these lands to local or private ownership, with local management of natural resources.

POLYCENTRIC

By polycentric I mean that these movements have many leaders or centers of leadership, and that these many leaders are not ultimately directed or commanded through a chain of command under a central leader. The leaders, like the segments, are not organized in a hierarchy; they are "heterarchic." They do not have a commander in chief. There is no one person who can claim to speak for the movement as a whole, any more than there is one group that represents the movement.

Initially we termed these movements polycephalous, or "many headed," because the movements we studied in the 1960s had many leaders, and these were not organized in a hierarchical chain of command. We changed the term to polycentric because movement participants since the 1960s often claim to have no leaders and are dismayed when a situational leader appears to be translating inspiration and influence into command. But whatever the attitude toward leadership, social movements do have multiple centers of leadership.

While the press often picks out an individual to feature and quote, in reality it is rare for one person to be acknowledged by participants as the movement leader. Movement leaders are more likely to be charismatic than bureaucratic. People become leaders chiefly by inspiring and influencing others rather than by being chosen for their political or organizational skills. This leadership is usually situational, as leaders arise to cope with particular situations or episodic challenges in the life of a movement. Leaders must continue to prove their worth and are often challenged by rivals.

In the 1970s, Amory Lovins rose to prominence as an international leader in the movement to resist nuclear energy and to promote solar energy alternatives. In the late 1980s, Dave Foreman became famous as the leader of direct action to stop logging of old growth forests, and Petra Kelly captured attention as the leader of the West German Greens and an international exponent of environmentalism. In the 1990s, Ron Arnold, Alan Gottlieb, and Charles Cushman were recognized as founders and leaders of the Wise Use/property rights movement.

Although movement participants share common views, they also disagree. Different leaders reflect these disagreements. In the environmental movement, many believe that economic growth is incompatible with saving the environment; others think that growth and development can be sustained through efficient and benign or "soft" technologies. Some believe militant direct action is necessary to force change, and others want to work peacefully within established rules. Even in the relatively small Redwood Summer, some wanted to protest the U.S. action in the Persian Gulf War and attack big corporations, while others wanted to focus on stopping logging.

There is no one person or group able to make decisions that are binding upon all or even most of the participants in a movement. This makes negotiation and settlement difficult, if not impossible. Temporary leaders of a specific protest action may be able to reach agreement on concessions that will end the action, but they have no power to prevent anyone from launching new protests.

Leaders must carefully balance the need to demonstrate personal strength and ability with the need to heed the desire for democratic participation that drives many of their fellows. It is this desire that helps motivate movement participants to challenge established orders, and that is nourished in confrontation with such orders. Social movement groups are likely to try to make decisions by having everyone agree or consent, and leaders must learn how to work within this often long and laborious process. Thus a leader must often act as no more than a "first among equals," or *primus inter pares*. In the United States at any rate, such leaders find it useful to be known as someone who has risen up from the proverbial "grassroots," through ability and hard work, and who can speak for the people.

NETWORKED

The diverse groups of a movement are not isolated from each other. Instead, they form an integrated network or reticulate structure through nonhierarchical social linkages among their participants and through the understandings, identities, and opponents these participants share. Networking enables movement participants to exchange information and ideas and to coordinate participation in joint action.

Networks do not have a defined limit but rather expand or contract as groups interact or part ways.

Movement participants are not only linked internally, but with other movements whose participants share attitudes and values. Through these links, a movement can draw material support, recruit new supporters, and expand coordination for joint action. In the 1970s, protesters of nuclear and fossil fuel power plants and related electrical power lines formed alliances with organizations working to make and market solar and renewable energy technologies. In the 1980s, environmentalists established relationships with feminist, labor, and civil rights organizations to overcome their image as urban elites. In the 1990s, Wise Use members sought to expand beyond their rural and suburban bases in the West. They have joined with urban groups in eastern cities to oppose rent control laws and allied with the political and religious right to elect conservatives to public office.[3]

Linkages

Personal relationships connect participants in different groups through kinship, marriage, friendship, neighborliness, and other associations. Even if groups split, the personal connections remain. Often an individual will participate in more than one group. Leaders are particularly active in networking. Indeed, one way to become and remain a leader is to recruit participants and to link groups together, that is, to become a node connecting many groups.

Traveling evangelists and other visitors provide living links in the movement network. They carry information across the network, from group to group, and build personal relationships with those they visit. Many of the people recognized by the media as leaders of a movement are more accurately understood as its evangelists. Thus, while they may lead segments of a movement, directing the actions of its participants, they play a more significant role as those who evangelize across the movement as a whole. In a general sense, evangelists are those who zealously spread the ideology of any movement, promoting its ideas, reinforcing the beliefs of participants, exhorting them to

---

[3]Ron Arnold, personal communication, 1998.

action, and helping them recruit newcomers and form groups, raise funds, and mobilize against opponents. Many are recognized leaders, who draw crowds when they visit different places. Some are ordinary participants who write about their travels and visits in movement newsletters. In the late 1960s and early 1970s, students often traveled across the United States and western Europe by navigating movement networks, helped by local people who gave them contacts elsewhere.

It is not only those famed as evangelists who travel movement networks, visiting its various segments. Ordinary participants do this as well, and in the process also help to vitalize the networks and carry ideas to and fro across them, including by writing trip reports in movement newsletters. In the late 1960s and early 1970s, students involved in what was called the "counterculture," traveled across the United States and to Western Europe by navigating networks of what they sometimes called "affinity groups." They would be helped by people at local centers of information, called "switchboards." These would tell them where they could find affinity groups to live with for a time and where they might find new switchboards and affinity groups when they moved on. From the 1960s through the 1990s, people moved along such networks to participate in demonstrations or meetings launched by movement segments.

Large gatherings for conventions, conferences, workshops, "teach-ins," and demonstrations allow participants to learn and share ideas, and to act on them. Through participation in gatherings, people not only learn movement ideology and demonstrate their commitment to the cause, they also make or reestablish relationships with each other. Conversely, mobilizing people to attend gatherings through their local groups reinforces movement linkages. In recent years, a growing way of gathering and linking an array of diverse environmental, rights, and labor activists has been to call these activists to protest at meetings of world leaders and managers held to address global economic and resource issues, from global trade to climatic change and genetic engineering.

Communications technologies, such as the telephone, radio and television talk shows, letters, newsletters, and membership magazines, allow individuals to extend their reach far beyond their own group.

Since the mid-1980s, email and the Internet have been added to this repertoire. Individuals and groups reinforce and extend their relationships, consult with each other, and share information and interpretations. This helps them to coordinate their actions and act jointly, even over long distances. Email and the Internet provide the main channels through which people have been mobilized to protest at the global trade and resource management meetings mentioned above. People and groups, such as the Independent Media Center (IMC or indymedia), have been organizing in what itself has become a grassroots movement to produce and distribute reports via the Internet and other media about such protests (Stringer, 2000).

The web is aptly named. With the advent of the Internet, movement participants are now organizing "cyberconferences," or "virtual conferences," and exchanging information and ideas through email and web sites. Many environment/ecology groups have web sites, as do the Center for the Defense of Free Enterprise, a main coordinating node in the Wise Use network, and the Heartland Institute, a Wise Use think tank working to counter environmentalist positions on environmental risks such as global warming.[4]

Linkages and Information Sharing

Movement participants use these personal relationships, traveling evangelists, gatherings, and multimedia communication technologies to share the information that enables them to act in concert.

By the late 1980s, among the most important types of information shared by participants in both the environmentalist movement or the Wise Use/property rights movement were about the activities, ideas, "leaders," and organization of the other. One function of the conferences held by each was, indeed, to communicate this information and to arouse their fellows to expose and combat their opposition. It is interesting that each movement attacks the legitimacy of the other by claiming that the other is not really a popular movement, but instead a conspiracy. Thus, Wise Use members listen to speakers and read handouts detailing the activities of "ecogroups," showing how

---

[4]Web site addresses are listed at the end of this chapter.

these groups are "trashing the economy" and promoting "socialist" agendas, "neopaganism" or one world government under the guise of environmental protection. Environmentalists listen to speakers and read handouts claiming that the Wise Use movement is not a grass-roots organization, as its leaders claim, but instead the creature of a few manipulators and "slick foxes" fronting for big and polluting industry, and oil, mining, timber, and other resource development interests.

INTEGRATING FACTORS

The segments in social movement networks are also integrated by what they share or hold in common. These include a shared opposition and ideology. These factors complement each other and help constitute the culture of the movement.

Shared Opposition

The recognition or perception of an external opposition helps diverse movement groups to unite and to expand. A movement grows with the strength of its opposition, much as a kite flies against the wind. Opposition creates a sense of solidarity, an "us" against "them." In many instances, movement participants see their cause as a small and heroic David against the Goliath of the establishment. As "underdogs," they must put aside their differences and work together.

When movements face countermovements, such as the environmental versus the Wise Use movement, each wages a propaganda war against the other, using the threat of one to mobilize the other. Environmentalists warn that opponents are growing in power; their leaders "want to destroy the environmental movement" (Western States Center, 1993, p. 1), and their hidden supporters are the industrialists and developers who wish to exploit the environment for narrow economic gain. Wise Use theorists Arnold and Gottlieb argue in public addresses, in publications (Arnold and Gottlieb, 1993, pp. 53–77), and online (www.cdfe.org) that the environmental movement has close ties to government agencies and big foundations, a powerful combination requiring committed and united counteraction by the "citizen groups" sharing Wise Use ideas.

Both of these movements regularly research and "expose" each other. Wise Use member Charles Cushman alerts his National Inholders Association and Multiple-Use Land subscribers to environmentalist threats through his occasional newsletters. Barry Clausen, a private investigator hired by timber, mining, and ranching interests to investigate Earth First! wrote *Walking on the Edge: How I Infiltrated Earth First!* to expose it as an "ecoterrorist" organization that threatens the lives of loggers and miners and dupes ordinary environmentalists (Clausen, 1994). Dave Mazza, a professional investigator and environmental activist, wrote about the connection between Wise Use and the Christian Right movements (Mazza, 1993), while Carl Deal published *The Greenpeace Guide to Anti-Environmental Organizations* (1993). William Burke surveyed Wise Use activities in New England (1992), and the Wilderness Society commissioned MacWilliams Cosgrove Snider, a media and political communications firm, to study and report on the capabilities and limitations of the Wise Use movement (1993). Arnold and Gottlieb countered by claiming that the research was funded by a few private foundations not to advance understanding but rather to destroy the movement.

Since at least the 1990s, an increasingly broad array of environmental, rights, social justice, farm, and labor activists, as well as anticapitalist anarchists, have worked in various ways to define multinational corporations and international banking, trade, and economic-development organizations as threats to human welfare and environmental health, because of their pursuit of global economic integration and growth. These activists promulgate their ideas about these global threats through personal contact, print media, and especially the Internet.[5] Thus informed, the activists use major worldwide meetings of officials of the international organizations as forums to gather in protest and publicly communicate the threats they perceive. Their often militant demonstrations force responses from police and local governments, which then provide new opposition against which they can converge. One noted example took place in Seattle, Washington, from late November to early December 1999 at a meeting of the World

---

[5]See http://indymedia.org and http://www.zmag.org/globalism/globalecon.htm.

Trade Organization (see De Armond, 2000, and Chapter Seven of this volume).

Shared Ideology

Movement ideology operates on two levels. All participants share basic beliefs or core themes, which are sometimes articulated as slogans or aphorisms. The ecology movement used such concepts as ecosystem, interdependence, limited resources, renewable resources, spaceship earth, and no-growth economy. Wise Use has employed the concept of balance—the harmonizing of economy *and* ecology. At another level is a myriad of different interpretations and emphases on these themes. Disagreement may generate splits, but shared beliefs contribute to a sense of participating in a single movement. Sometimes these unifying tenets become master concepts that shape the discourse not only of movements, but of society as a whole. Once ecology passed into the popular lexicon the prefix "eco" became widely used to give new meaning to other words (e.g., ecofeminism). Sometimes beliefs or slogans change over time. Between the 1960s and the 1990s *ecology* became *environmentalism*, and after the Green movement became popular in Germany in the late 1970s and early 1980s, "green" became a synonym for both of these.

Core beliefs can be shared because they are ambiguous and flexible, and they vary locally because they can be changed situationally. In 1972, biologist and early environmental evangelist René Dubos coined the term "thinking globally, acting locally" to warn that programs to protect the global environment cannot easily be translated everywhere into local actions but must be tailored to suit local ecological, economic, and cultural conditions (Dubos, 1981). By the 1980s, environmentalists had given the phrase multiple meanings. Some used it to encourage people to act locally on environmental problems in expectation that actions would combine to produce desirable global results. Some used it to imply that global exigencies override local ones. Some used it to claim that local actions serving local causes helped meet the challenge of global poverty and pollution (Gerlach, 1991).

In their efforts to build alliances across movements, activists search for common interests and seek to express these through an encompassing ideology. For instance, environmentalists have advanced the theme of environmental justice to support collaboration with civil and human rights groups and probably also to overcome the criticism that environmentalism is a movement of the affluent. Thus, in January 2001 the Sierra Club announced on its web site that it is working with the Earth Day 2001 Network and Amnesty International to "highlight human rights abuses and environmental destruction associated with fossil fuel extraction." According to the announcement, these and other groups will work in a campaign

> to help the world's marginalized and powerless communities defend themselves against the alliances of multinational corporations with undemocratic and repressive regimes.[6]

Participants in the disparate movements protesting the environmental and social costs of economic globalization are also eschewing the idea that they are *against* globalization in favor of the idea that they are *for* global justice.

It is through their production and use of such ideas and symbols, and their ongoing efforts to reshape them to meet their evolving interests and changing challenges, that participants in movements help coordinate the actions of their various groups and collaborate with other movements. It is this that helps make it possible for groups to share leaders and evangelists, to coalesce temporarily for specific actions, to maintain a flow of financial and material resources through both bureaucratic and nonbureaucratic channels, to identify and organize against external opposition, and to unite in common purpose.

ADAPTIVE FUNCTIONS

The type of organization we here describe as SPIN has often been labeled disorganized, poorly organized, loosely organized, or underdeveloped—and thus it has been denigrated or criticized not only by opponents or observers but at one time by movement participants. A

---

[6]From the web site for the Sierra Club's human rights campaign.

common assessment has been that this type of organization as well as the movements themselves represent lower stages in organizational or cultural evolution. It is said that in time, groups or societies organized so loosely will evolve to become centralized bureaucracies or states, because centralized bureaucracies are more efficient, more adapted, more advanced. Our argument against this assessment is that SPINs exhibit a number of properties that are adaptive under certain conditions of turbulence.

The SPIN style of organization supports rapid organizational growth in the face of strong opposition, inspires personal commitment, and flexibly adapts to rapidly changing conditions. It is highly adaptive for the following reasons.

1. It prevents effective suppression by the authorities and the opposition. To the extent that local groups are autonomous and self-sufficient, some are likely to survive the destruction of others. This is also true of leaders; some will survive and even become more active and radical when others are removed, retired, or co-opted. For every group or leader eliminated, new ones arise, making movements look like the many-headed Hydra of mythology. It is difficult to predict and control the behavior of the movement by controlling only some of these components. In the 1960s and 1970s, authorities used the metaphor of trying to grab Jell-O to portray their difficulties in investigating and controlling a variety of protest movements. In 2001, an FBI agent used the same metaphor to describe efforts even to find members of the Earth Liberation Front.[7]

Even with suppression, burnout causes casualties. Having multiple groups limits the consequences of burnout. During the energy conflicts of the 1970s, when one group of power line protesters despaired of stopping line construction, another group took up the challenge.

2. Factionalism and schism aid the penetration of the movement into a variety of social niches. Factionalism along lines of preexisting socioeconomic or cultural cleavages supplies recruits from a wide range of backgrounds, classes, and interests. Groups can be formed in many different sectors or communities. Redwood Summer and Earth First!

---

[7]David S. Jackson, "When ELF Comes Calling," *Time*, January 15, 2001, p. 35.

recruited young adults who could afford to take personal risks. The Environmental Protection Information Center attracted Californians, mostly white, from the middle and upper classes, whose politics were more moderate than that of other environmental groups. Many Native American tribes are also organizing to protect their rights to natural resources. Some hunters of waterfowl and big game have allied themselves with nature conservancy and other environmental organizations to obtain and protect wetlands and other habitats for the wildlife they appreciate in different ways (Gerlach, 1995).

Wise Use is also diverse. It hosts groups of ranchers and farmers, loggers, miners, recreational vehicle users, land developers, other property rights advocates, and also hunters and anglers. While some Wise Use groups attract libertarians and free market advocates, others attract religious fundamentalists worrying that environmentalism is a type of neopaganism.

3. Multiplicity of groups permits division of labor and adaptation to circumstances. The greater the differentiation of groups, the more likely the movement is able to offer something for every sympathizer to do to further the movement's goals. In the environmental movement, some groups take direct physical action to prevent loggers from cutting down redwoods in Northern California or red and white pines in Minnesota, and other groups work with lawyers and public relations specialists to persuade courts and legislatures to block this logging. In Minnesota in 1997 and 1998, a group of Earth First! activists asked the founder of another group, Earth Protector, to use legal action to complement their direct demonstrations against logging in the Superior National Forest (Grow, 1998). In Northern California, opposition to logging old growth redwoods in the privately held Headwaters Forest has effectively included both the direct actions of Earth First! and the legal and legislative operations coordinated by the Environmental Protection Information Center. Wise Use has many segments: corporate and industry interests who contribute legal and financial resources, issue entrepreneurs who act as information clearinghouses and mobilizers of public responses, and individuals who worry that their way of life is threatened by environmental regulation (Switzer, 1996, 1997).

4. Segmentary, polycentric, and networked organization contributes to system reliability. Failure of one part does not necessarily harm the other parts since these are not connected (Landau, 1969). Instead, groups learn from failures and are free to disavow parts of the movement that fail. Just as one movement group is ready and able to take over the functions of another when it is no longer viable, so can a group disavow another if the latter's actions put the former at risk, or copy another if its actions prove successful.

5. Competition between groups leads to escalation of effort. When one group or leader attracts more attention than another, the latter often steps up its activities to regain prominence. When a movement group threatens established institutions, they may respond by negotiating with a more moderate group, making gains for the movement and building outposts in the established order. Often, the more threatening group accuses the dealmakers of selling out. This may motivate the latter to renew its militancy or to demand more from the establishment. The process repeats, opening new fronts while consolidating old gains. In the 1970s, a leader of a Minnesota branch of a mainline environmental organization, the Isaac Walton League, urged legislators to pass a bill to establish the Boundary Waters as a Canoe Area Wilderness and exclude motorized travel by saying that if the "kids" do not see that they can protect the environment by working within the system, they will join radical groups that act more "on emotion." In an interview, he said that he was motivated to lead this legislative action to prove wrong the young ecology activist who called him an "Uncle Tom" on conservation efforts.

6. SPIN organization facilitates trial-and-error learning through selective disavowal and emulation. Movement groups challenge established orders and conventional culture both in the ideas they espouse and in the tactics they use to promote these ideas. Through trial and error come social and cultural forms that prove to be successful and adaptive. Because the groups are connected in a network of social relationships and information flows, knowledge about successes and failures flows rapidly from one group to another. While some environmental groups alienated loggers and millers, others sought to work with these loggers and millers by arguing that it was the big corporations who had depleted the forests of harvestable trees to make short-

term profits. When the environmentalist call for no-growth was rejected by people because it appeared to threaten their economic opportunity and well-being, some environmental groups proposed the alternative idea of "sustainable development" through appropriate technology and resource management. When people worried about the consequences of not building more nuclear or fossil fuel energy facilities, proponents of solar energy took this as an opportunity to make and market solar technologies.

7. SPIN promotes striving, innovation, and entrepreneurial experimentation in generating and implementing sociocultural change. Environmental groups led the way in promoting the conservation of resources and recycling of waste. They developed new approaches to teaching about ecology; involved children and adults in monitoring water quality in lakes and rivers; helped to persuade government, foundations, and private firms to institutionalize new approaches; pushed industries to use less-polluting and more-resource-efficient technologies; and pushed government to legislate environmental protection. By militantly resisting fossil and nuclear technologies and promoting the use of solar energy technologies, environmental groups have encouraged both government and private industry to rethink the future of nuclear energy (Gerlach, 1978, 1979; Gerlach and Eide, 1978). Since the 1980s, environmental groups across the world have taken the lead in warning governments and public bodies about the causes and risks of global climatic and other environmental change, and have done much to promote agreements among nations to reduce and control emissions into the atmosphere and oceans and other great bodies of water. Wise Use has both complemented and challenged the environmentalist agenda. Its demands for inclusion in decisionmaking have helped open the process and fostered a debate over how to balance environmental protection and economic development and established the idea that development should be sustainable not only ecologically, but also socially and politically (Gerlach and Bengston, 1994).

CONCLUSION

Social movements that are segmentary, polycentric, and networked have a very effective form of organization. In particular, this form helps its participants to challenge and change the established order and to survive overwhelming opposition. It makes the movement difficult to suppress; affords maximum penetration of and recruitment from different socioeconomic and subcultural groups; contributes to system reliability through redundancy, duplication, and overlap; maximizes adaptive variation through diversity of participants and purposes; and encourages social innovation and problem solving. SPINs may well be the organizational form of the global future, the one best suited to reconcile the need to manage globally and locally, comprehensively and democratically, for the common good as well as individual interest, institutionalizing ecological and economic interdependence as well as ethnolocal independence.

BIBLIOGRAPHY

Arnold, Ron, and Alan Gottlieb. 1993. *Trashing the Economy: How Runaway Environmentalism Is Wrecking America*. Bellevue, Wash.: Distributed by Merril Press.

Burke, William Kevin. 1992. *The Scent of Opportunity: A Survey of the Wise Use/Property Rights Movement in New England*. Cambridge, Mass.: Political Research Associates, December.

Clausen, Barry, with Dana Rae Pomeroy. 1994. *Walking on the Edge: How I Infiltrated Earth First!* Olympia: Washington Contract Loggers Association, distributed by Merril Press.

Cushman, Charles. 1980s. *National Inholders Association (NIA), Multiple-Use Land Alliance (MULTA) Newsletter*. Bellevue, Wash.: Occasionally published since the early 1980s.

De Armond, Paul, *Netwar in the Emerald City*, Public Good Project (Bellingham, Wash.), February 2000, www.nwcitizen.com/publicgood/reports/wto.

Deal, Carl. 1993. *The Greenpeace Guide to Anti-Environmental Organizations*. Berkeley, Calif.: Odonian Press.

Dubos, René. 1981. "Think Globally, Act Locally," in *Celebrations of Life*. New York: McGraw-Hill.

Environmental Protection Information Center (EPIC). 1997. *Wild California, A Newsletter of the Environmental Protection Information Center,* Spring, p. 2. (P.O. Box 397, Garberville, Calif. 95542).

Freeman Jo, and Victoria Johnson, eds. 1999. *Waves of Protest: Social Movements Since the Sixties,* Lanham, Mass.: Rowman and Littlefield.

Gerlach, Luther P. 1971. "Movements of Revolutionary Change: Some Structural Characteristics." *American Behavioral Scientist* 14: 812–836. Abridged version in Jo Freeman, ed. 1983. *Social Movements of the Sixties and Seventies,* New York: Longman.

Gerlach, Luther P. 1978. "The Great Energy Standoff." *Natural History* 87 (January).

Gerlach, Luther P. 1979. "Energy Wars and Social Change," in *Predicting Sociocultural Change,* Susan Abbot and John van Willigen, eds. Southern Anthropological Society Proceedings #13. Athens: University of Georgia Press.

Gerlach, Luther P. 1991. "Global Thinking, Local Acting: Movements to Save the Planet: Evaluation Review," special issue, *Managing the Global Commons* 15, no. I (February).

Gerlach, Luther P. 1995. "Innovations in Cooperation: The North American Waterfowl Management Plan," *Resolution of Water Quantity and Quality Conflicts,* Ariel Dinar and Edna Loehman, eds. Westport, Conn., Greenwood Publishing, pp. 337–353.

Gerlach, Luther P., and David Bengston. 1994. "If Ecosystem Management Is the Solution, What Is the Problem?" *Journal of Forestry* 92, no. 8 (August): 18–21.

Gerlach, Luther P., and Paul Eide. 1978. *Grassroots Energy,* 16-mm 27-minute, sound, color film. University of Minnesota Media Resources. Distributed by Penn State University Film.

Gerlach, Luther P., and Virginia H. Hine. 1970. *People, Power, Change: Movements of Social Transformation.* Indianapolis: Bobbs-Merrill.

Gerlach, Luther P., and Virginia H. Hine. 1973. *Lifeway Leap: The Dynamics of Change in America.* Minneapolis: University of Minnesota Press.

Grow, Doug. 1998. "For Him, It's All Passion, No Profit: Activist Trying to Save Red Pines While Looking for a Place to Live." *Star Tribune*, January 5, Metro/State, Minneapolis, p. B2.

Landau, Martin. 1969. "Redundancy, Rationality, and the Problem of Duplication and Overlap." *Public Administration Review* 24 (July/August): 346–358.

MacWilliams Cosgrove Snider (media, strategy, and political communications firm). 1993. *Report on the Wise Use Movement* (authors anonymous). Clearinghouse on Environmental Advocacy and Research, Center for Resource Economics (1718 Connecticut Ave., N.W. #300, Washington, D.C., 20009) for the Wilderness Society.

Mazza, Dave. 1993. *God, Land and Politics: The Wise Use and Christian Right Connection in 1992 Oregon Politics.* The Wise Use Public Exposure Project: Western States Center, 522 S.W. 5th Ave., Suite #1390, Portland, OR 97204; Montana AFL-CIO, P.O. Box 1176, Helena, Mont., 59624.

National Inholders Association (NIA). 1991. *Multiple-Use Land Alliance (MULTA)*, newsletter, Bellevue, Wash., January 21, p. 1.

Pickett, Karen. 1990. "Redwood Summer." *Earth First! Journal* 11, no. 1 (November 1): 8.

Stringer, Tish. *Molecular Decentralization*, unpublished manuscript, December 2000.

Switzer, Jacqueline Vaughn. 1996. "Women and Wise Use: The Other Side of Environmental Activism." Paper delivered at the annual meeting of the Western Political Science Association, San Francisco, March 14–16.

Switzer, Jacqueline Vaughn. 1997. *Green Backlash: The History and Politics of Environmental Opposition in the U.S.* Boulder, Colo., Lynne Rienner.

Western States Center. 1993. "Inside the 1993 Wise Use Leadership Project." *Western Horizons*, newsletter of the Wise Use Public Exposure Project's Grassroots Information Network, Vol. 1, No. 3, special issue, September. Western States Center, 522 S.W. 5th Ave., Suite

1390, Portland, Oreg., 97204, in collaboration with the Montana State AFL-CIO, P.O. Box 1176, Helena, Mont., 59624.

Web Addresses

Center for the Defense of Free Enterprise (www.cdfe.org)

Earth First! (www.envirolink.org/orgs/ef)

The Environmental Protection Information Center (www.igc.org/epic/)

Environment '97 (www.environment97.org/framed/village/index.html)

The Heartland Institute (www.heartland.org)

Independent Media Center (http://indymedia.org)

Natural Resources Defense Council (www.nrdc.org/field/enashrae/html)

Protest.Net (www.protest.net)

Public Good Project (http://nwcitizen.com/publicgood)

Sierra Club (www.sierraclub.org/human-rights)

Z Net (www.zmag.org/Globalism/GlobalEcon.html)

Chapter Ten

WHAT NEXT FOR NETWORKS AND NETWARS?

David Ronfeldt and John Arquilla

Editors' abstract. As with other new modes of conflict, the practice of netwar is ahead of theory. In this concluding chapter, we suggest how the theory of netwar may be improved by drawing upon academic perspectives on networks, especially those devoted to organizational network analysis. Meanwhile, strategists and policymakers in Washington, and elsewhere, have begun to discern the dark side of the network phenomenon, especially among terrorist and criminal organizations. But they still have much work to do to harness the bright side, by formulating strategies that will enable state and civil-society actors to work together better.

THE SPREAD OF NETWORK FORMS OF ORGANIZATION

The deep dynamic guiding our analysis is that the information revolution favors the rise of network forms of organization. The network appears to be the next major form of organization—long after tribes, hierarchies, and markets—to come into its own to redefine societies, and in so doing, the nature of conflict and cooperation. As noted in the introductory chapter, the term *netwar* calls attention to the prospect that network-based conflict and crime will be major phenomena in the years ahead. The chapters in this volume provide early evidence for this.

Changes for the Better

The rise of networks is bringing many changes for the better. Some hold out the promise of reshaping specific sectors of society, as in writings about the promises of "electronic democracy," "networked corporations," "global civil society," and even "network-centric warfare."[1] Other likely effects are broader and portend the reshaping of societies as a whole, such that writers herald the coming of "the network society," "the network age," and even the redefinition of "nations as networks."[2] In addition, key academic studies of globalization revolve around observations about the growth of global networks and their interconnection with networks at local levels of society.[3] Many writings are speculative, but others, particularly in the business world, are usually quite practical, inquiring into exactly what kinds of network structures and processes work, and which do not.[4]

At a grand theoretical level, age-old ideas about life as a "great chain of being" or as a progression of nested hierarchies are giving way to new ideas that networks are the key to understanding all of life. Here, theorists argue that hierarchies or networks (or markets, for that matter) are mankind's finest form of organization, and that one or the other design underlies essentially all order in the world. In the social sciences, for example, some key 1960s writings about general systems theory (e.g., Bertalanffy, 1968) and social complexity (e.g., Simon, 1969) took stances lauding the roles of hierarchy in many areas of life. But since the 1970s, and especially in the 1990s, ideas have come slowly to the fore that networks are the crucial design. Thus, it is said that "most real systems are mixtures of hierarchies and networks" (Pagels, 1989, p. 51; also La Porte, 1975), and that "the web of life consists of networks within networks," not hierarchies (Capra, 1996, p. 35; also

---

[1]The literatures on each of these concepts is, by now, quite large, except for "network-centric warfare," whose main source is Cebrowski and Garstka (1998). Some writers (e.g., Florini, 2000) prefer the term "transnational civil society" over "global civil society."

[2]See Kelly (1994) and Lipnack and Stamps (1994) on "the network age," Castells (1996) and Kumon (1992) on "the network society," and Dertouzos (1997) on "networks as nations."

[3]See Held and McGrew (2000), esp. Ch. 2 (excerpted from a 1999 book by David Held, Anthony McGrew, David Goldblatt, and Jonathan Perraton), and Ch. 11 (from a 1997 paper by Michael Mann). Also see Rosenau (1990) and Nye and Donahue (2000).

[4]The *Harvard Business Review* is a fine source of business-oriented references, e.g., Evans and Wurster (1997) and Coyne and Dye (1998), which address banking networks, and Jacques (1990), which provides a classic defense of the importance of hierarchy in corporate structures.

Kelly, 1994). So many advances are under way in the study of complex networks that

> In the longer run, network thinking will become essential to all branches of science as we struggle to interpret the data pouring in from neurobiology, genomics, ecology, finance, and the World-Wide Web (Strogatz, 2001, p. 275).

The Dark Side

Most people might hope for the emergence of a new form of organization to be led by "good guys" who do "the right thing" and grow stronger because of it. But history does not support this contention. The cutting edge in the early rise of a new form may be found equally among malcontents, ne'er-do-wells, and clever opportunists eager to take advantage of new ways to maneuver, exploit, and dominate. Many centuries ago, for example, the rise of hierarchical forms of organization, which displaced traditional, consultative, tribal forms, was initially attended, in parts of the world, by the appearance of ferocious chieftains bent on military conquest and of violent secret societies run according to rank—long before the hierarchical form matured through the institutionalization of states, empires, and professional administrative and bureaucratic systems. In like manner, the early spread of the market form, only a few centuries ago, was accompanied by a spawn of usurers, pirates, smugglers, and monopolists, all seeking to elude state controls over their earnings and enterprises.[5]

Why should this pattern not be repeated in an age of networks? There appears to be a subtle, dialectical interplay between the bright and dark sides in the rise of a new form of organization. The bright-side actors may be so deeply embedded in and constrained by a society's established forms of organization that many have difficulty becoming the early innovators and adopters of a new form. In contrast, nimble bad guys may have a freer, easier time acting as the cutting edge—and reacting to them may be what eventually spurs the good guys to innovate.

---

[5]Adapted from Ronfeldt (1996).

The spread of the network form and its technologies is clearly bringing some new risks and dangers. It can be used to generate threats to freedom and privacy. New methods for surveillance, monitoring, and tracking are being developed; and the uproars over the intelligence systems "Echelon," "Semantic Forests," and "Carnivore" manifest what will surely be enduring concerns. Critical national infrastructures for power, telecommunications, and transportation, as well as crucial commercial databases and information systems for finance and health, remain vulnerable to computer hackers and cyberterrorists. Furthermore, a growing "digital divide" between information "haves" and "have-nots" portends a new set of social inequities. All this places new strains on the world's democracies. Even worse is the possibility that information-age dictatorships will arise in parts of the world, based on the skillful exploitation of the new technologies for purposes of political command and control.

Ambivalent Dynamics of Netwar

As this volume shows, netwar, in all its varieties, is spreading across the conflict spectrum. Instances abound among violent terrorists, ethnonationalists, criminals, and ideological fanatics who are anathema to U.S. security interests and policies. At the same time, many militant yet mainly peaceable social netwars are being waged around the world by democratic opponents of authoritarian regimes and by protestors against various risky government and corporate policies—and many of these people may well be agents of positive change, even though in some cases their ideas and actions may seem contrary to particular U.S. interests and policies.

In other words, netwar is an ambivalent mode of conflict—it has a dual nature. While it should not be expected that the dystopian trends associated with the dark side of netwar will prevail in the years ahead, they will surely contend, sometimes bitterly, with the forces of the bright side.

Netwar is not likely to be a passing fancy. As the information revolution spreads and deepens around the world, instances of netwar will cascade across the spectrum of conflict and crime. So will the sophistication and the arsenal of techniques that different groups can muster. At present, the rise of netwar extends from the fact that the world system is in a turbulent, susceptible transition from the modern era, whose climax was reached at the end of the cold war, to a new era that is yet to be aptly named. Netwar, because of its dependence on networks, is facilitated by the radical increases in global and transnational connectivity, as well as from the growing opportunities for increased connectivity in another sense—the ability of "outsiders" and "insiders" to gain access to each other, and even for insiders to be secreted within an organization or sector of society.[6] All this means that netwar is not a transitional phenomenon; it will likely be a permanent aspect of the new era.

WHEN IS A NETWORK REALLY AN ORGANIZATIONAL NETWORK?[7]

Netwar rests on the dynamics of networks. Yet, what does the term "network" mean? Discussions about networks are proliferating, and three usages are in play, with clear distinctions rarely drawn among them. One common usage refers to communications grids and circuits—as though networking were a technological phenomenon, such that placing a set of actors (military units, for example) atop a grid would make them a network. This is a limited usage; we have spoken about its pitfalls in this and earlier studies, and thus will not dwell further upon it here.

In two other prominent usages, the term refers either to social networks or to organizational networks (or to a conflation of both). But social and organizational networks are somewhat different organisms. This is what needs discussion here, because the difference is a significant issue for theory and practice, affecting how best to think about the dynamics of netwar. The field of network analysis, writ large, has been dominated by social network analysis, but organizational network analysis can be even more helpful for understanding the nature of netwar.

---

[6]The success of Otpor ("Resistance") in overthrowing the regime of Slobodan Milosevic in Serbia is an example of a combined insider-outsider strategy (Cohen, 2000).

[7]Stephen Borgatti and Peter Monge deserve special words of thanks for the informal review comments and significant criticisms they provided regarding this section.

Our main point is that netwar (and also counternetwar) is principally an organizational dynamic, even though it requires appropriate social and technological dynamics to work well. But our deeper point is that there is still much work to be done to clarify the meaning of "network" and come up with better, easier methods of analysis for policymakers and strategists. Both the social and organizational schools can contribute to this—but in different ways, because they have different tendencies.

Social Network Analysis[8]

Social network analysis is an important academic specialty pursued by a relatively small number of anthropologists, sociologists, and organization theorists. It has grown in influence for several decades. Generally speaking, their view—see a book like *Networks and Organizations*, or *Social Structures: A Network Approach*, or *Social Network Analysis*, or the web site of the International Network for Social Network Analysis (INSNA)—holds that all social relationships, including all social organizations, can and should be analyzed as networks: that is, as sets of actors (nodes) and ties (links) whose relationships have a patterned structure.[9]

Social network analysis traces many of its modern roots back to efforts, decades ago, to develop sociograms and directed graphs to chart the ties among different actors in particular contexts—what gradually became known as a network. Later, some social network analysts, along with social psychologists and organizational sociologists who studied what were then called organization-sets, observed that networks often come in several basic shapes (or topologies): notably, *chain* or line networks, where the members are linked in a row and communications must flow through an adjacent actor before getting to the next; *hub*, star, or wheel networks, where members are tied to a central node and must go through it to communicate with each other; and *all-channel* or fully connected or full-matrix networks, where ev-

---

[8]Some of this subsection is verbatim from Ronfeldt (2000).

[9]The references are, respectively to books by Nohria and Eccles (1992); Wellman and Berkowitz (1997); and Wasserman and Faust (1994). The INSNA's web site is at www.heinz.cmu.edu/project/INSNA/.

eryone is connected to and can communicate directly with everyone else (from Evan, 1972).[10] Other shapes have also been identified (e.g., grids and lattices, as well as center/periphery networks and clique networks[11]); so have combinations and hybrids, as in sprawling networks with myriad nodes linked in various ways that are sometimes called "spider's web" networks. Moreover, any particular network may itself be embedded within surrounding networks. Yet, few social network analysts say much about such typologies; their concern is usually to let the data sets speak for themselves.

Classic studies concern topics like friendship cliques among school children, interlocking memberships in corporate boards, job search and occupational mobility patterns that depend on personal connections, partnerships among business firms, and even the structure of the world economic and political system. When a social network analyst studies a primitive tribe, a hierarchical bureaucracy, or a market system, he or she searches for the formal and informal networks that undergird it and emphasizes their roles in making that social organization or system work the way it does (e.g., as in Granovetter, 1985).

In this view, power and influence depend less on one's personal attributes (e.g., resources, attitudes, behaviors) than on one's interpersonal relations—the location and character of one's ties in and to the network. The "unit of analysis" is not so much the individual as it is the network in which the individual is embedded. Not unlike complexity theorists, social network analysts view a network as a systemic whole that is greater than and different from its parts. An essential aim is to show how the properties of the parts are defined by their networked interactions, and how a network itself functions to create opportunities or constraints for the individuals in it.

Many social network analysts stress the importance of location: as in whether an actor's power and prestige stem from his "centrality" in a

---

[10]More complicated designs may be laid out, depending on how many nodes and variations in ties are taken into account. While we appreciate the simplicity of the three designs mentioned here, a more complex depiction of networks composed of three to five persons appears in Shaw (1976), which uses the term "comcon" instead of "all-channel."

[11]For discussions of center/periphery and clique networks, see writings by Stephen Borgatti posted at www.analytictech.com/borgatti/.

network, or whether he has greater autonomy and potential power if he is located at a "structural hole"[12] (a kind of "nonredundant" location that can provide an opening or bridge to an actor in a nearby network). Other analysts stress the importance of the links between actors: whether the ties are strong (tightly coupled) or weak (loosely coupled), and what difference this may make for acquiring and acting on information about what is happening in and around the network.[13] Other questions may be asked about the overall "connectedness" of a network, and the degrees of "reciprocity" and "mutuality" that characterize flows and exchanges within it.

For social network analysts, then, what is keenly interesting about individuals is not their "human capital" (personal properties) but their "social capital" (interpersonal or relational properties). Social networks are often said to be built out of social capital. Many—the ones that tend to be favored in a society, such as business partnerships—thrive when mutual respect and trust are high. But the cohesion and operation of other social networks—such as illicit ones for access to drugs and prostitution—may not require much respect or trust.

Social network analyses tend to be intricately methodological, placing a premium on mathematical modeling and visualization techniques.[14] Although there are exceptions related to measures of efficiency and effectiveness, these analyses are generally not normative or prescriptive, in the sense of observing that one kind of network structure may be better than another for a particular activity, such as a business alliance or a social movement. Moreover, these analyses are not evolutionary, in the sense of observing that the network may be a distinct form of organization, one that is now coming into its

---

[12]Term from Burt (1992). See also his chapter in Nohria and Eccles (1992), and his writings posted at http://gsbwww.uchicago.edu/fac/ronald.burt/research/. The "structural hole" concept is quite prominent in the literature about social network analysis. Meanwhile, a somewhat similar, equally interesting concept is the "small world network" being developed separately by mathematicians. See footnote 19.

[13]Granovetter (1973) is the classic reference about strong versus weak ties; see Perrow (1979) about tightly versus loosely coupled systems.

[14]For a fascinating discussion of the history of visualization techniques, see Freeman (2000).

own. For many social network analysts, the network is the mother of all forms, and the world amounts to a network of networks.

Organizational Network Analysis[15]

Organizational network analysts—or, since this phrase is not widely used, analysts who use network perspectives for studying organizational forms—utilize many of the methods and measures developed for social network analysis. But their approach is quite different—many of them view the network as a distinct form of organization, one that is gaining strength as a result of advances in communications. Also, many of them think that network forms of organization have advantages over other (e.g., hierarchical) forms, such as flexibility, adaptability, and speed of response. For social network analysts, almost any set of nodes (actors) that have ties amounts to a network. But for organizational analysts, that is not quite enough. They might ask, for example, whether the actors recognize that they are participating in a particular network, and whether they are committed to operating as a network.

This literature arises mainly in the fields of organizational and economic sociology, and in business schools. There are various accounts as to who, in recent decades, first called attention to the emergence of networked organizational designs. But most accounts credit an early business-oriented analysis (Burns and Stalker, 1961) that distinguished between *mechanistic* (hierarchical, bureaucratic) and *organic* (networked, though still stratified) management systems. The organic form was deemed more suited to dealing with rapidly changing conditions and unforeseen contingencies, because it has "a network structure of control, authority, and communication" along with a "lateral rather than vertical direction of communication" (p. 121).

Nonetheless, and despite other insightful efforts to call attention to network forms of organization (e.g., Perrow, 1979; Miles and Snow, 1986), decades passed before a school of thinking began to cohere.

---

[15]The discussion here, like the one in the prior subsection, is selective and pointed. For broader, thorough discussions of the various literatures on organizational forms and organizational network analysis, see Monge and Contractor (2001) and Monge and Fulk (1999).

One seminal paper in particular (Powell, 1990) looked beyond informal social networks to argue that formal organizational networks were gaining strength, especially in the business world, as a distinct design—distinct in particular from the "hierarchies and markets" that economic transaction theorists, some other organizational economists, and economic sociologists were accustomed to emphasizing.

> [T]he familiar market-hierarchy continuum does not do justice to the notion of network forms of organization. . . . [S]uch an arrangement is neither a market transaction nor a hierarchical governance structure, but a separate, different mode of exchange, one with its own logic, a network (Powell, 1990, pp. 296, 301).

But this new thinking remained focused mostly on innovative approaches to economic organization and business competition.[16] Moreover, definitional issues remained (and still do) as to precisely what is and is not a network form of organization; often, a definition that may be appropriate in the business world might not apply well in other contexts, such as for analyzing networked social movements.

Since the early 1990s, the literature on networks has grown immensely. Yet, the distinctions between the social and organizational approaches to analysis remain sources of academic debate. An important effort to bridge the debate (Nohria and Eccles, 1992) focused on inquiring "whether 'network' referred to certain characteristics of any organization or whether it referred to a particular form of organization" (p. vii). The question was left unresolved—a lead-off author claimed the pro-form view was largely rhetorical, while the concluding authors implied the academic debate was less significant than the fact that business strategists were developing and applying the new form.[17] In contrast, a later effort by a set of scholars who believe the network is a distinct form of organization (DeSanctis and Fulk, 1999)

---

[16]For example, Miles and Snow (1992) discuss why network organizations in the business world may fail rather than succeed; and Kumar and Dissel (1996) discuss interorganizational business systems whose topologies correspond to chain, hub, or all-channel networks. Also see references in footnote 4.

[17]In that volume, Perrow (1992) sounds a new note when he concludes that the large, fully integrated firms so characteristic of American life may have eroding effects on civil society—and the growth of small firm networks may have revitalizing effects.

ends by noting how much work remains to be done to clarify this phenomenon and its relation to the advances in communications technology. A key task is to create better typologies, since the study of organizational forms still "tends to be dominated by such dichotic concepts as market versus hierarchy or bureaucratic versus post-bureaucratic" (p. 498).

Lately, these unsettled debates over how to think about networks have affected major writings about where societies as a whole may be headed. Consider, for example, this treatment in Francis Fukuyama's *The Great Disruption* (1999), which does not view networks as a distinctive form of organization that is newly on the rise:

> If we understand a network not as a type of formal organization, but as *social capital*, we will have much better insight into what a network's economic function really is. By this view, a network is a moral relationship of trust: A network is a group of individual agents who share *informal* norms or values beyond those necessary for ordinary market transactions. The norms and values encompassed under this definition can extend from the simple norm of reciprocity shared between two friends to the complex value systems created by organized religions (Fukuyama, 1999, p. 199, italics in original).

This is different from the view espoused by Manuel Castells in *The Rise of the Network Society* (1996). He recognizes, in a manner not unlike Fukuyama, the importance that values and norms play in the performance of networks and other forms of organization. Yet, his deeper point is that networks are spreading and gaining strength as a distinct form of organization:

> Our exploration of emergent social structures across domains of human activity and experience leads to an overarching conclusion: as a historical trend, dominant functions and processes in the information age are increasingly organized around networks. Networks constitute the new social morphology of our societies . . . . While the networking form of social organization has existed in other times and spaces, the new information technology paradigm provides the material basis for its pervasive expansion throughout the entire social structure (Castells, 1996, p. 469).

Fukuyama's view reflects mainly the social network approach to analysis, Castells's the organizational approach—and his view is more tied to the influence of the information revolution. Our own view is decidedly in the latter camp (Arquilla and Ronfeldt, 1996, 2000; Ronfeldt, 1992, 1996); but that is not the main point here. The point is that these debates are far from settled; they will persist for years. Meanwhile, where netwar is the object of concern—as in assessing the degree to which an adversary is or is not a netwar actor, and how well it is designed for particular strategies and tactics—the analyst should be steeped in the organizational as much as the social approach. Organizational design is the decisive factor (even when the actors are individuals).

Against this backdrop, good progress at network analysis is being made by anthropologists, sociologists, and political scientists who are studying the growing roles of organizational networks in social movements. Their definitions of "network" have not always improved on prior ones. For example, a pathbreaking study of transnational advocacy movements (Keck and Sikkink, 1998) defines networks rather vaguely as "forms of organization characterized by voluntary, reciprocal, and horizontal patterns of communication and exchange" (p. 8). But their full discussion considers all the organizational, doctrinal, technological, and social dynamics that an effective social movement—and netwar actor—requires.

As noted in Chapter One of this volume, one of the earliest studies to point in this direction was about SPIN (segmented, polycentric, ideologically integrated network) movements in the 1960s. This concept, though rarely noticed by scholars in either the social or the organizational school, remains relevant to understanding the theory and practice of netwar—which is why this volume includes Chapter Nine by Luther Gerlach, updating and summarizing his views about SPIN dynamics. While he has focused the SPIN concept on social movements in the United States, the concept also illuminates dynamics that are under development in various terrorist, criminal, ethnonationalist, and fundamentalist networks around the world.

Furthermore, complexity theorists in the hard and social sciences—theorists interested in discerning common principles to explain "the architecture of complexity" across all natural and human systems—

are delving into the structures and dynamics of biological, ecological, and social systems where networks are the organizing principle (e.g., see Strogatz, 2001). Of the many orderly patterns they have found, one seems particularly worth mentioning here. Many such systems feature a small number of highly connected nodes acting as hubs, along with a large number of less connected nodes—a pattern that proves resilient to systemic shocks, unless a key hub is disrupted or destroyed.[18] This apparently resembles a well-structured, multihub "spider's web" network, or a set of interconnected center/periphery networks. Also, this is the kind of pattern—one or more actors as key hubs, around which are arrayed a large number of actors linked to the hubs but less so to each other, yet with frequent all-channel information-sharing across all actors—that was seen in the social netwars in Seattle and in Mexico. It may also characterize some sprawling terrorist and criminal networks.

WHAT MAKES A NETWORK EFFECTIVE, BESIDES ORGANIZATION?[19]

What holds a network together? What makes it function effectively? The answers involve much more than the organizational aspects emphasized above. While there is no standard methodology for analyzing network forms of organization, our familiarity with the theoretical literature and with the practices seen among netwar actors indicates that the design and performance of such networks depend on what happens across five levels of analysis (which are also levels of practice):[20]

---

[18]George Johnson, "First Cells, Then Species, Now the Web," *The New York Times*, December 26, 2000, pp. M1, M2, provides an overview and relates how this pattern may reflect a mathematical "power law" that is of interest to complexity theorists.

[19]Some of the text in this section is from our earlier books (see Arquilla and Ronfeldt, 1996, 2000). What is analytically new here is the addition of the "narrative level" to the scope of analysis.

[20]This assumes that there are enough actors and resources to organize a network in the first place. Otherwise we would have to specify a recruitment and resource level as part of what makes a network strong and effective.

- Organizational level—its organizational design
- Narrative level—the story being told
- Doctrinal level—the collaborative strategies and methods
- Technological level—the information systems
- Social level—the personal ties that assure loyalty and trust.

The strength of a network, perhaps especially the all-channel design, depends on its functioning well across all five levels. The strongest networks will be those in which the organizational design is sustained by a winning story and a well-defined doctrine, and in which all this is layered atop advanced communications systems and rests on strong personal and social ties at the base. Each level, and the overall design, may benefit from redundancy and diversity. Each level's characteristics are likely to affect those of the other levels.

These are not idle academic issues. Getting a network form "right"—like getting a hierarchical or market form "right"—can be a delicate enterprise. For practitioners trying to organize a new network or adjust one that already exists, various options may merit consideration—and their assessment should assure that all the organizational, narrative, doctrinal, technological, and social levels are well-designed and integrated.

This applies to netwar and counternetwar actors across the spectrum. However, our discussion emphasizes evidence from social netwar actors, mainly activist NGOs, because they have been more open and expressive than have terrorist, criminal, and other violent, secretive actors. The discussion draws on some of the cases presented in earlier chapters, but also affords an opportunity to bring in other recent examples.

Each of these levels of analysis deserves more elaboration than we give here. Our goals are to get people to think in these terms and point the way, even though we cannot pretend to offer final methodological guidance.

Organizational Level

To what extent is an actor, or set of actors, organized as a network? And what does that network look like? This is the top level—the starting point—for assessing the extent to which an actor, or set of actors, may be designed for netwar.

Nowadays, many writings about terrorists, criminals, and activists observe that one grouping or another is organized as a network. But the analyst should be able to specify more than simply that. Among other things, assessment at this level should include showing exactly what type of network design is being used, whether and how members may act autonomously, where leadership resides and/or is distributed, and whether and how hierarchical dynamics may be mixed in with the network dynamics.

As noted earlier, networks come in three major typologies: chain, hub, and all-channel. There are also complex combinations and hybrids involving myriad nodes and links—as in "spider webs," as well as in center/periphery and clique networks. There are also designs that amount to hybrids of networks and hierarchies. In many cases, an important aspect may be the variety of "structural holes" and "bridges" that exist within and between networks—and whether "short cuts" exist that allow distant actors to connect with only a few hops across intermediates, as in a "small world network."[21] Henry Mintzberg (1981) suggests that short cuts may be facilitated by the rise of "mutual adjustment" practices in cross-disciplinary teams. He notes this in the context of business organizations, where the "adjustment phenomenon" will break down "line and staff as well as a number of other distinctions" (p. 5).

Netwar analysts writing for policymakers and strategists should be able to identify and portray the details of a network's structure—as well as they traditionally do when charting an adversary's leadership structures, especially for analyzing terrorist and criminal groups.

---

[21]See Burt (1992, and his web site) on "structural holes" and "bridges," and Watts (1999) and Strogatz (2001) on "small world networks." Watts and Strogatz approach the study of complex networks as mathematicians.

In an archetypal netwar, the units are likely to resemble an array of dispersed, internetted nodes set to act as an all-channel network. Recent cases of social netwar by activist NGOs against state and corporate actors—e.g., the series of campaigns known as J18, N30, A16, etc.—show the activists forming into open, all-channel, and multihub designs whose strength depends on free-flowing discussion and information sharing. The chapters on Burma, Mexico, and the Battle of Seattle substantiate this.

In addition, the International Campaign to Ban Landmines (ICBL) is the prime case of a social netwar developed by NGO activists whose network eventually included some government officials, in a campaign that one prominent organizer, Jody Williams, called "a new model of diplomacy" for putting pressure on the United States and other recalcitrant governments:

> It proves that civil society and governments do not have to see themselves as adversaries. It demonstrates that small and middle powers can work together with civil society and address humanitarian concerns with breathtaking speed. It shows that such a partnership is a new kind of "superpower" in the post-Cold War world. . . . For the first time, smaller and middle-sized powers had not yielded ground to intense pressure from a superpower to weaken the treaty to accommodate the policies of that one country (Jody Williams, *1997 Nobel Lecture,* December 10, 1997, www.wagingpeace.org/articles/ nobel_lecture_97_williams.html).

This campaign had no central headquarters or bureaucracy. Instead, it had a netwar design—a pattern of constant, open communication and coordination among a network of national campaigns that worked independently but coordinated constantly with each other on behalf of their common goal (also see Williams and Goose, 1998).

Such flatness and openness may be impossible for terrorist, criminal, and other violent netwar actors who depend on stealth and secrecy; cellular networks and/or hierarchies may be imperative for them, along with hybrids of hierarchies and networks. Consider the Earth Liberation Front (ELF), a radical environmental group of unclear origins. The ELF may in fact have only a small core of true believers who commit its most violent acts, such as arson and vandalism at new construction sites in naturally wild landscapes (e.g., Long Island, New York). But according to ELF publicist, Craig Rosebraugh, the ELF consists of a "series of cells across the country with no chain of command and no membership roll." It is held together mainly by a shared ideology and philosophy. "There's no central leadership where they can go and knock off the top guy and it will be defunct."[22] In other words, the ELF is allegedly built around "autonomous cells" that are entirely underground. This is different from the "leaderless resistance" doctrine discussed later, which requires a mix of aboveground and underground groups. This is also different from those terrorist networks discussed in Chapter Two that are characterized by horizontal coordination among semiautonomous groups.

In netwar, leadership remains important, even though the protagonists may make every effort to have a leaderless design. One way to accomplish this is to have many leaders diffused throughout the network who try to act in coordination, without central control or a hierarchy. This can create coordination problems—a typical weakness of network designs—but, as often noted, it can also obviate counterleadership targeting. Perhaps a more significant, less noted point is that the kind of leader who may be most important for the development and conduct of a netwar is not the "great man" or the administrative leadership that people are accustomed to seeing, but rather the doctrinal leadership—the individual or set of individuals who, far from acting as commander, is in charge of shaping the flow of communications, the "story" expressing the netwar, and the doctrine guiding its strategy and tactics.

We often posit that it may take networks to fight networks. Yet, government interagency designs for waging counternetwar against terrorists, criminals, and other violent, law-breaking adversaries will have to be built around hybrids of hierarchies and networks. Governments cannot, and should not, attempt to do away with all hierar-

---

[22]From Dan Barry and Al Baker, "Getting the Message from 'Eco-Terrorists': Mystery Group Takes Its Campaign East," *The New York Times*, January 8, 2001, A15. The ELF sometimes operates in alliance with the Animal Liberation Front. See the web site at www.earthliberationfront.com.

chy.[23] Earlier chapters, especially the ones on dealing with terrorists, criminals, and gangs, expanded on this point.

Narrative Level

Why have the members assumed a network form? Why do they remain in that form? Networks, like other forms of organization, are held together by the narratives, or stories, that people tell.[24] The kind of successful narratives that we have in mind are not simply rhetoric—not simply a "line" with "spin" that is "scripted" for manipulative ends. Instead, these narratives provide a grounded expression of people's experiences, interests, and values.[25] First of all, stories express a sense of identity and belonging—who "we" are, why we have come together, and what makes us different from "them." Second, stories communicate a sense of cause, purpose, and mission. They express aims and methods as well as cultural dispositions—what "we" believe in, and what we mean to do, and how.

The right story can thus help keep people connected in a network whose looseness makes it difficult to prevent defection. The right story line can also help create bridges across different networks. The

---

[23]We have previously discussed the need for attention to hybrids of hierarchies and networks, most recently with regard to military swarming (Arquilla and Ronfeldt, 2000). Yet, the idea that such hybrids are a normal feature of social life has figured in a substream of academic writings for decades. In an exemplary volume from the 1970s (La Porte, 1975), the authors maintain that few social activities have structures that look like a "tree" (hierarchy) or a "full matrix" (an all-channel network). Most have "semilattice" structures—they resemble a set of oddly interconnected hierarchies and networks.

[24]Because we want to encourage a new turn of mind, we discuss this as the narrative level, in keeping with our sense that "whose story wins" is a vital aspect of netwars of all types. We could have also presented this level of analysis in a more traditional light, as a cultural, ideological, and/or political level, but the concepts of "narratives" and "stories" seem equally useful and more dynamic for capturing how people actually communicate with each other.

[25]This has been a strong theme of American radical activist organizers, from early pre-netwar ones like Saul Alinsky to contemporary strategists like Gene Sharp.

right story can also generate a perception that a movement has a winning momentum, that time is on its side.[26]

Doctrinal and other leaders may play crucial roles in designing winning stories and building organizational cultures around them. This has long been recognized for executives in corporate systems.[27] It is also true for netwar actors.

All the netwar actors examined in this volume engage in narrative assurance, and use old and new media to do so. All are very sensitive about the stories they use to hold a network together and attract external audiences. For terrorists, the stories tend to herald heroic deeds, for criminals their adventures in greed, and for social activists their campaigns to meet human needs. If it sounds odd to cast criminals this way, note that Colombian (not to mention Mexican and other) drug traffickers have no problem viewing and presenting themselves in a positive light as archnationalists who do good for their communities, for example through financial donations to churches, hospitals, and schools, as well as through legitimate investments in sagging local economies.

On this point, Manuel Castells (1998, pp. 196–201) discusses cartel behavior in Colombia to underscore his thesis (p. 197) about "the importance of cultural identity in the constitution, functioning, and strategies of criminal networks."

> The attachment of drug traffickers to their country, and to their regions of origin, goes beyond strategic calculation. They were/are deeply rooted in their cultures, traditions, and regional societies. Not only have they shared their wealth with their cities, and invested a

---

[26]This, of course, is true for earlier modes of conflict too. Modern guerrilla wars placed very strong emphasis on winning by convincing an opponent that an implacable insurgent movement can never be decisively defeated. In counterinsurgency, similar efforts are made to win the "hearts and minds" of indigenous peoples.

[27]According to a classic of organization theory (Schein, 1985, p . 2), "there is a possibility . . . that the only thing of real importance that leaders do is to create and manage culture." According to Bran Ferren, former Walt Disney Imagineering executive, "The core component of leadership is storytelling, how to articulate a vision and communicate it to people around you to help accomplish the mission" (see Tony Perry, "Navy Takes a Scene Out of Hollywood," *Los Angeles Times*, November 27, 2000, pp. C1, C5, on Ferren's design of a new command center for a Navy command ship).

significant amount (but not most) of their fortune in their country, but they have also revived local cultures, rebuilt rural life, strongly affirmed their religious feelings, and their beliefs in local saints and miracles, supported musical folklore (and were rewarded with laudatory songs from Colombian bards), made Colombian football teams (traditionally poor) the pride of the nation, and revitalized the dormant economies and social scenes of Medellin and Cali—until bombs and machine guns disturbed their joy (p. 199).

In the abstract, his points might apply as well to some leading terrorist groups in the Middle East.

Writings about social activism are especially keen about the narrative level. Keck and Sikkink (1998, citing Deborah Stone) observe that it is crucial for social campaigns to follow the lines of a "strategic portrayal" based on a "causal story." Rutherford (1999) relates the growth of the ICBL to the story it choose to tell: "By controlling the agenda—what was to be discussed and how—the ICBL established the context of the landmine debate as humanitarian rather than military." Also, Otpor ("Resistance"), the netwar-like underground movement to overthrow Milosevic and democratize Serbia, adopted a doctrine of nonviolence, not simply because that was the ethical thing to do, but because it would help provoke the regime into resorting to force in ways that would undermine its authority and give Otpor the high ground regarding whose story should win (Cohen, 2000).

Military campaigns also depend on whose story wins. For example, the highly networked Chechens won their military campaign against Russia during the 1994–1996 war—and they also won the battle of the story, portraying themselves as plucky freedom fighters ridding their land of the last vestiges of a tottering, evil old empire. But in the second war, beginning in 1999, the Russians not only improved their own ability to fight in small, dispersed, networked units, but also mobilized Russian society, including many organizations that opposed the first war in Chechnya, by portraying this second round as a war against terrorism. This story, advanced in the wake of urban bombings in Russia in 1999, even played well in the industrialized West, which has given the Russians a free hand in Chechnya this time, with no threats to withhold new loans because of what might be going on in the transcaucasus region.

In the current Intifadah, both the Palestinians and the Israelis have waged an ever-shifting "battle of the story." The Palestinians have depicted the Israelis as having abrogated the Oslo Accord, while the Israelis have depicted Arafat and his advisers as unwilling to make any—even reasonable—concessions. Moreover, the Palestinians have portrayed the Israelis as using excessive force—although this thrust is vitiated by the Palestinians own violent acts. Meanwhile in cyberspace, both sides have reached out successfully to their ethnic diasporas, for moral as well as financial support. Both have also successfully encouraged distributed-denial-of-service (DDOS) attacks on each other's information systems—the Israelis going so far as to provide a web site for encouraging average Israeli citizens to join the cause by downloading and using various computer attack tools. The Palestinians have used a narrative-level twist on this—they have invoked a "cyber jihad" against Israel, which has resulted in much participation in the cyberspace aspects of this conflict by Muslims from Morocco to Pakistan. Hizbollah in particular has articulated a strategy that includes both computerized swarming attacks on Israeli information infrastructures and selective attacks on commercial firms doing business with Israel.[28]

Disinformation, misrepresentation, and outright lying are eternal downsides that should not be overlooked at the narrative level. Some actors may be unscrupulously cunning about the story lines they unfold in the media.[29] Nonetheless, many of the major trends of the information age—e.g., the continued growth of global media of all types, the proliferation of sensors and surveillance devices, the strengthening of global civil society—imply that the world will become ever more transparent. This may well be a mixed blessing, but it should be to the advantage of democratic state and nonstate actors who thrive on openness (Florini, 1998; Brin, 1998).

---

[28]Lee Hockstader, "Pings and E-Arrows Fly in Mideast Cyber-War," *Washington Post Foreign Service*, October 27, 2000. Carmen Gentile, "Israeli Hackers Vow to Defend," *Wired News*, November 15, 2000.

[29]Gowing (1998) provides a distressing account of how well-meaning but naïve and presumptuous humanitarian NGOs were outmaneuvered by Rwandan officials and their allies in the battle for the control and manipulation of information in the Great Lakes region of Africa in the mid 1990s. Rothkopf (1999), among others, warns about the advent of "the disinformation age," although his examples are not from netwars.

As this occurs, a premium will be placed on using public diplomacy to advance one's messages. Jamie Metzl (1999, pp. 178, 191) explains that

> the struggle to affect important developments across the globe is increasingly an information struggle. Without winning the struggle to define the interpretation of state actions, the physical acts themselves become less effective. . . . [T]he culture of foreign policy must change from one that along with protecting secrets and conducting secret negotiations recognizes that openness—achieved through the development of broad information networks and multiple temporary mini-alliances with both state and nonstate actors—will be the key to foreign policy success.

This may give presumably weaker actors, like NGOs intent on social netwar, a soft-power edge in dealing with presumably stronger actors, like states. As Martin Libicki (1999–2000, p. 41) argues,

> The globalization of perception—the ability of everyone to know what is happening in minute detail around the world and the increasing tendency to care about it—is another way that the small can fend off the large.

Many approaches are being developed for analyzing the narrative level—for example, by scholars who study soft power, political discourse, narrative paradigms, story modeling, agenda setting, metaphors, frames, messages, and/or perspective-making. Some approaches reflect established social-science efforts to understand psychology, propaganda, ideology, and the media, and, in the field of political science, to develop a norm-oriented "constructivist" paradigm as an alternative to the dominant "neorealist" paradigm.[30] Other approaches reflect the rise of "postmodernism" in academia (as in the writings of Pierre Bourdieu, Jacques Derrida, Michel Foucault,

---

[30]Standard sources on neorealism include a range of writings by Kenneth Waltz and John Mearshimer in particular. The literature on constructivism is much more recent and less settled but revolves mainly around writings by Emanuel Adler, Peter Katzenstein, Terrence Hopf, and Alexander Wendt, among others. An interesting effort to split the difference, by focusing on how people argue their stories, is Risse (2000). Our own interest in the narrative level stems in part from our work on the concept of "noopolitik" (Arquilla and Ronfeldt, 1999, and Ronfeldt and Arquilla, 2000).

and Gilles Deleuze and Felix Guattari). All of them show the importance of this level of analysis and practice.

Doctrinal Level

What doctrines exist for making best use of the network form of organization? This level of analysis is very important for explaining what enables the members to operate strategically and tactically, without necessarily having to resort to a central command or leader. The performance of the multihub and all-channel designs in particular may depend on the existence of shared principles and practices that span all nodes and to which the members subscribe in a deep way. Such a set of guiding principles and practices—a doctrine—can enable them to be "all of one mind" even though they are dispersed and devoted to different tasks. It can provide a central ideational, strategic, and operational coherence that allows for tactical decentralization. Overall, this is a looser approach to decisionmaking and operations than traditionally found in right- or left-wing movements—compare this approach, for example, to Mao Zedong's maxim that "command must be centralized for strategic purposes and decentralized for tactical purposes."

So far, two doctrinal practices seem particularly apt for netwar actors. One is to organize and present a network in a way that is as "leaderless" as possible, by having no single leader who stands out, by having (or appearing to have) multiple leaders, and by using consultative and consensus-building mechanisms for decisionmaking.[31] This principle is quite evident in several cases in this book. The second is to use swarming strategies and tactics by having myriad small units that are normally kept dispersed turn to converge on a target from all directions, conduct an attack, and then redisperse to prepare for the next operation. This second principle—swarming—has not been explicitly espoused or adopted by the actors we have looked at, but it is implicitly there, awaiting refinement in many of them—from Middle Eastern terrorists seeking to enter the United States from different di-

---

[31]Commonly recognized downsides are the possibilities that no decision is made, that unaccountable ones are made, or that a network will lack a "center of gravity."

rections in order to converge on a bombing target, to NGO activists who swarmed into Mexico in 1994 and Seattle in 1999.

An example of the first principle is the doctrine of "leaderless resistance" elaborated by right-wing extremist Louis Beam. This doctrine downplays hierarchy in favor of organizing networks of "phantom cells." It reveals a belief that the more a movement conforms to a networked organizational style, the more robust it will be defensively, and the more flexible offensively:

> Utilizing the Leaderless Resistance concept, all individuals and groups operate independently of each other, and never report to a central headquarters or single leader for direction or instruction. . . . [P]articipants in a program of Leaderless Resistance through Phantom Cell or individual action must know exactly what they are doing, and exactly how to do it. . . . Organs of information distribution such as newspapers, leaflets, computers, etc., which are widely available to all, keep each person informed of events, allowing for a planned response that will take many variations. No one need issue an order to anyone (Beam, 1992).

The underground element of Beam's doctrine originally called for four types of secretive, decentralized cells: command, combat, support, and communiqué cells. Each should consist of about eight "minutemen" and have its own leader. But late in the 1990s, practice diverged from this doctrine, allowing "lone wolves" to instigate violent acts, like bombings, seemingly on their own initiative.

The leaderless resistance doctrine has permeated far right circles in the United States (see Burghardt, 1995a, 1995b; Stern, 1996).[32] In addition, it has reached hate groups in Germany, some of which are stockpiling weapons and explosives and posting death lists on web sites.

> "What we are seeing is a very worrying trend in the organization of far right groups with a view to committing terrorism," says Graeme

---

[32]According to Paul de Armond, many far rightists may now regard leaderless resistance as a backward step, since it means that they should not, indeed cannot, organize a mass party and be very public about their leaders and aims. See Barkun (1997) for further discussion of leaderless resistance.

Atkinson, European editor of the anti-fascist magazine *Searchlight*. "They are talking about creating a 'leaderless resistance' of terrorist cells—and of ensuring the creation of liberated zones, with foreigners driven out from rural areas and smaller towns" (Martin A. Lee, "Neo-Nazism: It's Not Just in Germany's Beer Halls Anymore," *Los Angeles Times*, December 31, 2000, p. M2).

By itself, a tenet like leaderless resistance is only a partial step toward having a doctrine for netwar. What operational behavior may in fact be most effective for small, dispersed, mobile forces that are joined in networks? The short answer is swarming (for elaboration, see Arquilla and Ronfeldt, 1997, 2000). If the optimal organizational form for netwar is the dispersed network, the corresponding doctrine must surely consist of swarming. Swarming may well become the key mode of conflict in the information age. But swarming doctrines and strategies have barely begun to emerge for the conduct of terrorist, criminal, and social conflicts.

In this volume, the Zapatista and Seattle cases show swarming in action. Today, one of the most sophisticated doctrines for social netwar comes from the Direct Action Network (DAN), which arose from a coalition of activists dedicated to using nonviolent direct action and civil disobedience to halt the WTO meeting in Seattle. Its approach to netwar epitomizes swarming ideas. Participants are asked to organize, at their own choice, into small (5–20 people) "affinity groups"— "self-sufficient, small, autonomous teams of people who share certain principles, goals, interests, plans or other similarities that enable them to work together well."[33] Each group decides for itself what actions its members will undertake, ranging from street theater to risking arrest.[34] Where groups operate in proximity to each other, they are further organized into "clusters"—but there may also be "flying groups" that move about according to where needed. Different people in each group take up different functions (e.g., police liaison), but every effort is made to make the point that no group has a single lead-

---

[33]See DAN's web site, www.directactionnetwork.org/. It is the source of the observations and quotations in the paragraph.

[34]One role in an affinity group might be police liaison, but it carries the risk that this person would be perceived as a group leader, when in fact the group does not have a leader per se, making all decisions through consensus.

er. All this is coordinated at spokescouncil meetings where each group sends a representative and decisions are reached through democratic consultation and consensus (in yet another approach to leaderlessness).

This approach generated unusual flexibility, mobility, and resource sharing in the Battle of Seattle. It is discussed at length in Chapter Seven, but here is another eyewitness account:

> In practice, this form of organization meant that groups could move and react with great flexibility during the blockade. If a call went out for more people at a certain location, an affinity group could assess the numbers holding the line where they were and choose whether or not to move. When faced with tear gas, pepper spray, rubber bullets and horses, groups and individuals could assess their own ability to withstand the brutality. As a result, blockade lines held in the face of incredible police violence. When one group of people was finally swept away by gas and clubs, another would move in to take their place. Yet there was also room for those of us in the middle-aged, bad lungs/bad backs affinity group to hold lines in areas that were relatively peaceful, to interact and dialogue with the delegates we turned back, and to support the labor march that brought tens of thousands through the area at midday. No centralized leader could have coordinated the scene in the midst of the chaos, and none was needed—the organic, autonomous organization we had proved far more powerful and effective. No authoritarian figure could have compelled people to hold a blockade line while being tear gassed— but empowered people free to make their own decisions did choose to do that (Starhawk, *How We Really Shut Down the WTO*, December 1999, www.reclaiming.org/starhawk/wto.html).

This is very much a netwar doctrine. It is not quite an explicit swarming doctrine—but almost.

An unusually loose netwar design—one that is eminently leaderless yet manages to organize a large crowd for a rather chaotic, linear kind of swarming—is found in the pro-bicycle, anti-car protest movement known as Critical Mass (CM) in the San Francisco Bay area. Since its inception in 1992, CM's bicycle activists (sometimes numbering 2,000) have converged on the last Friday of every month from around the Bay area to disrupt traffic at peak hours along a chosen route.

They slow and block traffic, while handing out pamphlets about pollution and other detriments of the automobile culture. CM riders are proud of their lack of formal organization and leadership and constitute what they call a "xerocracy," which amounts to governance by distributing copies of an idea online or on the scene, say for a ride route, and letting a vote by the assembled decide. A key doctrinal tenet is "organized coincidence," by which "CM rides simply 'materialize' every month even though there are no leaders or organizational sponsorships." This way, "No one need take responsibility but everyone can take credit."

The aim is to ride en masse. The preference may be for "keeping Mass" (riding in a single, large, spread-out mass), but for safety or other reasons a ride may splinter into "minimasses" (multiple, dense small groups). Group decisionmaking about when and where to alter the route of a ride may occur on the fly, as a function of "dynamic street smarts" among the bicyclists up front. A "buddy system" is used to watch out for each other within a mass. Whistle signals are used for some command and control (e.g., stop, go, turn). "Cell phone contact" is used for communications between minimasses, which is particularly helpful if riders want to regroup splinters into a single mass. Tactics during a ride may include "corking" an intersection and "swarming" around a lone car. For much of the 1990s, there were tendencies for confrontation—if not by the riders then by police who came to "escort" and "herd" them. But by 1999, CM became "a ride dominated by creative self-governance and celebratory experimentation—with little or no ill will, and an eye out for avoiding confrontation."[35]

In netwars, swarming often appears not only in real-life actions but also through measures in cyberspace. Aspirations for a leaderless swarming doctrine, beginning with a rationale and a capability for

---

[35]Sources are Dylan Bennett and Gretchen Giles, "Spokes Persons: Bicyclists See Transportation As Critical," *Sonoma County Independent*, April 3–9, 1997, www.metroactive.com/papers/sonoma/04.03.97/bikes-9714.html; *Critical Mass*, undated brochure, http://danenet.wicip.org/bcp/cm.html; Joel Pomerantz, *A San Francisco Critical Mass Glossary: 7 Years of Building a Culture & Learning Lessons, As Reflected in Our Terminology,"* September 1999, http://bok.net/~jig/CM/glossary.html; and Joel Pomerantz, *A Few Comments on Critical Mass: A Brief Introduction to the Critical Mass Glossary,* October 1999, http://bok.net/~jig/CM/glossaryintro.html.

"electronic civil disobedience," show up among hacktivists who advocate the usage of online tools to flood (i.e., overwhelm) a target's computer systems, email inboxes, and web sites, thereby disrupting and even defacing them (see Wray, 1998). Virtually anybody can log into one of these tools and, with a few commands, mount an automated DDOS attack. For example, a device called FloodNet, developed by a collectivity named the Electronic Disturbance Theater (EDT), has been used since the late 1990s against government and corporate sites in Mexico, the Middle East, Europe, and the United States (e.g., against Etoys). Hacktivists associated with the EDT would like to create a new device named SWARM (after our writings), in order to move "digital Zapatismo" beyond the initial emphasis on FloodNet and create new kinds of "electronic pulse systems" for militant activism.[36]

A newer device, called Tribal FloodNet, evidently programmed by a German hacker named Mixter, is technically more powerful. It can enable a lone anonymous individual to mount a far more massive DDOS attack than is the case with FloodNet, which requires publicly announced mass participation (a virtual sit-in) to function well. Tribal FloodNet gained notoriety for its usage in shutting down Yahoo! and other U.S. sites early in 2000. But since then, the contrast between the two systems has led to an ideological controversy. Hacktivist proponents of FloodNet—not only in the EDT, but also in the Electrohippies and, to a lesser extent, the Cult of the Dead Cow—prefer to assert "the presence of a global group of people gathering to bear witness to a wrong." They criticize the Tribal version for being undemocratic and secretive.[37]

---

[36]Interested readers should visit www.nyu.edu/projects/wray/ and related web sites.

[37]From Stefan Krempl, "Computerized Resistance After the Big Flood: Email Interview with Ricardo Dominguez," *Telepolis* (European online magazine), February 16, 2000, www.heise.de/tp/english/inhalt/te/5801/1.html; and Carrie Kirby, "Hacking with a Conscience Is a New Trend," *San Francisco Chronicle*, November 20, 2000, posted at www.sfgate.com/cgi-bin/article.cgi?file=/chronicle/archive/2000/11/20/BU121645.DTL. Also see the web sites of the EDT, the Electrohippies, and the Cult of the Dead Cow.

Technological Infrastructure

What is the pattern of, and capacity for, information and communications flows within an organizational network? What technologies support them? How well do they suit the organizational design, as well as the narrative and doctrinal levels? The new information and communications technologies are crucial for enabling network forms of organization and doctrine. An ample, blossoming literature speaks to this (e.g., DeSanctis and Fulk, 1999). Indeed, the higher the bandwidth and the more dispersed the means of transmission, reception, storage, and retrieval, the better the prospects for success with network-style organization. The multihub and all-channel designs in particular depend on having a capacity—an infrastructure—for the dense communication of functional information. Current advances in peer-to-peer computing (as seen with Napster, Publius, and FreeNet) may give netwar actors an even greater technological edge in the future.[38]

Yet, as noted in Chapter One, netwar can be waged without necessarily having access to the Internet and other advanced technologies. This level may mix old and new, low- and high-tech capabilities. Human couriers and face-to-face meetings may still remain essential, especially for secretive actors like terrorists and criminals.

Many of the chapters in this volume speak to these points. Additional evidence comes from other interesting cases of netwar. Consider the development of the ICBL. Its protagonists got the movement off the ground in the early 1990s by relying mainly on telephones and faxes. They did not turn to the Internet until the mid 1990s, using it first for internal communication and later to send information to outside actors and to the media. Thus, it is "romanticized gobbledygook" that the Internet was essential for the ICBL's early efforts—email and web technologies were not widely used until late in the development of the campaign, and even then usage remained quite limited, rarely including government officials. Nonetheless, the late turn to the new

---

[38]We thank Bob Anderson of RAND for pointing out the importance of peer-to-peer computing. He observes that peer-to-peer computing can enable its users to prevent censorship of documents, provide anonymity for users, remove any single point of failure or control, efficiently store and distribute documents, and provide plausible deniability for node operators.

technologies did improve communication and coordination and helped the ICBL create, and present to the world, a sense that it was a close-knit community on the move, with an important story for the world to hear. A leading academic analyst of the ICBL's use of technology, Ken Rutherford (1999)[39] concludes,

> One of the most significant aspects of the ICBL case is that it shows how NGO coalitions can use communications technologies in order to increase their opportunities for success in changing state behavior. It highlights the importance of how NGOs might be able to address security and social issues that states have thus far proven unable to manage. . . . [T]he role of communications technologies in future international NGO coalitions will be more important than they were in the landmine case.

That is in the case of a well-organized movement. But the new technologies can also have a catalyzing effect for the rapid, unexpected emergence of a spontaneous protest movement. Evidence for this— and for the further spread of the netwar phenomenon—appeared during a wild week in Britain in September 2000, when about 2,000 picketing protesters, alarmed by soaring gasoline prices, quickly organized into dispersed bands that blocked fuel deliveries to local gas stations. The protestors were brought together by cell phones, CB radios, in-cab fax machines, and email via laptop computers. They had no particular leader, and their coordinating center constantly shifted its location. Will Hutton, director general of Britain's Industrial Society (a probusiness group), called it "a very 21st-century crisis made possible by information technology":

> Old organizational forms have been succeeded by a new conception, the network. . . . Using mobile phones, people with no experience of protest were able to coalesce around common aims while never actually meeting.[40]

---

[39]Rutherford (1999), with original text corrected via email correspondence. Also see Williams and Goose (1998, esp. pp. 22–25).

[40]Alexander MacLeod, "Call to Picket Finds New Ring in Britain's Fuel Crisis," *The Christian Science Monitor,* September 19, 2000. MacLeod notes that recent commercial practices increased Britain's vulnerability to this social netwar: Many tanker drivers were freelancers, with no contractual obligations to the oil companies; and many gas stations operated under a "just-in-time" delivery system, keeping few reserves in place.

An earlier example of the use of advanced communications in support of a protest movement can be found in the Polish Solidarity movement of the 1980s. In the wake of the imposition of martial law, mass arrests and some brutality, Solidarity had difficulties keeping its members mobilized and informed. The United States, which was actively trying to undermine communist rule, went to great lengths to provide the movement with sophisticated communications equipment that could not easily be monitored or located. The new gear re-empowered the movement, giving it the ability to once again mount strikes and demonstrations that repeatedly took the government (and the KGB) by surprise.[41]

Social Underpinnings

The full functioning of a network also depends on how well, and in what ways, the members are personally known and connected to each other. This is the classic level of social network analysis, where strong personal ties, often ones that rest on friendship and bonding experiences, ensure high degrees of trust and loyalty. To function well, networks may require higher degrees of interpersonal trust than do other approaches to organization, like hierarchies. This traditional level of theory and practice remains important in the information age.

In this book, the chapters on terrorist, criminal, and gang organizations referred to the importance of kinship, be that of blood or brotherhood. Meanwhile, news about Osama bin Laden and his network, al-Qaeda (The Base), continue to reveal his, and its, dependence on personal relationships he formed over the years with "Afghan Arabs" from Egypt and elsewhere who were committed to anti-U.S. terrorism and Islamic fundamentalism. In what is tantamount to a classic pat-

---

[41]Schweizer (1994) details the CIA's sending of advanced communications devices to Solidarity, and notes (p. 146) that "the administration also wanted the underground fully equipped with fax machines, computers, advanced printing equipment, and more." Woodward (1987, p. 66) observes that these secure lines of communication were also used to maintain contact with the CIA, which often gave Solidarity early warning of the military regime's planned "sweeps" for activists and leaders.

tern of clan-like behavior, his son married the daughter of his long-time aide and likely successor, Abu Hoffs al-Masri, in January 2001.[42]

The chapters on activist netwars also noted that personal friendships and bonding experiences often lie behind the successful formation and functioning of solidarity and affinity groups. And once again, the case of the ICBL speaks to the significance of this level, when organizer Jody Williams treats trust as the social bedrock of the campaign:

> It's making sure, even though everybody was independent to do it their own way, they cared enough to keep us all informed so that we all had the power of the smoke-and-mirrors illusion of this huge machinery. . . . And it was, again, the follow up, the constant communication, the building of trust. Trust, trust, trust. The most important element in political work. Once you blow trust, you've blown it all. It's hard to rebuild.[43]

The tendency in some circles to view networks as amounting to configurations of social capital and trust is helpful for analyzing this level. But there are other important concepts as well, notably about people forming "communities of practice" (Brown and Duguid, 2000), "communities of knowing," and "epistemic communities" (Haas, 1992). In a sense, all these concepts reflect the ancient, vital necessity of belonging to a family, clan, or tribe and associating one's identity with it.

Meanwhile, the traditions of social network analysis and economic transaction analysis warn against the risks of having participants who are "free riders" or lack a personal commitment to teamwork. Indeed, compared to tribal/clan and hierarchical forms of organization, networks have more difficulty instilling, and enforcing, a sense of personal identity with and loyalty to the network. This is one of the key weaknesses of the network form—one that may affect counternetwar

---

[42]See the three-part series of articles in *The New York Times* on "Holy Warriors," beginning with Stephen Engelberg, "One Man and a Global Web of Influence," *The New York Times*, January 14, 2001, pp. A1, A12–A13.

[43]From the discussion following the speech by Jody Williams, *International Organization in the International Campaign to Ban Landmines*, at a gathering of recipients of the Nobel Peace Prize, University of Virginia, November 6, 1998, www.virginia.edu/ nobel/transcript/jwilliams.html.

designs as well. It extends partly from the fact that networks are often thought to lack a "center of gravity" as an organization.

THE PRACTICE OF NETWAR (AND COUNTERNETWAR)

Netwar actors that are strong at all five levels are, and will be, very strong indeed. Netwar works—and it is working for all types: good guys and bad guys, civil and uncivil actors. So far, all have done quite well, generally, in their various confrontations with nation states. A significant question, then, is whether one or the other type could predominate in the future? Will NGOs proselytizing for human rights and high ethical standards reshape the world and its statecraft? Or will violent terrorists, criminals, and ethnonationalists have greater impact—in a dark way? Or will all types move ahead in tandem?

Growing Recognition of Netwar's Dark Face

Practice has been outrunning theory in one area after another where netwar is taking hold. Most commentaries and case studies about organizational networks (and networked organizations) have concerned competitive developments in the business world. However, the year 2000 brought an advance in U.S. government thinking about networking trends among our adversaries, and in the consideration of new options for dealing with them. Government- and military-related research institutes paid the most attention (e.g., see Copeland, 2000),[44] but high-level offices and officials were not lagging far behind.

The first landmark was the annual report, *Patterns of Global Terrorism: 1999*, released by the U.S. State Department's Office of the Coordinator for Counterterrorism in April 2000. It provided the strongest statement yet about networking trends:

> U.S. counterterrorist policies are tailored to combat what we believe to be the shifting trends in terrorism. One trend is the shift from well-organized, localized groups supported by state sponsors to loosely organized, international networks of terrorists. Such a net-

---

[44]In Copeland (2000), see especially the statements by James Rosenau and Steven Metz.

work supported the failed attempt to smuggle explosives material and detonating devices into Seattle in December. With the decrease of state funding, these loosely networked individuals and groups have turned increasingly to other sources of funding, including private sponsorship, narcotrafficking, crime, and illegal trade.[45]

By December 2000, observation of this trend—and of the links growing between crime and terrorism—became even more pronounced in the report of a U.S. interagency group on global crime. While noting that most criminal organizations remain hierarchical—they still have leaders and subordinates—the *International Crime Threat Assessment* found that:

> International criminal networks—including traditional organized crime groups and drug-trafficking organizations—have taken advantage of the dramatic changes in technology, world politics, and the global economy to become more sophisticated and flexible in their operations. They have extensive worldwide networks and infrastructure to support their criminal operations . . . . Much more than in the past, criminal organizations are networking and cooperating with one another, enabling them to merge expertise and to broaden the scope of their activities. Rather than treat each other as rivals, many criminal organizations are sharing information, services, resources, and market access according to the principle of comparative advantage.[46]

Also in December, a forecasting report with a 15-year outlook—*Global Trends 2015*—was produced by the National Intelligence Council, based largely on conferences sponsored by the Central Intelligence Agency for consulting nongovernment experts.[47] The report often us-

---

[45]From the "Introduction" to *Patterns of Global Terrorism: 1999*, Department of State Publication 10687, Office of the Secretary of State, Office of the Coordinator for Counterterrorism, released April 2000, www.state.gov/www/global/terrorism/1999report/1999index.html.

[46]From U.S. Government Interagency Working Group, *International Crime Threat Assessment*, December 2000, Chapter 1, www.whitehouse.gov/WH/EOP/NSC/html/documents/pub45270/pub45270chap1.html#4.

[47]National Intelligence Council, *Global Trends 2015: A Dialogue About the Future with Nongovernment Experts*, NIC 2000-02, Central Intelligence Agency, December 2000, www.cia.gov/cia/publications/globaltrends2015/index.html.

es the word "network" and observes that the world and many of its actors, activities, and infrastructures are ever more networked. Nonetheless, network dynamics appear more in a background than a foreground role—the report does not do much to illuminate network dynamics. Moreover, where this future outlook highlights the growing power and presence of networked nonstate actors of all varieties, it mostly plays up the perils of terrorists, criminals, and other possible adversaries, along with the challenges that activist NGOs may pose for states. The report has little to say about the promising opportunities for a world in which civil-society actors continue to gain strength through networking and where states may learn to communicate, coordinate, and act conjointly with them to address legitimate matters of mutual concern, from democracy to security.

Nationalism, Globalism, and the Two Faces of Netwar

Which face of netwar predominates will depend on the kind of world that takes shape. The key story lines of the 20th century have come to an end. Imperialism, for example, has been virtually extirpated. Over half the world's landmass was under colonial control in 1900,[48] but only a few tiny colonies are left now. The world's major totalitarianisms are also passé. Fascism has gone from being the preferred form of governance among half the great powers and many lesser states in the 1930s, to near extinction today. Communism has moved from being a world threat in 1950, to a mere shadow of itself at the turn of the millennium.

The major old force that remains strongly in play at the dawn of the 21st century is nationalism, particularly its violence-prone ethnonationalist variety. A good measure of the continuing power of nationalism, and of the attractiveness of the state as a form of organization and a focus of nationalist loyalty, is the number of states in existence. When the United Nations was organized after World War II, almost every nation in the world joined, for a total of 54 members. Half a century later, membership has more than tripled, and is closing in on 200.

---

[48]See Lenin (1916, p. 76), whose breakdown showed 90 percent of Africa under colonial control in 1900, 60 percent of Asia, all of Polynesia and Australia, and nearly a third of the Americas.

People without state status want it—and will often engage in terrorist actions to pursue it. Indeed, the majority of terrorist groups, for a long time, arose from nationalist motivations (Hoffman, 1998).[49]

Playing against the old, persistent, often divisive force of nationalism is the new, more unifying force of globalism. It is, to an extent, a reincarnation of the 19th century "Manchester Creed," which held that the growth of industry and trade would create a unified, peaceful world governed by a harmony of interests (see Carr, 1939, pp. 41–62). But today's concept of globalization has many new elements and dynamics, particularly in its deemphasis of the state and its association with the information revolution.[50]

Both nationalism and globalism will continue to coexist, much as the Manchester Creed coexisted with classic power politics.[51] Both will continue to galvanize all kinds of netwars around the world. While many of the violent terrorist, criminal, and ethnic netwars have mainly nationalist origins and objectives, most social netwars have strong globalist dimensions. Thus, the two forces in play in today's world—nationalism and globalism—mirror significant aspects of the two faces of netwar. This is worth pointing out, partly because many current discussions about networked actors and information-age conflict treat them as being mainly the products of globalization, and downplay the enduring significance of nationalism. However, it is important to note that some "dark netwarriors" (e.g., criminal networks) have little or no nationalist motivations.

An eventual question is whether a new "harmony of interests" based on the rise of global civil-society actors relying on soft power will erode the dominance of hard-power, nation-state politics. To some extent, developments in the theory and practice of netwar will affect both these world tendencies. That is, learning better how to build net-

---

[49]Hoffman (1998) notes that religion is also a rising force behind terrorism.

[50]See Held and McGrew (2000), esp. Ch. 2 (excerpted from a 1999 book by David Held, Anthony McGrew, David Goldblatt, and Jonathan Perraton), and Ch. 11 (from a 1997 paper by Michael Mann). Also see Rosenau (1990) and Nye and Donahue (2000).

[51]In the 19th century, the notion of a harmony of interests seemed to predominate over realpolitik—at least from the fall of Napoleon in 1815 to the onset of the social revolutions of 1848, and even, though falteringly, until the onset of World War I. The 20th century, on the other hand, seems to have been mainly the child of realpolitik.

works against crime and terror may tamp down some of the problems that attend ethno- and hypernationalism. Also, states that learn to nurture nonstate civil-society actors may help reduce some of the "demand" for terror, and some of the quests to create ever more nation states. Whichever path unfolds, it will be one in which netwar will surely be found at every turn.

The duality of Janus, first discussed in our introductory chapter, is reintroduced here. According to a modern interpretation by Arthur Koestler (1978), Janus symbolizes the eternal human tension between the need for individual self-assertion and the progress that comes with integration into larger, ultimately global groupings. When kept in equilibrium, in a system allowing individual striving but encouraging connectedness to the world as a whole, the bright face of this dual spirit moves ahead. Today, that tendency is represented by activist NGOs waging social netwar on behalf of human rights and political democracy; they aim to integrate the world around a model of civil society based on common, worldwide values. But "under unfavourable conditions, the equilibrium is upset, with dire consequences" (p. 58).[52] Trouble, for Koestler writing in the 1970s, arises especially when the individual is suborned in a totalitarian society—he gives the examples of Stalinist excesses, Nazi atrocities, and the infamous Milgram "authority experiments" of the 1950s. The modern-day netwar equivalent corresponds to the dark-side terrorists, criminals, and ethnonationalists who pursue self-assertion for narrow purposes.

Two Axes of Strategy

The chapters on terrorist, criminal, and gang networks ended with observations and recommendations for strengthening counternetwar. The chapters on the social netwars—Burma, Mexico, and Seattle—did not end this way, although they mentioned the countermeasures taken by the Burmese and Mexican governments and the City of Seattle. Instead, these latter case studies implied that social netwar could pressure authoritarian regimes to become democratic and impel democracies to become more responsive and transparent.

---

[52]Koestler (1978) does not adequately consider the kind of disequilibrium in which a refusal to connect with the world as a whole may lead to mischief.

In other words, netwar is not a uniformly adverse phenomenon that can, or should, always be countered. It is not necessarily a mode of conflict that always gets in the way of government aims.

States have a range of plausible strategies for dealing with networked nonstate actors. Which strategies are pursued can make a difference in whether the dark or the bright face of netwar predominates. The dark face—with its terrorists, criminals, and virulent ethnonationalists—must be countered by the United States and its allies. But, at times and in particular places, social netwar may complement a government's strategies. Who may benefit from which face depends on what government is being discussed.

In a basic sense, strategy is the methodical art of relating ends and means to deal with other actors. We view the general field of alternatives for strategists as consisting of two axes: one based on military and economic hard power, the other on idea-based soft power (see Figure 10.1). The principal axis for most strategists, and the easy one to describe, is the hard-power one—ranging from active opposition at one pole to material support at the other. In today's parlance, this axis runs from containment and deterrence at one end to engagement and partnership at the other. This axis, for example, permeates most U.S. discussions about China today.

But that is not the only axis. Strategists also think along the lines of an axis for soft-power strategies, where using military and/or economic means to oppose or support another actor is deliberately avoided. At one extreme, the soft-power axis means thoroughly shunning another actor, perhaps because of being disappointed in it, or deploring its behavior without wanting to take active measures against it, or even in the hope of arousing it to behave more positively. At the other pole, this axis consists of trying to influence an actor's behavior, rather indirectly, by holding out a set of values, norms, and standards—"dos" and "don'ts," and hopes and fears—that should determine whether or not one may end up materially favoring or opposing that actor in the future. This might be viewed as the "shining beacon on the hill" ap-

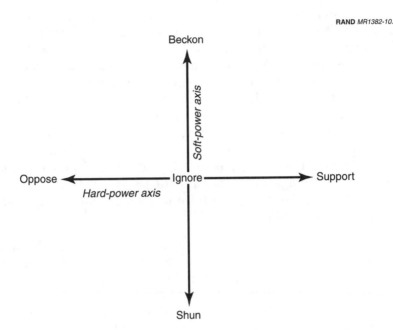

RAND *MR1382-10.*

Figure 10.1—The Two Axes of Strategy

proach to strategy.[53] The midpoint of this axis—and of the hard-power axis, too—is the origin point, where no action at all is taken, perhaps because of having little or uncertain interest in an actor.

These dual axes frame the range of alternative strategies that states use in dealing with each other. Over time, the United States has used them all, often in hybrid blends. For example, during the cold war era,

---

[53]This unexpectedly paraphrases President Reagan, whose national security strategy articulated in June of 1981 called for the spread of American values, creating a new dimension of American power. He wanted to encourage the world to see, in the American example, "a shining city on a hill." As Reagan observed in his farewell address to the nation (given January 11, 1989):

> I've spoken of the shining city all my political life, but I don't know if I ever quite communicated what I saw when I said it. In my mind it was a tall, proud city built on rocks stronger than oceans, windswept, God-blessed, and teeming with people of all kinds living in harmony and peace; a city with free ports that hummed with commerce and creativity (Hannaford, 1998, p. 278).

U.S. strategy revolved mainly around the hard-power axis, with emphasis on containing the Soviet Union and strengthening the NATO alliance. Lines were drawn around the world; actors were obliged to take sides. In today's loose, multipolar world, however, the soft-power axis is more in play. It is now feasible just to shun some states that once required rising degrees of containment (e.g., Cuba). Much of U.S. strategy is now more intent on using soft-power measures to exposit our standards and to attract a target (e.g., like Vietnam) into affiliation with us. Meanwhile, some states, such as Mexico and Canada, have long been subjected to a broad array of alternative strategies—depending on the times and the issues, the United States has ignored and beckoned, supported and even cautiously opposed our neighbors on occasion.

Nonstate actors of all types—especially the kinds of civil and uncivil actors analyzed in this volume—are now so powerful around the world that they cannot be dismissed by national security strategists. As strategists increasingly turn to address them, particularly the ones intent on netwar, this dual-axis perspective on strategy seems likely to frame the options usefully, with each having different implications for the future of netwar.

Each strategy has its merits, but also its costs and risks. For example, trying to stamp out criminal networks—the preferred strategy of the international community today—entails a heavy investment, including the cost involved in trying to achieve a level of cooperation among nations sufficient to deny the criminals (or terrorists, for that matter) any useful "safe havens." Choosing this strategy presupposes that the balance of forces between states and these networks still runs heavily in favor of the former, and that firm action must be taken before criminal networks grow beyond control. For some dictatorships, of course, the target networks are not the criminal ones, but rather the local and transnational NGOs that aim to expand civil society and promote democracy.

A strategy of neglect is quite characteristic of many states' approaches to NGOs—basically ignoring them but also allowing them to grow, to engage state actors, sometimes even to pressure states into action (e.g., as in the antipersonnel landmine campaign and the effort to establish an international criminal court). This strategy holds out the prospect of keeping the various costs of dealing with nonstate civil-society actors to a minimum, by responding to them only when necessary. It also reserves states' options, either to act directly against NGOs at some future point, or to turn to actively embrace them. A preference for this strategy may be based on an assumption that state power still dwarfs the energy and efficacy of nonstate actors; but it differs from the previous strategy in the belief that this gap in relative power seems unlikely to be narrowed anytime soon. For some states, this pattern of behavior may also apply to criminal networks in their midst.

Alternatively, states could pursue a "beacon" approach, by proclaiming standards that will determine whether active opposition or support becomes the eventual recourse. This approach holds great promise for the United States, which has often practiced it without being analytically explicit about it. It is an expression of what, in another writing, we term "noopolitik" (Arquilla and Ronfeldt, 1999). And actually it has been a regular practice of human-rights and other NGOs, more than of U.S. policymakers and strategists. George Kennan's life and work offer exemplary forms of both axes in practice—from his blueprint for active, hard-power containment during the cold war (see his famous "long telegram") to his call to rely principally on passive, soft-power ideals and values in the new era.[54] For an example elsewhere, one could note that Colombia's government has been resorting to aspects of this strategy—i.e., shifting from a principally hard to a soft-power approach—in its newest efforts to deal with the guerrilla organizations that control much of the national territory.

Finally, states could actively embrace and nurture favorable nonstate actors and their networks, encouraging their growth, enhancing their potency, and working with them in a coordinated manner. This may prove a boon to statecraft, when the goals of both coincide. But the risk of such a strategy is that states might unwittingly assist in the creation of a new, networked fabric of global society that may, in the end, be strong enough to constrain states when there are conflicts of inter-

---

[54]Kennan (1996, p. 282) puts it concisely, noting that what we call the "beacon" strategy "would be a policy that would seek the possibilities for service to morality primarily in our own behavior, not in our judgment of others."

est. This may well be an acceptable risk; but it is one that has to be thoroughly assessed.[55] As we look around the world today, we see little sustained embrace of networks of civil-society actors, and only faint hints that some states may be reaching out to transnational criminal and terror networks.[56]

Individual state strategies toward nonstate networks have in practice tended to feature some mixing and blending of these approaches. The United States, in particular, has pursued confrontation against criminal and terror networks, while trying to ignore NGOs when their aims conflict with government policy (e.g., as in the anti-landmine movement and the international criminal court initiative). With regard to the Intifadah waged by the Palestinians, American strategy can be characterized as active support for the "rights" of the Palestinians (not to mention Israeli rights), mixed with "shunning" those who are associated with violent acts—on both sides.

Much more can and should be done to shift to a strategy of both cultivating and cooperating with NGOs. Since U.S. policymakers have tended to emphasize the threats posed by emerging nonstate ac-

---

[55]We have related in other writings (e.g., Arquilla and Ronfeldt, 1999) our own view that states will remain the paramount actors in the international system. As a result of the information revolution, nonstate actors will continue to gain strength and influence; and this will lead to changes in the nature of the state—but not its "withering away." What will occur is a transformation, where some states will emerge stronger than ever because of a capacity to work conjointly with NGOs and other civil nonstate actors. As this process unfolds, there will be a rebalancing of relations among state, market, and civil-society actors around the world—in ways that favor "noopolitik" over realpolitik.

[56]For example, Afghanistan's Taliban government, while it refuses to extradite Osama bin Laden, shows little sign of protecting him out of self-interest. Rather, its position seems to stem from a sense of obligation to a heroic fighter in the war against the Soviet occupation of Afghanistan in the 1980s. In Colombia, far from embracing criminal networks, the government is imperiled by them. The only unambiguously clear example of a state reaching out to a nonstate organization thought to engage in terrorist attacks is that of Iran and Hizbollah, which operates out of southern Lebanon and recently drove the Israeli Defense Forces out of Lebanon, after two decades of occupation. Finally, there are some signs that China is cooperating on some levels with certain criminal networks—modern-day pirates in particular—but the evidence is scant at best.

tors,[57] it is not hard to see how the potential opportunities of engaging and helping to build a global civil society may have been overlooked so far. But the cost of inattention to this issue is already substantial (e.g., political opprobrium suffered because of lack of U.S. support for the antipersonnel landmine ban), and will grow.

Learning not only to live but also to work with NGOs to create new governance schemes for addressing social problems is becoming the cutting edge of policy and strategy.[58] It would seem advisable for the United States to take the lead at this—possibly in connection with newly emerging concepts about "information engagement." However, the states that may be more willing to engage NGOs may well be the ones that possess less hard power and are less interested in competitive realpolitik. Sweden, a good friend to nonstate actors, has not been in a shooting war for 200 years. So perhaps the "beaconing" and nurturing strategies toward nonstate actors that we have articulated will have to diffuse from the periphery of the world political system to its core actors—slowly and over time—if the greater powers cannot advance the process themselves.

This concluding discussion could no doubt be made more thorough and nuanced. But, brief and selective as it is, it serves to underscore what we think is the important point: The rise of netwar and its many early successes imply the need for statecraft to adjust to—perhaps be transformed by—these civil and uncivil manifestations of the information revolution. Most central concepts about national security are over half a century old now. Containment, mutual deterrence, coercive diplomacy, all seem ever less relevant to the types of challenges confronting nation states. Netwar—with its emphasis on empowering

---

[57]See the discussion above about the recently released *Global Trends 2015* report (National Intelligence Council, 2000) which focuses to a large extent on the rise of networked criminal and terrorist organizations, while spending very little time on the opportunities that may arise from working with and supporting nonstate civil-society actors.

[58]A growing literature has begun to identify lessons and options for states and NGOs to work together. Recent sources we consulted include Florini (2000), Reinicke (1999–2000), Gerlach, Palmer, and Stringer (2000), and Simmons (1998); Fukuyama and Wagner (2000) for a RAND research perspective; Chayes, Chayes, and Raach (1997) on conflict management situations; Metzl (1996) and Tuijl (1999) on human rights issues; and Carothers (1999–2000) and Clark, Friedman, and Hochstetler (1998) for cautionary observations about expecting too much from global civil society.

dispersed small groups, its reliance on the power of the story, and its suitability to leaderless networks adept at swarming—should call forth a strategic renaissance among those who would either employ it or oppose it. This conceptual rebirth, if allowed to thrive, will undoubtedly take us all far from the old paradigms. Deterrence and coercion will not disappear entirely as tools of statecraft; but, more and more often, suasion will have to be tried, as our understanding of the limited usefulness of force grows ever clearer.

BIBLIOGRAPHY

Arquilla, John, and David Ronfeldt, *The Advent of Netwar,* Santa Monica, Calif.: RAND, MR-789-OSD, 1996.

Arquilla, John, and David Ronfeldt, *The Emergence of Noopolitik: Toward an American Information Strategy,* Santa Monica, Calif.: RAND, MR-1033-OSD, 1999.

Arquilla, John, and David Ronfeldt, *Swarming and the Future of Conflict,* Santa Monica, Calif.: RAND, DB-311-OSD, 2000.

Arquilla, John, and David Ronfeldt, eds., *In Athena's Camp: Preparing for Conflict in the Information Age,* Santa Monica, Calif.: RAND, MR-880-OSD/RC, 1997.

Barkun, Michael, *Religion and the Racist Right: The Origins of the Christian Identity Movement,* Revised edition, Chapel Hill, N.C.: University of North Carolina Press, 1997.

Beam, Louis, "Leaderless Resistance," *The Seditionist,* Issue 12, February 1992 (text can sometimes be located at www.louisbeam.com/ leaderless.htm).

Bertalanffy, Ludwig von, *General Systems Theory: Foundations, Development, Applications* (revised edition), New York: George Braziller, 1968.

Brin, David, *The Transparent Society: Will Technology Force Us to Choose Between Privacy and Freedom?* Reading, Mass.: Addison-Wesley, 1998.

Brown, John Seely, and Paul Duguid, *The Social Life of Information*, Boston: Harvard Business School Press, 2000.

Burghardt, Tom, *Leaderless Resistance and the Oklahoma City Bombing*, San Francisco: Bay Area Coalition for Our Reproductive Rights (BACORR), April 1995a.

Burghardt, Tom, *Dialectics of Terror: A National Directory of the Direct Action Anti-Abortion Movement and Their Allies*, San Francisco: Bay Area Coalition for Our Reproductive Rights (BACORR), October 1995b.

Burns, Tom, and G. M. Stalker, *The Management of Innovation*, London: Tavistock, 1961.

Burt, Ronald S., *Structural Holes: The Social Structure of Competition*, Cambridge, Mass.: Harvard University Press, 1992.

Capra, Fritjof, *The Web of Life: A New Scientific Understanding of Living Systems*, New York: Anchor Books, 1996.

Carothers, Thomas, "Civil Society," *Foreign Policy*, No. 117, Winter 1999–2000, pp. 18–29.

Carr, Edward Hallett, *The Twenty Years' Crisis, 1919–1939*, London: Macmillan, 1939.

Castells, Manuel, *The Information Age: Economy, Society and Culture*, Vol. I, *The Rise of the Network Society*, Malden, Mass.: Blackwell Publishers, 1996.

Castells, Manuel, *The Information Age: Economy, Society and Culture*, Vol. III, *End of the Millennium*, Malden, Mass.: Blackwell Publishers, 1998.

Cebrowski, Vice Admiral Arthur, and John Garstka, "Network-Centric Warfare," *Proceedings of the United States Naval Institute*, Vol. 24, No. 1, January 1998, pp. 28–35.

Chayes, Antonia Handler, Abram Chayes, and George Raach, "Beyond Reform: Restructuring for More Effective Conflict Intervention," *Global Governance*, Vol. 3, No. 2, May–August 1997, pp. 117–145.

Clark, Ann Marie, Elisabeth J. Friedman, and Kathryn Hochstetler, "The Sovereign Limits of Global Civil Society: A Comparison of NGO Participation in UN World Conferences on the Environment, Human Rights, and Women," *World Politics*, Vol. 51, No. 1, October 1998, pp. 1–35.

Cohen, Roger, "Who Really Brought Down Milosevic," *The New York Times Magazine*, November 26, 2000, pp. 43–47, 118, 148.

Copeland, Thomas E., ed., *The Information Revolution and National Security*, Carlisle, Pa.: Strategic Studies Institute, U.S. Army War College, 2000.

Coyne, Kevin P., and Renée Dye, "The Competitive Dynamics of Network-Based Businesses," *Harvard Business Review*, January–February 1998, pp. 99–109.

Dertouzos, Michael, *What Will Be: How the New World of Information Will Change Our Lives*, San Francisco: HarperCollins, 1997.

DeSanctis, Gerardine, and Janet Fulk, eds., *Shaping Organizational Form: Communication, Connection, and Community*, Thousand Oaks, Calif.: Sage Publications, 1999.

Evan, William M., "An Organization-Set Model of Interorganizational Relations," in Matthew Tuite, Roger Chisholm, and Michael Radnor, eds., *Interorganizational Decisionmaking*, Chicago: Aldine Publishing Company, 1972, pp. 181–200.

Evans, Philip B., and Thomas S. Wurster, "Strategy and the New Economics of Information," *Harvard Business Review*, September–October 1997, pp. 71–82.

Florini, Ann, "The End of Secrecy," *Foreign Policy*, No. 111, Summer 1998, pp. 50–63.

Florini, Ann M., ed., *The Third Force: The Rise of Transnational Civil Society*, Washington, D.C.: Carnegie Endowment for International Peace, 2000.

Freeman, Linton C., "Visualizing Social Networks," *Journal of Social Structure*, Vol. 1, No. 1, February 4, 2000, available only online, www.heinz.cmu.edu/project/INSNA/joss/vsn.html.

Fukuyama, Frank, *The Great Disruption: Human Nature and the Reconstitution of Social Order*, New York: The Free Press, 1999.

Fukuyama, Frank, and Caroline S. Wagner, *Information and Biological Revolutions: Global Governance Challenges—Summary of a Study Group*, Santa Monica, Calif.: RAND, MR-1139-DARPA, 2000.

Gerlach, Luther, "The Structure of Social Movements: Environmental Activism and Its Opponents," in Jo Freeman and Victoria Johnson, eds., *Waves of Protest: Social Movements Since the Sixties*, Lanham, Mass.: Rowman and Littlefield, 1999, pp. 85–98.

Gerlach, L. P., G. B. Palmer, and Tish Stringer, *Managing Global Change Through Segmentary and Polycentric Networks*, paper presented at the annual meeting of the American Anthropology Association, San Francisco, November 14, 2000, posted at www.unlv.edu/faculty/gbp/gerlach/managing2.htm.

Gowing, Nik, *New Challenges and Problems for Information Management in Complex Emergencies: Ominous Lessons from the Great Lakes and Eastern Zaire in Late 1996 and Early 1997*, paper prepared for a conference on Dispatches from Disaster Zones: The Reporting of Humanitarian Enterprises, London, May 27, 1998.

Granovetter, Mark, "The Strength of Weak Ties," *American Journal of Sociology*, Vol. 78, No. 6, May 1973, pp. 1360–1380.

Granovetter, Mark S., "Economic Action and Social Structure: The Problem of Embeddedness," *American Journal of Sociology*, Vol. 91, No. 3, November 1985, pp. 481–510.

Haas, Peter M., "Introduction: Epistemic Communities and International Policy Coordination," *International Organization*, Vol. 46, No. 1, Winter 1992, pp. 1–36.

Hannaford, Peter, *The Quotable Ronald Reagan*, Washington, D.C.: Regnery Publishing, Inc., 1998.

Held, David, and Anthony McGrew, eds., *The Global Transformations Reader: An Introduction to the Globalization Debate*, Malden, Mass.: Polity Press, Blackwell Publishers, 2000.

Hoffman, Bruce, *Inside Terrorism*, New York: Columbia University Press, 1998.

Jacques, Elliot, "In Praise of Hierarchy," *Harvard Business Review*, January–February 1990, pp. 127–133.

Keck, Margaret E., and Kathryn Sikkink, *Activists Beyond Borders: Advocacy Networks in International Politics*, Ithaca: Cornell University Press, 1998.

Kelly, Kevin, *Out of Control: The Rise of Neo-Biological Civilization*, New York: A William Patrick Book, Addison-Wesley Publishing Company, 1994.

Kennan, George, *At a Century's Ending*, New York: Morrow, 1996.

Koestler, Arthur, *Janus*, New York: Random House, 1978.

Kumar, Kuldeep, and Han G. van Dissel, "Sustainable Collaboration: Managing Conflict and Cooperation in Interorganizational Systems, *MIS Quarterly*, September 1996, pp. 279–300.

Kumon, Shumpei, "Japan as a Network Society," in Shumpei Kumon and Henry Rosovsky, eds., *The Political Economy of Japan*, Vol. 3, *Cultural and Social Dynamics*, Stanford, Calif.: Stanford University Press, 1992, pp. 109–141.

La Porte, Todd R., ed., *Organized Social Complexity: Challenge to Politics and Policy*, Princeton, N.J.: Princeton University Press, 1975.

Lenin, V. I., *Imperialism. The Highest Stage of Capitalism*, New York: International Publishers, [1916]1939.

Libicki, Martin, "Rethinking War: The Mouse's New Roar?" *Foreign Policy*, No. 117, Winter 1999–2000, pp. 30–43.

Lipnack, Jessica, and Jeffrey Stamps, *The Age of the Network*, New York: Wiley & Sons, 1994.

Metzl, Jamie Frederic, "Popular Diplomacy," *Daedalus*, Vol. 128, No. 2, Spring 1999, pp. 177–192.

Metzl, Jamie F., "Information Technology and Human Rights," *Human Rights Quarterly*, Vol. 18, No. 4, November 1996, pp. 705–746.

Miles, Raymond E., and Charles C. Snow, "Organizations: New Concepts for New Forms," *California Management Review*, Vol. 28, No. 3, Spring 1986, pp. 62–73.

Miles, Raymond E., and Charles C. Snow, "Causes of Failure in Network Organizations," *California Management Review*, Summer 1992, pp. 53–72.

Mintzberg, Henry, "Organizational Design: Fashion or Fit?" *Harvard Business Review*, January–February 1981.

Monge, Peter R., and Noshir S. Contractor, "Emergence of Communication Networks," in F. M. Jablin and L. L. Putnam, eds., *The New Handbook of Organizational Communication: Advances in Theory, Research, and Methods*, Thousand Oaks, Calif.: Sage, 2001, pp. 440–502.

Monge, Peter, and Janet Fulk, "Communication Technology for Global Network Organizations" in Gerardine DeSanctis and Janet Fulk, eds., *Shaping Organizational Form: Communication, Connection, and Community*, Thousand Oaks, Calif.: Sage Publications, 1999, pp. 71–100.

Nohria, Nitin, and Robert G. Eccles, eds., *Networks and Organizations: Structure, Form, and Action*, Boston: Harvard Business School Press, 1992.

Nye, Joseph S., and John D. Donahue, eds., *Governance in a Globalizing World*, Washington, D.C.: Brookings Institution Press, 2000.

Pagels, Heinz R., *The Dreams of Reason: The Computer and the Rise of the Sciences of Complexity*, New York: Bantam Books, 1989 (originally, Simon & Schuster, 1988).

Perrow, Charles, *Complex Organizations: A Critical Essay*, 2nd edition, Glenview, Ill.: Scott, Foresman and Company, 1979.

Perrow, Charles, "Small Firm Networks, in Nitin Nohria and Robert G. Eccles, eds., *Networks and Organizations: Structure, Form, and Action*, Boston: Harvard Business School Press, 1992, pp. 445–470.

Powell, Walter W., "Neither Market Nor Hierarchy: Network Forms of Organization," in Barry M. Staw and L. L. Cummings (ed.), *Research in Organizational Behavior: An Annual Series of Analytical Essays and Critical Reviews*, Vol. 12, Greenwich, Conn.: JAI Press Inc., 1990, pp. 295–336.

Reinicke, Wolfgang H., "The Other World Wide Web: Global Public Policy Networks," *Foreign Policy*, No. 117, Winter 1999–2000, pp. 44–57.

Risse, Thomas, "'Lets Argue!'—Communicative Action in World Politics," *International Organization*, Vol. 54, No. 1, Winter 2000, pp. 1–39.

Ronfeldt, David, "Cyberocracy Is Coming," *The Information Society*, Vol. 8, No. 4, 1992, pp. 243–296.

Ronfeldt, David, *Tribes, Institutions, Markets, Networks—A Framework About Societal Evolution*, Santa Monica, Calif.: RAND, P-7967, 1996.

Ronfeldt, David, "Social Science at 190 MPH on NASCAR's Biggest Superspeedways," *First Monday*, Feb 2000, http://firstmonday.org/issues/issue5_2/index.html.

Ronfeldt, David, and John Arquilla, "From Cyberspace to the Noosphere: Emergence of the Global Mind," *New Perspectives Quarterly*, Vol. 17, No. 1, Winter 2000, pp. 18–25.

Rosenau, James N., *Turbulence in World Politics: A Theory of Change and Continuity*, Princeton, N.J.: Princeton University Press, 1990.

Rothkopf, David J., "The Disinformation Age," *Foreign Policy*, No. 114, Spring 1999, pp. 83–96.

Rutherford, Ken, *The Landmine Ban and NGOs: The Role of Communications Technologies*, draft paper presented at a conference sponsored by The Nautilus Institute, San Francisco, December 1999, www.nautilus.org/info-policy/workshop/papers/rutherford.html.

Schein, Edgar, *Organizational Culture and Leadership*, Oxford: Jossey-Bass, 1985.

Schweizer, Peter, *Victory: The Reagan Administration's Secret Strategy That Hastened the Collapse of the Soviet Union*, New York: The Atlantic Monthly Press, 1994.

Shafritz, Jay M., and J. Steven Ott, *Classics of Organization Theory*, 4th edition, New York: Harcourt Brace, 1996.

Shaw, Marvin, *Group Dynamics: The Psychology of Small Group Behavior*, 2nd edition, New York: McGraw-Hill Book Company, 1976.

Simmons, P. J., "Learning to Live with NGOs," *Foreign Policy*, No. 112, Fall 1998, pp. 82–96.

Simon, Herbert A., "The Architecture of Complexity," (1962), in Herbert A. Simon, *The Sciences of the Artificial*, Cambridge, Mass.: The M.I.T. Press, 1969.

Stern, Kenneth, *A Force upon the Plain: The American Militia Movement and the Politics of Hate*, New York: Simon & Schuster, 1996.

Strogatz, Steven H., "Exploring Complex Networks," *Nature*, Vol. 410, March 8, 2001, pp. 268–276.

Tuijl, Peter van, "NGOS and Human Rights: Sources of Justice and Democracy," *Journal of International Affairs*, Vol. 52, No. 2, Spring 1999, pp. 493–512.

Wasserman, S., and K. Faust, *Social Network Analysis: Methods and Applications*, Cambridge, Mass.: Cambridge University Press, 1994.

Watts, Duncan J., *Small Worlds: The Dynamics of Networks Between Order and Randomness*, Princeton, N.J.: Princeton University Press, 1999.

Wellman, Barry, and S. D. Berkowitz, *Social Structures: A Network Approach* (updated edition), Greenwich, Conn.: JAI Press, 1997.

Williams, Jody, and Stephen Goose, "The International Campaign to Ban Landmines," in Maxwell A. Cameron, Robert J. Lawson, and Brian W. Tomlin, eds., *To Walk Without Fear: The Global Movement to Ban Landmines*, New York: Oxford University Press, 1998, pp. 20–47.

Woodward, Robert, *Veil: The Secret Wars of the CIA, 1981–1987*, New York: Simon and Schuster, 1987.

Wray, Stefan, *Electronic Civil Disobedience and the World Wide Web of Hacktivism: A Mapping of Extraparliamentarian Direct Action Net Politics*, paper for a conference on The World Wide Web and Contemporary Cultural Theory, Drake University (Des Moines, Ia.), November 1998, www.nyu.edu/projects/wray/wwwhack.html.

AFTERWORD (SEPTEMBER 2001):
THE SHARPENING FIGHT FOR THE FUTURE[1]

John Arquilla and David Ronfeldt

Theory has struck home with a vengeance. The United States must now cope with an archetypal terrorist netwar of the worst kind. The same technology that aids social activists and those desiring the good of all is also available to those with the darkest intentions, bent on destruction and driven by a rage reminiscent of the Middle Ages.

Soon after we put the finishing touches on this book, terrorists attacked New York and Washington. In doing so, they confirmed the warnings (in retrospect, too briefly stated) in Chapter Two that information-age terrorist organizations like al-Qaeda might pursue a war paradigm, developing capabilities to strike multiple targets from multiple directions, in swarming campaigns that extend beyond an incident or two.[2] And, as Chapter Two said was increasingly likely, these terrorists used Internet email and web sites for their communications, sometimes relying on encryption and steganography for security. The picture emerging of these terrorists' network(s), although still obscure, also substantiates the analysis in Chapter Three, which discusses how criminal and other networks have cores and peripheries, with members playing varied, specialized roles. Chapter Three also explains how to attack such networks and their financial and other operations. Moreover, al-Qaeda and its affiliates resemble the SPIN-type organization and dynamics illuminated in Chapter Nine. Finally,

---

[1]This is an expanded version of the "Coda" that ends the paper by Ronfeldt and Arquilla (2001) posted online at *First Monday* (http://firstmonday.org).

[2]The idea of terrorists developing a war paradigm is outlined more fully in Lesser et al. (1999) and in Arquilla, Ronfeldt, and Zanini (2000).

in Los Angeles, the terrorist events had the effect of mobilizing the innovative Terrorist Early Warning Group discussed in Chapter Four.[3] This book is suddenly much more pertinent than we had expected.[4]

If Osama bin Laden's al-Qaeda network is the principal adversary—as seems likely, although other possibilities, including sponsorship by a rogue state like Iraq, cannot be discarded yet—then it may prove useful to view the network from the perspective of the five levels of theory and practice we elucidate in Chapter Ten (organizational, narrative, doctrinal, technological, and social).[5] First, at the organizational level, we see a major confrontation between hierarchical/state and networked/nonstate actors. For the United States and its friends and allies, one challenge will be to learn to network better with each other. Some of this is already going on, in terms of intelligence sharing, but much more must be done to build a globally operational counterterror network. A particular challenge for the cumbersome American bureaucracy will be to encourage deep, all-channel networking among the military, law enforcement, and intelligence elements whose collaboration is crucial for achieving success. U.S. agencies have been headed in this direction for years—in the areas of counternarcotics as well as counterterrorism—but interagency rivalries and distrust have too often slowed progress.

Regarding al-Qaeda, the organizational challenge seems to lie in determining whether this network is a single hub designed around bin Laden. If this is the case, then his death or capture would signal its defeat. However, the more a terrorist network takes the form of a multi-

---

[3]We are grateful to Paul de Armond, author of Chapter Seven, for a September 12, 2001, email that spelled out the ways in which these terrorist attacks took advantage of netwar and swarming paradigms and noted that the U.S. response should include a skillful information strategy.

[4]Meanwhile, the literature on other aspects continues to expand. Additions we like include Kalathil and Boas (2001), Kapstein (2001), Metzl (2001), and Tarrow (2001)—all of which bear, in one respect or another, on the prospects for improving cooperation between governments and nongovernmental organizations. Also see "Special Issue on Mapping Globalization," *American Behavioral Scientist*, Vol. 44, No. 10, June 2001, edited by Eszter Hargittai and Miguel Angel Centeno and supported by the International Networks Archive (based at www.princeton.edu/~ina).

[5]Joel Garreau, "Disconnect the Dots," *Washington Post*, September 17, 2001, offers additional discussion, based on interviews with social network analysts, about how to attack a terrorist network.

hub "spider's web" design, with multiple centers and peripheries, the more redundant and resilient it will be—and the harder to defeat.[6] In a somewhat analogous vein, it is worthwhile to note that since Napster's activities were curtailed by legal action in the United States, more free music is being downloaded and shared by loose peer-to-peer networks. Also, note that, despite the dismantling of the powerful Medellín and Cali cartels during the 1990s, drug smuggling by a plethora of small organizations continues to flourish in Colombia. The risk is that small, more nimble networks may spring up as successors to a defeated large network.

Second, at the narrative level, there is the broad contention of Western liberal ideas about the spread of free markets, free peoples, and open societies versus Muslim convictions about the exploitative, invasive, demeaning nature of Western incursions into the Islamic world. To use Samuel Huntington's phrase, this conflict involves a "clash of civilizations." Also, at the narrative level it might be deemed a "time war" (term from Rifkin, 1987), in that this terrorist mindset is, in a sense, so tribal, medieval, absolutist, and messianic that it represents an effort to challenge the 21st century with 16th century (and earlier) ideals—as well as to ruin Americans' hopes about their future. Indeed, it may be advisable for U.S. strategy to approach this conflict more as a time war than as a clash of civilizations. Bin Laden is an Arab Muslim, but that is not the only context in which to view him. He resembles, in many respects, some of the more fanatical figures out of

---

[6]A study with inputs from various researchers, "Special Report: Al-Qaeda," *Jane's Intelligence Review*, August 2001, pp. 42–51, provides an extensive analysis of al-Qaeda's organizational structure, history, and activities. The analysis views al-Qaeda as a kind of "conglomerate," with both formal vertical and informal horizontal elements, making it a partial hybrid of hierarchical and network forms of organization.

Norman Cohn's *The Pursuit of the Millennium* (1961)[7] and Eric Hoffer's *The True Believer* (1951).[8] Bin Laden is not clinically "insane," but he and his appeal are culturally and temporally perverse.[9]

To this basic imagery, the United States has made a point of adding that these terrorist attacks were "acts of war" against not only America

---

[7]Consider this statement from Cohn (1961, pp. 314–315) about messianic religious fanaticism, known as chiliasm, that coursed through Europe in the Middle Ages:

> In the Middle Ages, the people for whom revolutionary Chiliasm had most appeal were neither peasants firmly integrated in the village and manor nor artisans firmly integrated in their guilds. The lot of such people might at times be one of poverty and oppression, and at other times be one of relative prosperity and independence; they might revolt or they might accept the situation; but they were not, on the whole, prone to follow some inspired *propheta* in a hectic pursuit of the Millennium. . . . Revolutionary Chiliasm drew its strength from the surplus population living on the margin of society—peasants without land or with too little land even for subsistence; journeymen and unskilled workers living under the continuous threat of unemployment; beggars and vagabonds . . . . These people lacked the material and emotional support afforded by traditional social groups; their kinship-groups had disintegrated and they were not effectively organized in village communities or in guilds; for them there existed no regular, institutionalized methods of voicing their grievances or pressing their claims. Instead, they waited for a *propheta* to bind them together in a group of their own—which would then emerge as a movement of a peculiar kind, driven on by a wild enthusiasm born of desperation.

[8]Consider this statement by Hoffer (1951) (from a Harper Perennial edition of Hoffer's book issued in 1989, pp. 11–12) about "true believers" who enter into radical mass movements:

> For men to plunge headlong into an undertaking of vast change, they must be intensely discontented yet not destitute, and they must have the feeling that by the possession of some potent doctrine, infallible leader or some new technique they have access to a source of irresistible power. They must have an extravagant conception of the prospects and potentialities of the future. Finally, they must be ignorant of the difficulties involved in their vast undertaking. . . .

> On the one hand, a mass movement . . . appeals not to those intent on bolstering and advancing a cherished self, but to those who crave to be rid of an unwanted self. A mass movement attracts and holds a following not because it can satisfy the desire for self-advancement, but because it can satisfy the passion for self-renunciation.

[9]A further comparison, drawn from Greek myth and tragedy, is that bin Laden aims to be the *Nemesis* of American *hubris*. This goddess of divine retribution is sent by Zeus to destroy mortals afflicted with this capital sin of pride, the pretension to be godlike. However, bin Laden may yet reveal that he has a "hubris-nemesis complex." For background, see Ronfeldt (1994).

but also against "the civilized world," and American public opinion has been quickly galvanized by the revival of the Pearl Harbor metaphor. Indeed, the disproportionate nature of the terrorists' use of force—including the mass murder of civilians—can only reinforce feelings of righteous indignation. Against this, the perpetrators are likely to exalt their own "holy war" imagery, which they will have trouble exploiting beyond the Islamic world—and they cannot do even that well as long as they remain concealed behind a veil of anonymity. But while the United States may have the edge in the "battle of the story" in much of the world, it will have to think deeply about how to keep that edge if U.S. forces are sent into action in any Middle Eastern countries. The development of the new field of "information strategy" is needed more than ever (see Arquilla and Ronfeldt, 1999, including the notion of creating "special media forces").

Third, in terms of doctrine, the al-Qaeda network seems to have a grasp of the nonlinear nature of the battlespace, and of the value of attack from multiple directions by dispersed small units. If this is indeed a war being waged by al-Qaeda, its first campaign was no doubt the bombing of the Khobar Towers in Saudi Arabia in 1996, followed by a sharp shift to Africa with the embassy bombings of 1998. In between, and since, there have been a number of other skirmishes in far-flung locales, with some smaller attacks succeeding, and others apparently having been prevented by good intelligence. Thus, bin Laden and his cohorts appear to have developed a swarm-like doctrine that features a campaign of episodic, pulsing attacks by various nodes of his network—at locations sprawled across global time and space where he has advantages for seizing the initiative, stealthily.[10]

Against this doctrine, the United States has seemingly little to pose, as yet. Some defensive efforts to increase "force protection" have been pursued, and missile strikes in Afghanistan and the Sudan in 1998 suggest that the offensive part of U.S. doctrine is based on aging notions of strategic bombardment. Needless to say, if our ideas about netwar, swarming, and the future of conflict are on the mark, the

---

[10]For recent additions to the theoretical literature, see Johnson (2001) on "swarm logic," and Bonabeau and Meyer (2001) on "swarm intelligence." Swarming may benefit from advances in "peer-to-peer computing." On this, see Oram (2001).

former is not likely to be a winning approach; a whole new doctrine based on small-unit swarming concepts should be developed. It is possible that the notion of "counterleadership targeting" will continue to be featured—this was tried against Moammar Qaddafi in 1986, Saddam Hussein in 1991, Mohamed Aidid in 1993, and against bin Laden himself in 1998. Every effort to date has failed,[11] but that may not keep the United States from trying yet again, as this seems a part of its doctrinal paradigm. Besides, if bin Laden is the only hub of the al-Qaeda network—possible, though unlikely—his death, capture, or extradition might turn the tide in this conflict.

Fourth, at the technological level, the United States possesses a vast array of very advanced systems, while al-Qaeda has relatively few—and has great and increasing reluctance to use advanced telecommunications because of the risks of detection and tracking. But this category cannot be analyzed quite so simply. The United States, for example, has extensive "national technical means" for gathering intelligence and targeting information—but perhaps only a small portion of these means have utility against dispersed, networked terrorists. Orbital assets—now the linchpins of American intelligence—may prove of little use against bin Laden. At the same time, al-Qaeda has access to commercial off-the-shelf technologies that may prove a boon to their operations.

Fifth, at the social level, this network features tight religious and kinship bonds among the terrorists, who share a tribal, clannish view of "us" versus "them." Al-Qaeda's edge in this dimension ties into its narrative level, with Islam being the pivot between the story of "holy war" against "infidels" and the network's ability to recruit and deploy hate-filled, death-bound strike forces who evince a singleness of mind and purpose. Against this, the United States faces a profound defensive challenge at the social level: How will the American people, despite the arousal of nationalism, react to the potential need to become a less open society in order to become more secure? If the Pearl Harbor metaphor—key to the American narrative dimension—holds up, and

---

[11]The Russians succeeded in killing Dzhokhar Dudayev during the first (1994–1996) Chechen War—apparently triangulating on him while he used a cell phone—but the networked Chechens did quite well in that war, even without their "leader."

if U.S. operations result in successful early counterstrikes, then there may be unusual public solidarity to sustain the "war against terrorism" at the social level. But something of a social divide may emerge between the United States and Europe over whether the response to the attack on America should be guided by a "war" or a "law enforcement" paradigm.

In summary, a netwar perspective on the various dimensions of the struggle with al-Qaeda—again, *if* this is indeed the key adversary, or one of the them—renders some interesting insights into both the context and conduct of this first major conflict of the new millennium. At present, bin Laden and al-Qaeda seem to hold advantages at the social and doctrinal levels, and apparently in the organizational domain as well. The United States and its allies probably hold only marginal advantages at the narrative and technological levels. In terms of strategy, there appears to be less room for al-Qaeda to improve. However, its sound doctrinal and solid social underpinnings might be further enhanced—and a vulnerability removed—if it moved further away from being a hub network revolving around bin Laden. Indeed, this may be an optimal strategy for al-Qaeda, since it is delimited from waging an open "battle of the story" at the narrative level, its one other apparent strategic option.

For the United States and its allies, there is much room for improvement—most of all at the organizational and doctrinal levels. Simply put, the West must start to build its own networks and must learn to swarm the enemy, in order to keep it on the run or pinned down until it can be destroyed. The United States and its allies must also seize the initiative—including by applying pressure on any states that harbor or sponsor terrorists. To be sure, the edge at the narrative level in the world at large must be maintained, but this should be achievable with an economy of effort. The crucial work needs to be done in developing an innovative concept of operations and building the right kinds of networks to carry off a swarming campaign against networked terrorists. Because, at its heart, netwar is far more about organization and doctrine than it is about technology. The outcomes of current and future netwars are bound to confirm this.

BIBLIOGRAPHY

Arquilla, John, and David Ronfeldt, *The Emergence of Noopolitik: Toward an American Information Strategy*, Santa Monica, Calif.: RAND, MR-1033-OSD, 1999.

Arquilla, John, David Ronfeldt, and Michele Zanini, "Information-Age Terrorism," *Current History*, Vol. 99, No. 636, April 2000, pp. 179–185.

Bonabeau, Eric, and Christopher Meyer, "Swarm Intelligence," *Harvard Business Review*, May 2001, pp. 107–114.

Cohn, Norman, *The Pursuit of the Millennium: Revolutionary Messianism in Medieval and Reformation Europe and Its Bearing on Modern Totalitarian Movements*, New York: Harper Torch Books, 1961.

Hoffer, Eric, *The True Believer: Thoughts on the Nature of Mass Movements*, New York: Harper & Row, 1951.

Johnson, Steven, *Emergence: The Connected Lives of Ants, Brains, Cities, and Software*, New York: Scribner, 2001.

Kalathil, Shanthi, and Taylor C. Boas, "The Internet and State Control in Authoritarian Regimes: China, Cuba, and the Counterrevolution," *First Monday*, August 2001, Vol. 6, No. 8, http://firstmonday.org/issues/issue6_8/kalathil/.

Kapstein, Ethan B., "The Corporate Ethics Crusade," *Foreign Affairs*, Vol. 80, No. 5, September/October 2001, pp. 105–119.

Lesser, Ian O., Bruce Hoffman, John Arquilla, David Ronfeldt, Michele Zanini, and Brian Jenkins, *Countering the New Terrorism*, Santa Monica, Calif.: RAND, MR-989-AF, 1999.

Metzl, Jamie F., "Network Diplomacy," *Georgetown Journal of International Affairs*, Winter/Spring 2001, p. 796.

Oram, Andy, ed., *Peer-to-Peer: Harnessing the Power of Disruptive Technologies*, O'Reilly & Associates, 2001.

Rifkin, Jeremy, *Time Wars: The Primary Conflict in Human History*, New York: Simon & Schuster, 1987.

Ronfeldt, David, *The Hubris-Nemesis Complex: A Concept for Leadership Analysis*, Santa Monica, Calif.: RAND, MR-461, 1994.

Ronfeldt, David, and John Arquilla, "Networks, Netwars, and the Fight for the Future," *First Monday*, October 2001, Vol. 6, No. 10, http://firstmonday.org/issue6_10/index.html.

Tarrow, Sidney, "Transnational Politics: Contention and Institutions in International Politics," *Annual Review of Political Science*, Vol. 4, 2001, pp. 1–20.

CONTRIBUTORS

Paul de Armond is director of the Public Good Project, a research and education network based in Washington State that studies militant movements.

Tiffany Danitz is a journalist and a staff writer for stateline.org, an on-line news service that covers politics in the state legislatures. Earlier, she wrote extensively about national and international politics as a staff writer for *Insight Magazine* and *The Washington Times*.

Dorothy Denning is professor of computer science at Georgetown University and author of *Cryptography and Data Security* and *Information Warfare and Security*.

Sean Edwards is a doctoral fellow at the RAND Graduate School and author of *Swarming on the Battlefield: Past, Present, Future*.

Luther Gerlach is professor emeritus of anthropology at the University of Minnesota. He is coauthor of *People, Power, Change: Movements of Social Transformation* and has written numerous articles on social movements and environmental risks.

Warren Strobel is a journalist who has worked at *The Washington Times, U.S. News and World Report*, and is currently with the *Knight Ridder News Service*. He has written widely about international affairs.

John Sullivan is a sergeant with the Los Angeles Sheriff's Department. A specialist in terrorism, conflict disaster, urban operations, and police studies, he is editor of *Transit Policing* and cofounder of the Terrorism Early Warning (TEW) Group.

Phil Williams is professor of international affairs at the University of Pittsburgh and director of the Ridgway Center for International and Security Studies. He is a leading authority on transnational criminal networks.

Michele Zanini is a doctoral fellow at the RAND Graduate School and has written about information-age terrorism, NATO strategy in the Balkans and Mediterranean, and European defense planning.

John Arquilla is associate professor of defense analysis at the Naval Postgraduate School and a consultant to RAND.

David Ronfeldt is a senior social scientist working in the International Policy and Security Group at RAND.

Their publications include *In Athena's Camp: Preparing for Conflict in the Information Age, The Zapatista "Social Netwar" in Mexico, The Emergence of Noopolitik,* and *Swarming and the Future of Conflict.*